Mathematical Finance: Theories and Tools

Mathematical Finance: Theories and Tools

Edited by
Scarlett Morgan

www.willfordpress.com

Published by Willford Press,
118-35 Queens Blvd., Suite 400,
Forest Hills, NY 11375, USA

ISBN: 978-1-64728-525-8

Cataloging-in-Publication Data

Mathematical finance : theories and tools / edited by Scarlett Morgan.
p. cm.
Includes bibliographical references and index.
ISBN 978-1-64728-525-8
1. Finance--Mathematical models. 2. Investments--Mathematics. 3. Business mathematics.
4. Business--Mathematical models. I. Morgan, Scarlett.
HG106 .M38 2023
332.015 195--dc23

For information on all Willford Press publications
visit our website at www.willfordpress.com

Contents

Preface

Mathematical finance refers to a branch of applied mathematics concerned with the application of mathematics and mathematical modeling for solving financial problems and modeling financial markets. It is also known as quantitative finance and financial mathematics. The field combines tools from probability, statistics and stochastic processes with economic theory. There are various applications of mathematical finance, including risk management, data mining, stock trading, econometrics, forecasting, inventory management, marketing, and investing strategies. Risk management is one of the major applications of mathematical finance that helps in the identification and management of financial risks. Mathematical finance is used to mine financial data, which helps to manage financial risks and reduce expenses by recognizing the anomalies and patterns in data. It is used across several industries, such as manufacturing, banking, retail and technology for making predictions based on data. This book traces the progress of mathematical finance and highlights some of its key theories, tools and applications. It aims to present researches that have transformed this discipline and aided its advancement. With state-of-the-art inputs by acclaimed experts of this field, this book targets students and professionals.

All of the data presented henceforth, was collaborated in the wake of recent advancements in the field. The aim of this book is to present the diversified developments from across the globe in a comprehensible manner. The opinions expressed in each chapter belong solely to the contributing authors. Their interpretations of the topics are the integral part of this book, which I have carefully compiled for a better understanding of the readers.

At the end, I would like to thank all those who dedicated their time and efforts for the successful completion of this book. I also wish to convey my gratitude towards my friends and family who supported me at every step.

Editor

1

Markov-Switching Stochastic Processes in an Active Trading Algorithm in the Main Latin-American Stock Markets

Oscar V. De la Torre-Torres [1], Evaristo Galeana-Figueroa [1,*] and José Álvarez-García [2]

[1] Faculty of Accounting and Management, Saint Nicholas and Hidalgo Michoacán State University (UMSNH), 58030 Morelia, Mexico; oscar.delatorre.torres@gmail.com

[2] Financial Economy and Accounting Department, Faculty of Business, Finance and Tourism, University of Extremadura, 10071 Cáceres, Spain; pepealvarez@unex.es

* Correspondence: egaleana@umich.mx

Abstract: In the present paper, we review the use of two-state, Generalized Auto Regressive Conditionally Heteroskedastic Markovian stochastic processes (MS-GARCH). These show the quantitative model of an active stock trading algorithm in the three main Latin-American stock markets (Brazil, Chile, and Mexico). By backtesting the performance of a U.S. dollar based investor, we found that the use of the Gaussian MS-GARCH leads, in the Brazilian market, to a better performance against a buy and hold strategy (BH). In addition, we found that the use of t-Student MS-ARCH models is preferable in the Chilean market. Lastly, in the Mexican case, we found that is better to use Gaussian time-fixed variance MS models. Their use leads to the best overall performance than the BH portfolio. Our results are of use for practitioners by the fact that MS-GARCH models could be part of quantitative and computer algorithms for active trading in these three stock markets.

Keywords: Markov-Switching; Markov-Switching GARCH; Markovian chain; algorithmic trading; active stock trading; active investment; Latin-American stock markets; computational finance

---

1. Introduction

One of the main issues to be addressed in the investment management industry is the proper statistical parameter estimation of the behavior of a return time series (r_t). The seminal work of Markowitz [1,2], the foundational ground of classical Financial Economics, suggests the need of two parameters for security selection: (1) the expected return of the investor and (2) the expected risk exposure. In order to measure the expected return, usually the arithmetic ($\mu = \sum_{t=1}^{T} r_t \cdot T^{-1}$), exponential, or conditional mean is used, and several approaches have been proposed for this purpose. As an example, there is a wide literature review [3–6] about the use of factor models in which the value of another variable (such as a market index, a statistical, a financial, or an Economic indicator) determines the value of the expected mean. Other extensions are the use of time series models such as the ARMA (Auto Regressive Moving Average) one.

$$\hat{r}_t = \alpha + \sum_{p=1}^{P} \lambda_p \cdot r_{t-p} + \sum_{q=1}^{Q} \phi_p \cdot \varepsilon_{t-p} + \varepsilon_t,\ \varepsilon_t = r_t - \hat{r}_t \tag{1}$$

In this model, it is not assumed that the stochastic process of a return time series (r_t) is determined by an stationary, normally distributed stochastic process ($r_t \sim N(\mu, \sigma_r^2)$) with time-fixed mean ($\mu$) and variance ($\sigma^2 = \sum \varepsilon_t^2 \cdot T^{-1}$). In this process, r_t depends on p past realizations or lagged values (first term in (1) or AR term) and q past realizations of the residuals (ε_{t-p}) of the MA model estimated with the second term in Equation (1).

Following the modelling of the expected return with either an arithmetic mean (μ_r) or an AutoRegressive Moving Average models-ARMA model, the novel proposal of Engle [7] and Bollerslev [8] lead to measure the risk exposure (variance) as a time-varying parameter. This led to the widespread use of the Generalized Auto Regressive Conditional Heteroscedasticity (GARCH) models.

$$\hat{\sigma}_t^2 = \sigma_0 + \sum_{p=1}^{P} \beta_p \cdot \varepsilon_t^2 + \sum_{q=1}^{Q} \gamma_p \cdot \sigma_{t-q}^2 \tag{2}$$

In the previous expression, the value of the variance $\hat{\sigma}_t^2$ at t, is determined by the past squared values of the residuals ($\varepsilon_t^2 = (r_t - \mu)^2$ *or* $(r_t - \hat{r}_t)^2$) of an arithmetic mean (μ_r) or an ARMA stochastic process such as Equation (1). The second term in Equation (2) is known as the ARCH term and measures the impact that p lagged (or past) realizations of ε_t^2 have in the actual variance (σ_t^2) value. The third one is a generalization term, known as the GARCH term. This term allows parsimony in the model (a lower number of ARCH terms).

Despite the great breakthroughs and potential uses that the ARMA and GARCH models allow in the financial industry, these have some limitations in their original proposals. They assume that their stochastic process is time-fixed and comes from a single regime. The parameters λ_p, ϕ_q, β_p, and γ_q are the same for all the time series (r_t). This assumption leads to a practical drawback. It is assumed that their values are the same during "normal" and "distress" time periods in the financial markets. If we estimate μ_r or (1) for a given financial time series (e.g., the historical returns of a stock index), it is assumed that the value of μ_r or the parameters in Equation (1) will be the same. This occurs during normal time periods (such as the 2003–2007 in the main stock markets) and also during distressing periods (years 2007 to 2009).

Given the observed performance in the stock markets, this is not true during these two-time intervals or regimes. During the first regime (a "normal" one or $s = 1$), the analyst could have measured an $\mu_{r,s=1}$ expected value and another $\mu_{r,s=2}$ one in the distressed interval ($s = 2$). This lead us to assume lower expected return values during the distressing time periods than during the normal ones ($\mu_{s=2} < \mu_{r,s=1}$). In a similar fashion, the analyst could expect that the volatility values (either with a time-fixed or GARCH variance) in the distress time period are higher than in the normal ones ($\sigma_{r,s=1}^2 < \sigma_{r,s=2}^2$).

Another drawback of the GARCH process in Equation (3) is the fact that the sum of $\sum_{p=1}^{P} \beta_p + \sum_{q=1}^{Q} \gamma_p$ leads to a concept known as "persistence." This means that, if $\sum_{p=1}^{P} \beta_p + \sum_{q=1}^{Q} \gamma_p \approx 1$, there could be misleading estimations of $\hat{\sigma}_t^2$ and the analyst could over or underestimate its value. One theoretical cause for this persistence, as some authors suggest [9–13], is the fact that the behavior of the stochastic comes from a multimodal probability density function (pdf). This means that the studied time series has $S - 1$ structural breaks, which leads to S regimes or states of nature in the behavior of r_t.

For this reason and departing from the seminal work of Hamilton [14–16], the expected return (μ) and risk (σ_r^2) can be modelled in an S-state of the Gaussian or t-Student stochastic process. This follows for an arithmetic mean (μ) or for the ARMA model.

$$r_t \sim \Phi\left(\mu_s, \sigma_s^2\right) \tag{3}$$

$$r_t \sim \mathrm{t}\left(\mu_s, \sigma_s^2, \nu_s\right) \tag{4}$$

$$r_t \sim \Phi\left(\alpha_s + \sum_{p=1}^{P} \beta_{p,s} \cdot r_{t-p} + \sum_{q=1}^{Q} \gamma_{p,s} \cdot \varepsilon_{t-1}, \sigma_s^2\right), \tag{5}$$

$$r_t \sim \mathrm{t}\left(\alpha_s + \sum_{p=1}^{P} \beta_{p,s} \cdot r_{t-p} + \sum_{q=1}^{Q} \gamma_{p,s} \cdot \varepsilon_{t-1}, \sigma_s^2, \nu_s\right) \tag{6}$$

For the sake of simplicity in our review, we will assume that the return time series of a given stock index follows the stochastic processes given in Equations (3) and (4). We will use the arithmetic mean as a location parameter by following the original practice of Markowitz [1,2]. This sets aside the use of ARMA models to search for the most appropriate ARMA equation. This, in order to avoid more complexity in our estimations. This leaves, for further research, the use of other time series models such as ARMA. This will lead to a parameter set (θ) of two to three parameters in each regime s in which ν_s are the degrees of freedom of the t-Student pdf for each regime.

$$\theta = \left\{\mu_s, \sigma_s^2, \nu_s\right\} \tag{7}$$

In order to estimate the parameter set θ, the analyst must know which realization in r_t are generated with the stochastic process of each regime. For the sake of simplicity in the exposition as one of our main assumptions, we will model the stochastic process of r_t with two regimes. One regime ($s = 1$) for the "normal" or "good performing" time periods and another ($s = 2$) for the "distress", "crisis", or "bad performing" ones. We used this assumption because we are interested in simplifying the analysis using the two previously mentioned regimes. Additionally, we want to use only two regimes because we want to extend the current literature on Markov-Switching in stock index trading. Practically all these references, as we will mention in the next section, test the use of only two regimes. Lastly, we use a two-regime context by the fact that two regimes, in terms of market volatility or distress, are more easy to follow by the common. It is easier to follow the idea of a low and high volatility (normal or distress) context than a low, high volatility and extra high volatility one.

Given this, the analyst or stock trader must know the regime at t. Unfortunately, the regime is not observable, given the time series as the only input data. For this reason, Hamilton [14–16] suggests that the unobserved regime s at t could also be modeled as a time series stochastic process in which the values of s_t are modelled through an s-state hidden Markovian chain. With this assumption, the analyst can estimate the smoothed probability ($\xi_{s,t}$) of being in the s regime at t, along with the transition probability matrix ($\mathbf{\Pi}$). This matrix summarizes the likelihood of changing from one regime $\mathrm{s} = \mathrm{i}$ at $\mathrm{t}-1$, to another $\mathrm{s} = \mathrm{j}$ at t. We present the two-regime representation of this matrix next:

$$\mathbf{\Pi} = \begin{bmatrix} \pi_{i,i} & \cdots & \pi_{j,i} \\ \vdots & \vdots & \vdots \\ \pi_{i,j} & \cdots & \pi_{j,j} \end{bmatrix}, \left(\pi_{i,j} = P(s_t = i | s_{t-1} = j, \theta, r_t)\right) \tag{8}$$

Now, the use of the Markov-Switching (MS) model extends the probability set (θ) with the addition of the s smoothed probabilities of being in each regime, along with the transition probability matrix $\mathbf{\Pi}$.

$$\theta = \left\{\mu_s, \sigma_s^2, \nu_s, \xi_{s,t}, \mathbf{\Pi}\right\} \tag{9}$$

As noted, the MS models in Equations (3)–(6) assume that the variance in each regime (σ_s^2) is time fixed. This issue has an important implication because, even in each regime, the volatility levels could be time varying. If this effect is not taken into account in estimating the parameter set, the analyst could have misleading estimations of the risk exposure (σ_s^2) and also the smoothed regime probabilities ($\xi_{s,t}$). For this reason and as a natural solution to the persistence effect in the GARCH model (2), an ARCH (Markov Switching Component ARCH Model-MS-ARCH) or GARCH (Markov-Switching GARCH Models-MS-GARCH) extension of the MS model was proposed [12,13,17].

$$\theta = \left\{\mu_s, \sigma_{s,t}^2, \nu_s, \xi_{s,t}, \mathbf{\Pi}\right\}, \sigma_{s,t}^2 = \sigma_{0,s} + \sum_{p=1}^{P} \beta_{p,s} \cdot \varepsilon_{t-p}^2 + \sum_{q=1}^{Q} \gamma_{p,s} \cdot \sigma_{t-q}^2 + \nu_t \tag{10}$$

Given the features of Markov-Switching models, there are many practical applications of these. The most common one is to characterize the performance of a time series. Additionally, these can be used to measure the contagion (spillover) effect between financial markets and economic indicators in a multiple regime context. Therefore, the characterization or spillover test can be made (in the context of our paper) in a "normal" or low volatility ($s = 1$) regime or in a "distress" or high volatility period ($s = 2$). From all the potential uses, we are interested in their use for stock trading decisions.

Given the development of the 2007–2009 Global financial system crisis and its corresponding spillover effect, several investors faced significant losses in their portfolio's value. Had they had a quantitative model to forecast the probability of being in a distress time period at $t+1$, they would have rebalanced their equity portfolio into a less risky one.

The rationale of using MS models for investing comes from the pioneering proposal of Brooks and Persand [18]. This work has been extended by Kritzman Page and Turkington [19], Hauptmann et al. [20], Engel, Wahl, and Zagst [21], and De la Torre, Galeana, and Álvarez-García [22]. From all these works, only the last one reviews the use of MS models in a developing stock market (Mexico). The other works study their use in the equity markets of Germany, Switzerland, the U.S., or the U.K.

In a parallel fashion, all the previous papers used the original MS, which is a time-fixed variance model of Hamilton [14,15]. Given this, little has been written about the use of MS-GARCH models in trading algorithms.

Departing from these two issues, we have the main motivation of this paper. This includes performing one of the first tests of using MS-GARCH models in trading algorithms, which are applied in Latin-American stock markets. This comes from a U.S. dollar (USD) based investor perspective. Our intention is to extend the scant literature related to the use of MS models in stock trading. This, to the Latin American case. In addition, we are interested in extending the related literature to the review of MS-GARCH models in stock trading algorithms.

Even if it is possible to incorporate the effect of asymmetric shocks in the time varying GARCH variance, we will present the results of symmetric GARCH variances, along with symmetric Gaussian and t-Student likelihood functions in each regime. We did this because the combination of the homogeneous (same marginal pdf in each regime) and heterogeneous, symmetric, and asymmetric regime probabilities MS-GARCH models could be complex. In addition, presenting the results of homogeneous and heterogeneous variances could be wider in each of the three indexes. In order to expose the benefits of Gaussian or t-Student MS-GARCH models in stock trading, we will assume homogeneous GARCH and regime pdfs. Additionally, we will assume symmetric effects of financial shocks in the GARCH processes. As we will mention in our conclusions, the extension of our results to asymmetric variances and regime pdfs along with heterogeneous regime pdfs and variance processes will be left as a task for further research.

In order to run our review, our simulations (back tests) were made in the three main MSCI Latin America (LATAM) indexes: the MSCI Brazil, Chile, and Mexico stock indexes. We selected these three indexes because these markets showed the Highest Latin America (LATAM) turnover figures of the World Federation of Exchanges [23]. Even if Argentina has one of the most important stock markets in the region, we omitted this country due to its fiscal and economic issues observed in the last three years of our review. In some periods, this country's stock market was classified as an emerging market and, in some others, as a frontier one [24]. Additionally, the impact of their fiscal and debt negotiation, along with the emerging support of the International Monetary Fund, could lead to a behavior that is notably different from the other three markets.

We simulated the performance that U.S. dollar USD-based investors would have had, had they used our trading algorithm (with MS-GARCH models as a quantitative method). This takes place from 1 January 2000 to 30 January 2019.

Our rationale is that the use of the simulated algorithm with MS-GARCH models leads to better performance against a "buy and hold" strategy in the three main LATAM stock markets.

Our results, as we will observe, could give theoretical support for the use of two-regime MS-GARCH models in stock trading especially if a given practitioner wants to make active trading in the three main Latin-American stock indexes. In addition, we are providing practical evidence about the benefits of MS or MS-GARCH models as the core of a trading algorithm. Lastly, we extend the literature by testing the benefits of time-varying variances in MS-GARCH models.

Given our theoretical and practical motivations, we structured the paper as follows. In Section 2, we discussed the current and most relevant literature about the use of MS and MS-GARCH models.

In this review, we made a special emphasis in their little reviewed use in stock trading decisions. In Section 3, we present our back tests' material and methods. In this case, we present an introductory review of the MS and MS-GARCH models and their use in the trading algorithm. In addition, in that section, we tested the goodness of fit of MS and MS-GARCH models for the regime characterization of the three simulated indexes along with the back test's pseudo algorithm. In Section 4, we make a review of our simulation results and, in Section 5, we present our conclusions and guidelines for further research.

2. Literature Review of the Use of MS-GARCH Models

Since the original proposal made by Hamilton [14–16], MS and MS-GARCH models have been useful to characterize the behavior of a time series in S separate regimes. As mentioned in the previous section, this model allows us to estimate their corresponding location (usually mean, μ_s) and scale (variance or standard deviation, σ_s) parameters. Additionally, by treating these regimes as an unobserved first order Markovian process, MS and MS-GARCH models are useful for estimating several output parameters. One of these is the probability $\xi_{s,t}$ of being in each regime at t and the transition probabilities $\pi_{i,j}$ given in Equation (5). Departing from the output parameter set θ of MS and MS-GARCH models, several potential applications were tested in economics and finance.

As examples of their applications in time series modelling and crisis spillover measurement, we can mention the work of Ang and Bekaert [25,26], Kritzman, Page and Turkington [19], Klein [27], Areal, Cortez and Silva [28], Zheng and Zuo [29], and Hauptmann et al. [20]. These papers determine the appropriateness of MS models to characterize the time series of developed financial stock markets (the U.S., the U.K., Germany, France, Switzerland, or Japan). They do this with the presence of either two or three regimes. In addition, the works of Alexander and Kaeck [30], Castellano and Scacia [31], and Ma, Deng, and Ho [32] study the benefit of using MS models in the time series of credit default swaps (CDS) and the contagion that their performance have in some other markets, such as the oil ones.

As a practical and useful application, MS models are currently used to predict the probability that the U.S. could be in recession [33]. This use is based on the original paper of Chauvet [34] who suggested performing a two-regime MS regression of the U.S. Economic cycles with four coincident indicators: (1) the non-farm payroll employment, (2) the index of industrial production, (3) the real personal income (without transfer payments), and (4) the real manufacturing and trade sales. The paper shows evidence about the benefit of MS models to forecast recession time periods. This is an analogue used to the warning system with which our simulated trader will decide whether to invest in a stock index in normal or distress periods.

For the specific application of MS models in Emerging financial stock markets, we can mention the papers of Ma, Deng, and Ho [32], Riedel and Sottile [35], and Riedel, Thuraisamy, and Wagner [36]. These three works test the use of MS models in emerging economies' CDS or credit markets. Similar to the previous authors who test the impact of CDS rates in developed stock markets, these papers also find a relation and spillover effect between the CDS markets of the studied economies and their corresponding stock or money markets.

In a similar fashion, the works of Zhao [37], Walid et al. [38], Walid and Duc Khuong [39], and Rotta and Valls-Pereira [40] discuss the link between the volatility of emerging countries' stock markets and their corresponding currency pairs against the U.S. dollar. In the three works, the authors found a link between these two types of markets in the BRIC (Brazil, Russia, India, and China) region with a more observable two-way contagion effect during the high volatility regime ($s = 2$).

Another interesting review of the MS model use is the spillover of currency crises between integrated regions such as the European one. This is done, given the impact of monetary policy changes. Among the papers that address this issue in developed markets, we found the work of Mouratidis et al. [41], Miles and Vijverberg [42], and Lopes and Nunes [43]. These authors found the change in the inflation or monetary policy has a significant impact on the probability of the distress regime of the studied currencies.

In the study of the previously mentioned issue in emerging economies, we found the papers of Kanas [44], Álvarez-Plata and Schrooten [45], Parikakis and Merika [46], Girdzijauskas [47], Dubinskas and Stungurienė [48], Kutty [49], and Ahmed et al. [50]. These authors found a close relation between the changes in expectations in the interest target rate and the performance of the studied emerging markets' currencies or stock markets. In addition, Dufrénot, Mignon, and Péguin-Feissolle [51] found evidence of a spillover effect from the distress probability in the U.S. stock markets to the LATAM ones. Similarly, the work of Sosa, Ortiz, and Cabello [52] shows strong proof that the influence of the stock markets in its corresponding foreign exchange (FX) rate is strong in Chile, Colombia, Mexico, and Peru.

Even if the works mentioned in the previous paragraphs do not have a direct relationship with our tests, these give strong support for the use of MS and MS-GARCH models against single regime stochastic processes. In addition, these works characterize the MS stochastic process to emerging stock market indexes, such as the ones studied herein.

For the more specific application of MS models in emerging stock markets, we found the work of Cabrera et al. [53] and De la Torre, Álvarez-García, and Galeana-Figueroa [54]. These authors tested the use of MS and MS-GARCH models to characterize the performance of the main indexes of Bolivia, Brazil, Chile, Colombia, Mexico, Peru, Venezuela, and the MSCI LATAM index. This occurs along with the review of the U.S. S&P500 index. In their results, these two works show strong evidence in favor of the characterization of these stock indexes with two or three regimes. The paper of Cabrera et al. [53] found that it is preferable to use MS-GARCH models to characterize the stock indexes of these countries (a result in line with ours in this paper). In addition, the authors search for evidence in favor of a potential LATAM common distress cycle, but only Chile and Mexico show this behavior with the U.S. This last paper is one of the main motivations of ours and led us to test the use of MS-GARCH models for trading decisions in these three LATAM stock indexes.

All the previous literature makes a punctual and elegant review of the use of MS and MS-GARCH models in finance or economics. Despite this, none of these deal directly with the subject of using these models for trading.

The first work that propose the use of Gaussian MS models for trading is the one of Brooks and Persand [18]. It is the main theoretical motivation and departing point in this paper. In their work, the authors suggest the estimation of an MS model in the equity/bond yield ratio of the U.K. FTSE100 index and the 10-year Gilt. The rationale of their investment strategy is a very straightforward one: to invest in the U.K. Gilt and the FTSE100, given the smoothed probabilities $\xi_{s,t}$ of being in each regime at t. Therefore, the investment level in each asset is given by the next expression.

$$\begin{bmatrix} \omega_{FTSE\ 100} \\ \omega_{10-year\ Gilt} \end{bmatrix} = \begin{bmatrix} \xi_{s=1,t} \\ \xi_{s=2,t} \end{bmatrix} \tag{11}$$

With these trading weights, it is expected to invest more in the equity market in the normal time regime and less in the "risk-free" gilt. If the probability of being in the distress period is higher, it is expected to invest more in the U.K. gilt and less in the FTSE100 index. With the assumption of no trading costs, the authors found strong proofs in favor of using the MS model for active trading. This occurs against a "buy and hold" investment strategy in the FTSE100 or the 10-year U.K. Gilts. This work, as previously mentioned, is the departing theoretical point of our research. The difference between this paper and our tests is that, instead of using the smoothed probabilities as weights, we used these to decide whether to invest fully in the stock index. That is, we decided whether to invest all the portfolio proceedings in the stock index during normal periods or to do it in the risk-free asset during distressful periods.

In a similar manner to Brooks and Persand [18], Ang and Bekaert [25,26] extend the use of MS models to estimate MS covariances. With these, the authors simulated the performance that a USD-based investor would have had, had she made an international diversification in Germany, Japan, the U.K., and a European index. In their results, the authors found strong proof of a correlation clustering during distressful periods and a reduction in the diversification benefits. Given this result, they simulated the performance of an optimal portfolio in which the used regime-specific covariance matrix change, given the probability of being in a normal or distress period. Their results suggest that it is preferable to use an MS portfolio strategy than a passive or buy and hold one. This paper motivates the current one by the fact that the authors prove that the practice of rebalancing a portfolio in distress or the crisis time periods leads to an over performance against a buy-and-hold strategy.

With a similar portfolio perspective and a similar theoretical motivation to the present one, Kritzman, Page, and Turkington [19] suggest using the Baum and Welch [55] algorithm to perform a more computationally efficient estimation of the MS model. These authors use their proposed method for estimating the expected return vector of a multi-asset portfolio diversified with equities, strategy indexes, and financial risk factors. From a starting base portfolio, they suggest a series of tilts, given the smoothed probability of a distress time period in the stock market. They also incorporated (in the tilt decision) the probability of recession and the likelihood of a high inflation episode. Their simulations also show the benefit of using MS models for active investment.

Following these works, we found ones of Hauptmann et al. [20] and Engel, Wahl, and Zagst [21]. In these papers, the authors develop a warning system for the S&P 500, the Eurostoxx 50, and the Nikkei 225 stock indexes. In their proposal, the authors estimate a Gaussian three-regime model for the stock index and then use several economic and financial risk factors to estimate a logit warning system. This led to estimating time-varying smoothed regime probabilities and time-varying transition probabilities. In their review, the authors tested the performance of the S&P 500 index with a similar "tilt" investment strategy as the one of Kritzman, Page, and Turkington [19]. Their results show, as appropriate, the use of their warning system.

In a parallel test, Engel, Wahl, and Zagst [21] simulated, without the impact of transaction costs, the performance of an equally weighted portfolio of the S&P500, the Eurostoxx 50, and the Nikkei 250. This is a similar investment strategy to Hauptmann et al. [20]. Compared to the case of the previous works, their simulated portfolio also outperformed the corresponding passive investment strategy.

Lastly, we found the work of De la Torre, Galeana, and Álvarez-García [22] that test the use of MS models in the U.S., U.K., Italian, and Mexican stock markets. These authors simulated an agent that invested in stocks (risk-free asset) during normal (distress) time period at $t+1$. With weekly simulations that include the impact of trading costs, they found that the use of Gaussian two-regime MS models leads to alpha generation. For the specific Italian case, they found that the accumulated return in the simulations was negative. Despite this outcome, the result is better than the observed one in the corresponding passive strategy. The present paper has the intention of extending the review in the Mexican market by using Gaussian and t-Student MS-GARCH models in the trading decision algorithm.

Given our literature review and, as noted, little has been written about the use of MS models for investment decisions. Related to this use, we found two type of reviews in developed stock markets.

1. A two-asset trading algorithm that determines in which asset to invest, given the probability of being in each regime at t or $t+1$.
2. A portfolio management system, in which some tilts in the investment weights are made, given the same probabilities.

From the previously mentioned strategies, we want to extend use of the first type of strategy. Given our literature review and research interest, we found no literature that test the benefits of a two-asset strategy in the main LATAM markets. This uses a U.S. dollar (USD) based investor perspective. Additionally, we found that only the work of Engel, Wahl, and Zagst [21] use time-varying MS-GARCH models and these are used in developed stock markets.

Given this, we tested the performance of a USD-based investor that followed the next trading rule in the three main Latin America stock markets.

1. To invest in the simulated stock market index if the investor expects to be in the low volatility ($s=1$) regime at $t+1$ or

2. To invest in the U.S. risk-free asset if the investor expects to be in the high volatility ($s = 2$) regime.

Once our theoretical motivations were mentioned, we presented a brief review of the rationale behind the MS-GARCH model. We do this section for introductory purposes and also to explain how we used the MS and MS-GARCH models in the investment strategy.

3. Backtest Materials and Methods

3.1. The MS-GARCH Model and Its Use in the Active Trading Algorithm

The MS-GARCH model that we will test herein has the functional form of (10). As Haas, Mitnik, and Paolella [13] suggest, the MS-GARCH models must be inferred once the residuals have been calculated from a conditional mean such as the one of an arithmetic mean μ or an ARMA model in r_t. Since this paper is the first test of the application of the MS-GARCH model for investment purposes in the main LATAM stock markets, we used the arithmetic mean μ of the returns r_t (we will leave the application of ARMA models for further research). With the residuals determined as $\varepsilon_t = r_t - \mu$, we used these to calculate the MS-GARCH model. We did this by assuming that the stochastic process can be modeled with a two-regime Markovian chain that can be filtered, at t, from the time series of r_t. This takes place with either a Gaussian or a standardized t-Student filtered probability $\xi_{s,t}$ function as follows, respectively:

$$\xi_{s,t} = \frac{1}{\sqrt{2\pi}\sigma_s} e^{-\frac{1}{2}\left(\frac{\varepsilon_t}{\sigma_s}\right)^2} \tag{12}$$

$$\xi_{s,t} = \frac{\Gamma\left(\frac{\nu_s+1}{2}\right)}{\sqrt{(\nu_s-2)\pi}\Gamma\left(\frac{\nu_s}{2}\right)}\left(1+\frac{\left(\frac{\varepsilon_t}{\sigma_s}\right)^2}{(\nu_s-2)}\right)^{-\frac{\nu_s+1}{2}} \tag{13}$$

In these expressions, ν_s are the degrees of freedom that determine the shape of the t-student pdf. These will lead us to the maximization of the next log-likelihood function, in which p_s is the stable probability (mixing law) of each regime s (please refer to Hamilton [14,16] for further reference).

$$L(r_t, \theta) = \sum\nolimits_t^T \ln\left(\sum\nolimits_{S=1}^{S} p_s \cdot \xi_{s,t}\right), \theta = [\sigma_s, \pi_s, \Pi] \tag{14}$$

In order to estimate the parameter vector θ of (10), we will use a maximum likelihood method with the Viterbi [56] algorithm as used in the MSGARCH R package [57]. R and this package are the core computational software of our back tests and trading algorithm. From the inferred parameters with Equations (12) to (14), we were interested in the filtered probability $\xi_{s,t}$ at t and the transition probability matrix Π. With these, we can estimate the probability $\xi_{s=1,t+1}$ of being in each regime at $t+1$ as follows.

$$\begin{bmatrix} \xi_{s=1,t+1} \\ \xi_{s=2,t+1} \end{bmatrix} = \Pi \begin{bmatrix} \xi_{s=1,t} \\ \xi_{s=2,t} \end{bmatrix} \tag{15}$$

Given the two forecasted probabilities of being in the regime s at t, an investor can determine if she expects to be in a high-volatility regime at $t+1$ with the next indicator function.

$$s_{t+1} = \begin{cases} 1\ if\ \xi_{s=2,t+1} \leq 0.5 \\ 0\ if\ \xi_{s=2,t+1} > 0.5 \end{cases} \tag{16}$$

We propose this regime filtering, given in Equations (12) or (13) by following the standard suggestions in the related econometrics literature [14–16,18–20,25]. Once the filtered probabilities (12) or (13) were estimated with the Viterbi algorithm, we inferred their smoothed values, given Kim's method [58].

A natural question of the MS-GARCH model use is: why is it necessary to compare the performance of an investor if using either a MS, MS-ARCH, or MS-GARCH variance in Equations (12) or (13)? Our position is that, given the time-fixed ARCH or GARCH scale parameter used in Equations (12) or (13), the estimated smoothed probability $\xi_{s,t}$ of each regime will differ. This result will also lead to a different precision level in the forecast of $\xi_{s,t+1}$. With these different values of $\xi_{s,t+1}$, it is expected that there will be different trading decisions in Equation (16) at t. Given this, we backtested six scenarios in which we changed the variance estimation method (time-fixed, ARCH, or GARCH) and the Log-Likelihood Function (LLF) function with Equations (12) or (13).

Given the regime expectation in Equation (16), the trading algorithm that we tested herein was translated to the next computational decision system at t:

1. To invest in the simulated stock market index if the investor has a regime expectation of $s_{t+1} = 1$.
2. To invest in the U.S. risk-free asset if the investor has a regime expectation of $s_{t+1} = 0$.

Given the use of MS and MS-GARCH models in this computational trading algorithm, we will proceed to describe the simulation process used.

3.2. *Data Description and Markov-Switching Tests*

In order to simulate the benefit of investing with the MS trading algorithm in the stock markets of interest, we will assume (as mentioned previously) that the investor is a USD-based one.

Given the description of the MS-GARCH models used in the investment strategy, we will use the USD historical weekly data of the MSCI [24,59] indexes that we present in Table 1. All the time series start at 6 June 1998 and end on 1 February 2019 (a total of 1079 weekly realizations). As we will mention later, we simulated the investment strategy from 1 January 2000 in order to use the previous weeks of data at t for estimation purposes.

Table 1. The Latin American stock indexes that will be used in simulating the investment strategy.

Refinitiv™ RIC®	Source	Index Name	Ticker in the Paper	Country
.dMIBR00000PUS	Refinitiv™ Eikon™-Xenith™	MSCI Brazil price index (USD)	MSCIBRLUSD	Brazil
.dMICL00000PUS	Refinitiv™ Eikon™-Xenith™	MSCI Chile price index (USD)	MSCICHLUSD	Chile
.dMIMX00000PUS	Refinitiv™ Eikon™-Xenith™	MSCI Mexico price index (USD)	MSCIMEXUSD	Mexico
UST3MT = RR	Refinitiv™ Eikon™-Xenith™	U.S. 3-month treasury bill rate	USTBILL	U.S.

Source: Thomson Reuters [60].

All the time series used in our simulations came from the databases of Refinitiv™ Eikon™-Xenith™ [60]. In order to estimate the MS-GARCH model, we transformed the original price (P_t) time series of the six stock market indexes into a return (r_t) one. We did this by using the continuous time return calculation. Once we have done this, we had six return time series with a total of 1078 weekly returns each (since January of 1998).

The Statistical performance of this full time series is shown in Table 2. Almost all the six indexes had similar minimum weekly returns in a single week and similar mean weekly returns (with the exception of the Chilean market that has the lowest mean value of 0.15%). In terms of standard deviation, the Chilean index could be interpreted as the safest one (with a weekly standard deviation of 4.9%) and the Brazilian and Mexican markets as the riskiest ones.

In the same table, we present the results of the main probability density function (pdf) fiting tests. These include the Gaussian and t-Student Kolmogorov-Smirnov test (a test more focused in the central segment of the pdf), the Anderson-Darling normality test (more focused in the tails), and the skewness and kurtosis levels for the Jarque-Bera one. Given the high kurtosis level, there is no evidence that shows the returns of these three indexes are either gaussian or t-Student. This suggests two possible explanations: (1) we should consider a longer tail pdf such as the Generalized Error Distribution or

(2) there is a presence of potential high kurtosis, given the fact that the time series should be modeled with two or more regimes. In this paper, we test the second explanations. Essentially, the return time series can be modelled with a Gaussian or t-Student two-regime pdf. Thus, we observe a high kurtosis level in the entire time series, but this high kurtosis is due to the presence of two regimes (low and high volatility) in the stochastic process. If we model this stochastic process with such a number of regimes (two), it is possible to identify that the pdf in each regime is gaussian or t-Student. For this reason, we will perform the next Markov-Switching model fit tests with information criterions.

Table 2. Statistical summary of the weekly returns of the three stock indexes studied here (values as percentages).

Index or Asset	Min	5% Quantile	Mean	Standard Deviation	95% Quantile	Max
MSCIBRLUSD	−28.1500	−8.1800	0.2300	5.1900	7.7800	29.2000
MSCICHLUSD	−29.2900	−4.7800	0.1500	3.2900	4.9000	21.0900
MSCIMEXUSD	−26.4200	−6.2000	0.2100	4.0100	5.8900	25.3000
Index or Asset	**K-S Gauss**	**K-S t Student**	**A-D Test**	**Skewness**	**Kurtosis**	**J-B Test**
MSCIBRLUSD	0.7766	0.0000	0.0000	−0.0118	3.8109	0.0000
MSCICHLUSD	1.9346	0.0000	0.0000	−0.6560	8.2142	0.0000
MSCIMEXUSD	0.0445	0.0000	0.0000	0.0578	6.0194	0.0000

Source: own calculations with data of Thomson Reuters [60].

Following the statistical summary in Table 2, we determined, with the Akaike [61] Information Criterion (AIC), if it is appropriate to model, with one or two-regimes, the stochastic processes of each index's return. With the AIC, we also reviewed if it is more appropriate to use a MS-ARCH, a MS-GARCH, or a constant variance MS model (no ARCH and GARCH terms). We also made these comparisons by using either a Gaussian or a t-Student pdf in the estimation process.

We used the AIC instead of other information criterions or other criteria as the Root Mean Squared Error (RMSE) for two reasons.

1. The MSGARCH R package [57] can estimate only a GARCH(1,1) variance process. The maximum lag in the ARCH and GARCH terms are 1. Given this, the use of the Schwarz [62] or the Hannan-Quinn [63] one, are not necessary because we only want to determine the most accurate MS or MS-GARCH model (an issue in which the AIC is better than the other two criterions). This computational restriction sets aside the trade-off between accuracy and parsimony in the model.
2. We prefer to use an information criterion such as the AIC because we want to test the goodness of fit of the estimated model. For this reason, it is preferable to use the AIC criterion that uses the LLF as input. The value of this function, along with the number of parameters and the length of the time series (information set), could give us more accurate fitting values than the sample RMSE. In addition, since we used the Viterbi [56] algorithm, one of the outputs of the estimation process is the LLF (the value to be optimized in the estimation process). Given this, it is natural for us to use the AIC as a fitting parameter than the RMSE.

Departing from this fitting criteria for our analysis, we present the first fitting results of each index in Table 3.

The reader can note that, in all the indexes, the two-regime Markov-Switching are more appropriate to model the stochastic process of each index. From Table 3, it is also noted that the Gaussian MS-GARCH model gives the best fit for the Brazilian and Mexican indexes and that the MS-ARCH (with no GARCH term) gives the best fit for the Chilean case. This test is made on the full time series since 6 June 1998 to 1 February 2019.

Since the previous test applies to the entire time series and, in order to check the robustness of these results, we ran a recursive estimation of each model in each index from 7 January 2000 to

1 February 2019 (996 weeks). For estimation purposes, we used 7 January 2000 as starting date (t = 1), with all the historical time series since January 1998.

Table 3. Summary of fitting the different MS models used as candidates for the stochastic process in each index.

Stochastic Process	MSCIBRLUSD	MSCICHLUSD	MSCIMEXUSD
Single-Gaussian	−3301.55	−4285.52	−3858.55
Single-t Student	−3433.74	−4439.53	−4038.77
MS-Gaussian	−3467.75	−4492.07	−4093.52
MS-t Student	−3468.11	−4488.84	−4098.02
MSARCH-Gaussian	−3459.81	−4494.7929[Best]	−4092.63
MSARCH-t Student	−3461.04	−4487.31	−4093.82
MSGARCH-Gaussian	−3473.4399[Best]	−4493.92	−4108.1490 [Best]
MSGARCH-t Student	−3470.60	−4482.84	−4102.71

Source: own calculations with data from Thomson Reuters [60].

Next, in each simulated date *t*, we added the realization of the corresponding week to the information set. This test was useful to determine if the results of Table 3 hold across the time and will be of use in the investment strategy simulation.

In Table 4, we summarized the observed mean value of the weekly AIC for each index in each simulated scenario. The Gaussian time-fixed variance MS model is the best fit, which is a result that contradicts the results of the full time series analysis of Table 3. This is an interesting result because it gives guidelines to note that the use of Markov-Switching models is preferable in the risk modelling of these three indexes. However, in a week-by-week basis, it is preferable to use an MS constant variance instead of an MS ARCH or GARCH one.

Table 4. Summary of the weekly recursive fitting analysis of the different MS models tested in each index.

Stochastic Process	MSCIBRLUSD	MSCICHLUSD	MSCIMEXUSD
Single-Gaussian	−1706.5086	−2266.615	−1993.9081
Single-t Student	−1767.9515	−2346.8044	−2083.8841
MS-Gaussian	−1783.3242 [Best]	−2372.2979[Best]	−2109.2868[Best]
MS-t Student	−1777.5791	−2366.1851	−2106.7913
MSARCH-Gaussian	−1774.6806	−2368.5946	−2104.4816
MSARCH-t Student	−1768.8776	−2359.9256	−2100.1547
MSGARCH-Gaussian	−1771.4864	−2365.4424	−2106.9333
MSGARCH-t Student	−1765.128	−2355.3302	−2099.0413

Source: own calculations with data from Thomson Reuters [60].

A natural question with this estimation method that could arise is: Why did we not use a rolling time window instead of a recursive (length increasing) one? We did not want to use a rolling time window because the estimation method of the MS and MS-GARCH is a Bayesian one. Essentially, we are inferring the parameters of a hidden Markovian process from the observed data (time series). The estimation process is similar to the Kalman filter, as Hamilton [14,16] observed. Our rationale is: if the information set is bigger (the time series is longer), our estimations of the parameter set (θ) are more accurate. In order to use a time rolling window, we must determine the appropriate window length (number of realizations in the time series) and this question is part of parallel but different research efforts. Given this, we prefer to a use recursive, increasing window of time in order to have a more complete information set in our parameter estimations.

Additionally, we used a recursive and increasing time window due to past distress or high volatility periods, such as the 2000, 2007, or 2011 ones, which could not be in the parameter estimation. This episode is important to keep in all the simulated dates. This, in order to train our algorithm and to properly estimate the smoothed and transition probabilities of each regime. If we use a rolling time window with a given fixed length, it is expected that one or more distress episodes are not taken into account in our estimations. This could lead to poorer distress regime probability forecasts ($\xi_{s=2,t+1}$).

As mentioned in the conclusions, we suggest reviewing the proper time window for estimating MS and MS-GARCH models in further research. This, in order to contribute to the practitioners' literature (risk management) and for potential application of rolling time windows vs. recursive time increasing ones.

Following the results in Table 4, we answered another question for practical applications: if a given risk or portfolio manager would have run a backtest and a model selection process, how many times would each of the tested MS model have been used?

In order to answer this question, we determined, in each week of our simulations, the model with the lowest BIC and make a count of this result. The results are summarized in Table 5.

Table 5. Summary of the frequency of using each simulated model in the 996 weeks of the recursive simulation.

Stochastic Process	MSCIBRLUSD	MSCICHLUSD	MSCIMEXUSD
Single-Gaussian	0	0	0
Single-t Student	256	68	184
MS-Gaussian	487	394	275
MS-t Student	0	0	0
MSARCH-Gaussian	0	511	0
MSARCH-t Student	0	0	0
MSGARCH-Gaussian	253	23	537
MSGARCH-t Student	0	0	0
Total of dates	996	996	996

Source: own calculations with data of Thomson Reuters [60].

The MS-Gaussian with constant variance was the most used model in the Brazilian and Chilean indexes, but, in the Mexican case, was the Gaussian MS-GARCH.

The results of Tables 3–5 suggest that, in all weeks, the investor or risk manager would have used different models through time, but the Gaussian with constant variance MS model is the most appropriate in two of the three indexes.

Even if we found evidence in favor of using Gaussian MS, MS-ARCH, or MS-GARCH models, we have a question to be answered: As noted and for risk management purposes, is it suitable to use Gaussian MS models but would the use of such models lead to the best performance in terms of investment decisions? In order to answer this, we used the six MS models and we simulated the performance that an investor would have had, had the investor used each MS or MS-GARCH model during all the simulated time periods.

In order to simulate the performance that an investor would have had each weak, had the trading algorithm been used described in the previous section, we simulated the performance of a USD 100,000.00 start value portfolio. In this portfolio, there are only two possible assets to select.

1. The 7 January 2000 base 100 value of the simulated country's stock index. A value that we will assume as the theoretical market price of a zero-tracking error Exchange Traded Fund or ETF (risky asset) of such a simulated MSCI index. This will be our risky asset for normal or good performing time periods.
2. The 7 January 2000 base 100 value of a theoretical index calculated with the weekly rate of the 3-month U.S. Treasury-Bill. This will be also considered as the theoretical price of a zero-tracking

error theoretical ETF that pays the Treasury-Bill rate. This will be our risk-free asset for distressful time periods.

In order to simplify our simulations, we did not incorporate the impact of trading costs. Additionally, the FX risk and market price impact are not considered herein.

With these parameters and assumptions, we executed each week the next pseudo-Algorithm 1:

Algorithm 1. The MS-GARCH based trading algorithm's pseudocode

For date 1 to 996:

1. To determine the current balance in the portfolio (cash balance + market value of holdings).
2. To execute the Markov-Switching model analysis in Equation (6) with either GARCH, ARCH, or constant variance (either with Gaussian or t-Student probability density function).
3. To calculate, by using Equation (12), the forecasted smoothed probability $\xi_{s=2,t+1}$ related to being in a distressful or bad-performing regime at $t+1$.
4. If $s_{t+1} = 0$ ($\xi_{s=2,t+1} > 0.5$), then:
    a. To invest in the risk-free asset (Treasury-Bill ETF)

Else:

    b. To invest in the risky asset (The simulated index ETF)

5. To price the value of the portfolio with a market-to-market (with closing market prices at t) procedure.

End

We present the results of the simulations made in each market in the next section.

4. Simulation Results Discussion

For exposition purposes, we want to discuss the performance of a buy-and-hold strategy in each of the simulated MSCI indexes. This along with the performance of a theoretical portfolio that would have paid the three-month Treasury bill (T-Bill). For this purpose, we present the summary in Table 6.

Table 6. Summary of the performance of passive or buy-and-hold investment strategy in each index.

Index or Asset	Accumulated Return	Mean Return	Return Standard Deviation	Max Drawdown
MSCIBRLUSD	172.1344 [9.4753]	0.1006	5.0575	−33.0558
MSCICHLUSD	145.205 [7.9929]	0.0901	3.2291	−34.662
MSCIMEXUSD	162.0699 [8.9213]	0.0968	3.8617	−30.6773
USTBILL	38.3854 [2.113]	0.0374	0.0393	—
Index or Asset	**CVaR (95%)**	**CVaR (98%)**	**Sharpe Ratio**	
MSCIBRLUSD	−12.4852	−16.7693	0.0042	
MSCICHLUSD	−7.5621	−10.6382	0.0076	
MSCIMEXUSD	−9.132	−12.4010	0.0128	

Source: own calculations with data from Thomson Reuters [60].

The U.S. treasury-bill (USTBILL) had the lowest accumulated return with an accumulated return of 38.38% in the 996 weeks of simulation. This leads to a yearly rate (annually based values in brackets from now on) of 2.11%. Following this asset, the Chilean stocks are the ones with the lowest performance at 145.20% (7.99%), which means the Brazilian are the ones with the highest return paid (172.13% or 9.47% yearly).

The weekly max drawdown values are also of interest (the worst negative return in a single week) and fluctuate in a narrow range of 30.67% to 34.66%. We also present the weekly CVaR values at 95%

and 98% of confidence (as potential loss measures) and the annual Sharpe ratio [6] as mean-variance efficiency metric. We present the Conditional Value at Risk [64] (CVaR) figures in order to highlight how the risk profile improves with the active trading activities. In order to estimate the CVaR figures, given the $\alpha = 1 -$ confidence level (95% or 98%), we used the expected shortfall method. In such methods, q is the quantile ($q \in [q, 1]$) of maximum loss, given the confidence level and the historical percentage price variation of the simulated portfolios.

$$CVaR = \frac{1}{1-\alpha} \int_{q=\alpha}^{1} \Phi(q, \mu, \sigma) d\Phi \tag{17}$$

The idea is to compare this buy-and-hold strategy CVaR values with the ones observed in the three simulated portfolios. As the reader can observe, we found improvements in the risk exposure levels such as CVAR (at 95% and 98%) and the Max Drawdown (the lowest observed percentage variation).

Related to the Sharpe ratio values in Table 6, we can observe that only the Mexican index pays the best risk-return relationship. Therefore, the passive investment in this index pays a 0.0128% of extra weekly return (above the USTBILL rate of 0.0374%) given an extra 1% risk exposure.

Table 6 shows that all the indexes paid a positive risk premium to the USD-based investor. Our position, as previously stated in the introduction, is that the investor can use MS, MS-ARCH, or MS-GARCH models to perform active trading activities by executing the investment strategy of Section 3.2 each week.

By following the alphabetical order followed in the previous tables, we present the results of the simulations applied in the Brazilian stock market in Table 7. All but the t-Student MS and MS-GARCH models lead to an outperformance (alpha generation) against the passive or buy-and-hold strategy depicted in Table 6. From all the simulated models, the Gaussian MS-GARCH model is the best to use in order to perform active investment activities in the Brazilian market.

Table 7. Performance summary of the Markov-Switching investment strategy applied in the Brazilian stock market (from a U.S. dollar-based investor perspective).

Markov-Switching Model Used	Accumulated Return	Mean Return	Return Standard Deviation	Max Drawdown
MS-Gaussian	299.8539 [16.5057]	0.1393	3.9772	−15.4177
MS-tStudent	161.3091 [8.8794]	0.0965	4.028	−19.8727
MSARCH-Gaussian	216.5051 [11.9177]	0.1158	3.9228	−15.4154
MSARCH-tStudent	289.6302 [15.9429]	0.1367	4.0232	−19.8827
MSGARCH-Gaussian	441.0209 [24.2764]	0.1697	3.9149	−15.4184
MSGARCH-tStudent	59.6986 [3.2862]	0.0470	3.1778	−19.8978
Markov-Switching Model Used	**CVaR (95%)**	**CVaR (98%)**	**Sharpe Ratio**	**Mean Risky Exposure**
MS-Gaussian	−9.3428	−11.3154	0.0113	97.36%
MS-tStudent	−9.7378	−12.1689	−0.0006	96.53%
MSARCH-Gaussian	−9.2274	−11.2619	0.0095	95.76%
MSARCH-tStudent	−9.5271	−11.7792	0.0198	96.14%
MSGARCH-Gaussian	−9.1400	−11.2266	0.0244	93.75%
MSGARCH-tStudent	−8.4226	−11.3701	−0.0005	89.09%

Source: own calculations with data from Thomson Reuters [60].

Another interesting result of Table 7 is the value of the max drawdown that these actively managed portfolios had. The worst return in a single week is the −19.88% observed in the portfolio that used the t-Student MS-ARCH model. This result is lower than the −33.0558% of Table 6 for the Brazilian index (buy-and-hold strategy). Similar improvements are noted in the CVaR and Sharpe ratio values. With the exception of the t-Student MS and MS-GARCH models, the Sharpe ratios are higher than the passive strategy (index).

The last performance results can be observed in the simulated performance (base 100 value at the beginning of the simulations) of the six portfolios in the Brazilian market. This is illustrated in Figure 1. The Gaussian time-fixed variance (MS-Gaussian) and the Gaussian MS-GARCH show a better performance than the naïve buy-and-hold strategy (blue area). In the specific case of the 2007–2009 financial crisis episode, the reader can note that nearly all the simulated portfolios invested in the risk-free asset. The reader can note this because the performance of the simulated portfolio is practically flat in these periods. Therefore, the use of MS and MS-GARCH models leads to a better performance than a passive buy-and-hold strategy. Additionally, as noted in other time periods, the Gaussian MS-GARCH model proved to be more effective to forecast the distress time periods, which allows us to have better performance results than the other five portfolios.

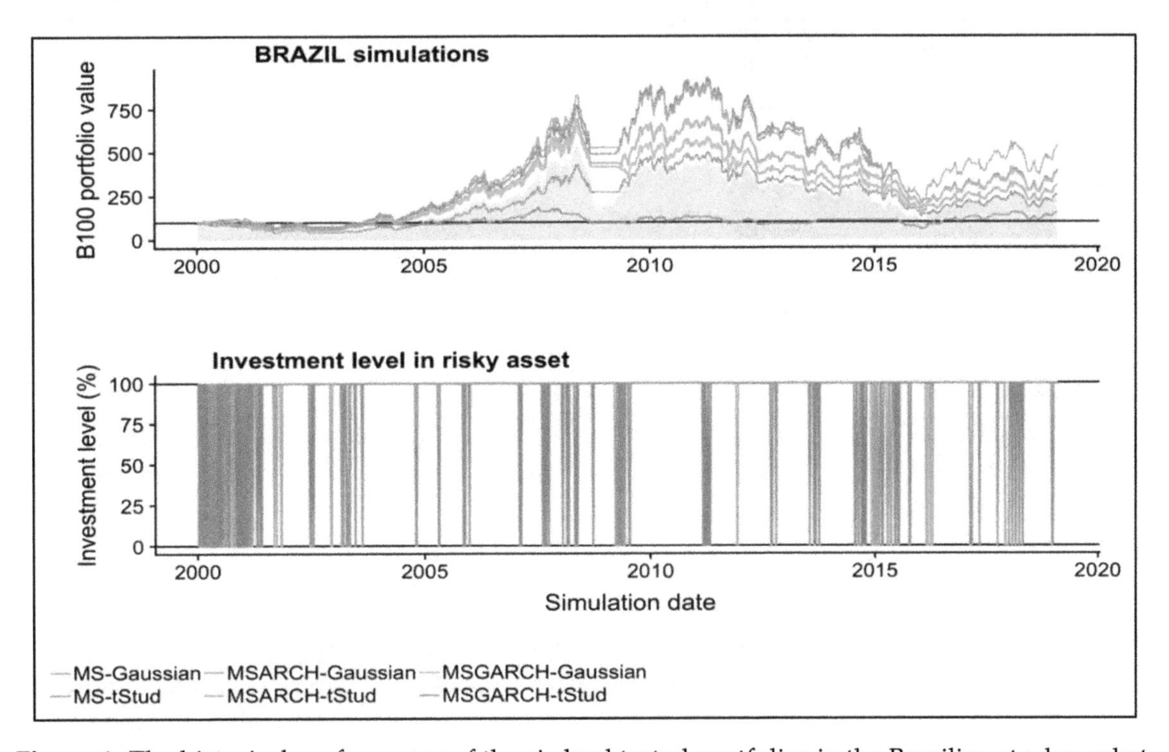

Figure 1. The historical performance of the six backtested portfolios in the Brazilian stock market.

In order to give support for the results we present, in the last column of Table 7 and in the bottom panel of Figure 1, the investment level in the stock index. As noted in that Table, the average observed investment level (mean risky exposure) in the Brazilian stock index are the third and second lowest levels. The average investment level in the Gaussian MS and MS-GARCH models are 97.36% and 93.75%, respectively. This summary of the investment level in the stock index gives an explanation about the performance improvements achieved with other scenarios. More specifically, the MS-GARCH (the best performing scenario) had significantly lower investment levels in the stock index during the distressing time periods. In addition, the use of Gaussian MS-GARCH models led to a better performance because the forecasted probability of the second regime ($\xi_{s=2,t}$) leads to more accurate sell decisions during small distress periods. At the end of 2014 and the beginning of 2015, (blue line) an appropriate sell sign is noted, which, during those weeks, allowed the simulated trader to reduce the loss value. In addition, in 2016, the buy decision made in the November 2016 weeks was earlier than the one observed in the other scenarios. Given this, the bottom panel of Figure 1 and the last column of Table 7 give proof about the outperformance of using Gaussian MS-GARCH models.

In a similar fashion to the Brazilian case, we present the results of the simulations in the Chilean market in Table 8. Similarly, for the Brazilian market, the t-Student MS and MS-ARCH models are the worst performers and lead to poorer performance than a passive or a buy-and-hold investment strategy. From all the six simulated portfolios, the one with a Gaussian MS-ARCH model is the one that showed the highest accumulated return (351.4477% or 19.3457% yearly). Furthermore, the improvements in the max drawdown, CVaR, and Sharpe ratio values against the observed ones in Table 6 are of interest.

Table 8. Performance summary of the Markov-Switching investment strategy applied in the Chilean stock market (from a U.S. dollar-based investor perspective).

Markov-Switching Model Used	Accumulated Return	Mean Return	Return Standard Deviation	Max Drawdown
MS-Gaussian	304.3449 [16.7529]	0.1404	2.35	−6.8837
MS-tStud	142.8021 [7.8607]	0.0892	2.3695	−10.0516
MSARCH-Gaussian	351.4477 [19.3457]	0.1515	2.2433	−6.8826
MSARCH-tStud	131.6221 [7.2453]	0.0844	2.344	−10.0567
MSGARCH-Gaussian	240.7838 [13.2542]	0.1232	2.2545	−14.8974
MSGARCH-tStud	296.9058 [16.3434]	0.1385	2.2653	−10.0541
Markov-Switching Model Used	**CVaR (95%)**	**CVaR (98%)**	**Sharpe Ratio**	**Mean Risky Exposure**
MS-Gaussian	−4.9655	−5.8791	0.013	96.97%
MS-tStud	−5.3301	−6.611	0.0085	96.96%
MSARCH-Gaussian	−4.8441	−5.8324	0.0314	95.96%
MSARCH-tStud	−5.3446	−6.6101	0.0099	95.74%
MSGARCH-Gaussian	−5.2099	−6.6896	0.019	89.12%
MSGARCH-tStud	−5.0197	−6.2081	0.0218	92.55%

Source: own calculations with data from Thomson Reuters [60].

In Figure 2, we present the historical performance of the six simulated portfolios of the Chilean market. The over performance in the active trading strategy is observable, especially in the Gaussian MS-ARCH and t-Student MS-GARCH cases. Practically all the MS and MS-GARCH models were very useful to forecast the 2007–2009 distress time period and, in October 2008, practically all were useful to forecast a rebalancing sign. This takes place in order to invest in the risk-free asset during this time period and to avoid the value drawdown that the buy-and-hold strategy had in this time period.

Even if this good performance is held for all the simulated portfolios, the use of Gaussian MS-ARCH models leads to a more sensitive forecast of the distress time period in other time intervals. This last result also led to a better portfolio performance, as noted in the March 2011 period in which this model forecasted the end of the distress time period. This allows the simulated trader to invest in the Chilean stock market during the corresponding upward movement. More specifically, as noted in the bottom panel of Figure 2 for this time period, the use of the Gaussian MS-ARCH model led to an earlier buy decision. This trade enhanced the performance of the simulatedportfolio. Despite this, no other new differences were observed in the six simulated portfolios. This is noted by the fact that the mean investment level in five of the six portfolios (including the best performer) are similar. This led us to conclude that, even if the use of the Gaussian MS-ARCH model leads to more precise investment decisions, this precision is less observable if we compared it with the Brazilian case.

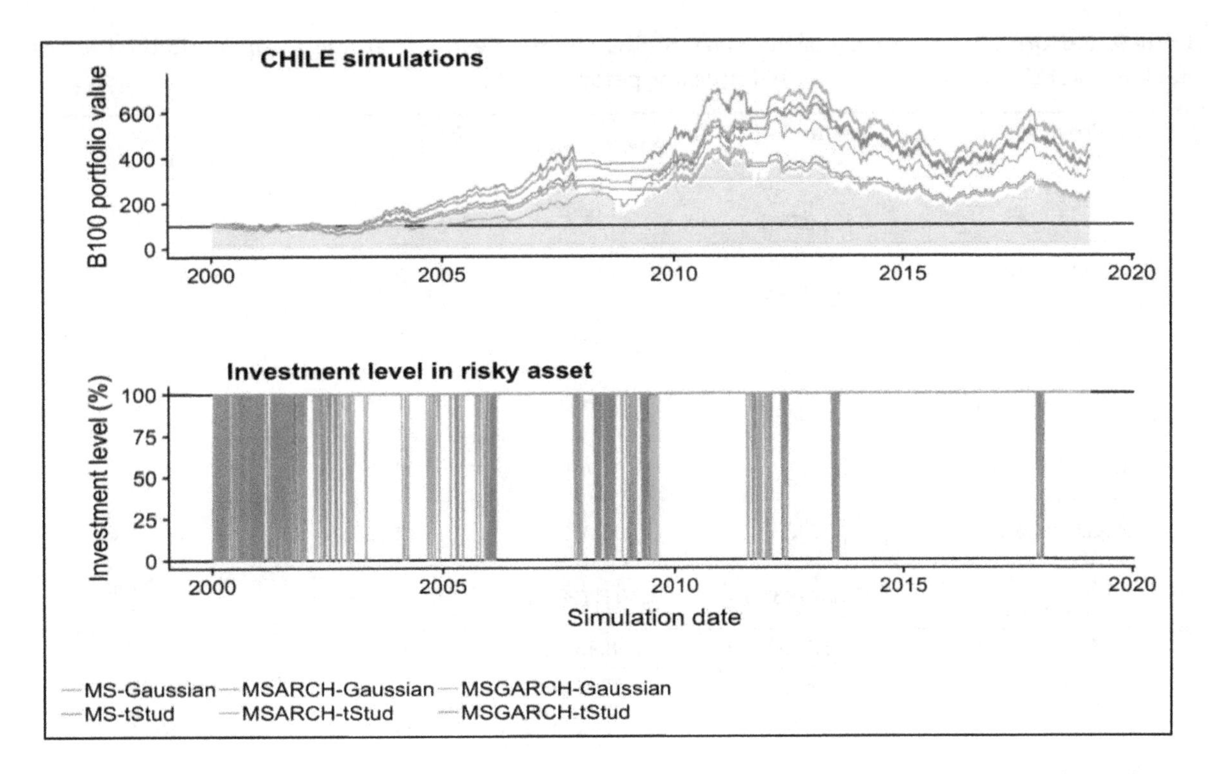

Figure 2. The historical performance of the six backtested portfolios in the Chilean stock market.

Lastly, we present the performance results of the simulations observed in the Mexican stock market in Table 9. In this last case, the worst performers are the portfolios that used the Gaussian and t-Student MS-ARCH or MS-GARCH models. Only the portfolios with the time-fixed variance generated alpha and, from these two, the Gaussian model is the one with the highest accumulated return (259.3627% or 14.2768% yearly). Furthermore, only these two models (the ones with constant variance) show improvements in the Sharpe ratio when compared with the passive investment strategy.

This last result can be observed in detail in the historical performance of Figure 3. As noted at the beginning of the simulated period, the two MS portfolios invested in the risk-free asset and they remained invested in it during the 2000 technology stock crisis. Given their less sensitive characteristic to switch the forecast of the smoothed probability of the distress regime, these two portfolios remained invested in the risk-free asset. This is a situation that explains why these two portfolios have a positive performance in the first part of the simulation chart.

In the 2007–2009 period, these two portfolios invested in the risk-free asset, but the Gaussian pdf MS one remained invested more time in it. This led to a better portfolio value during this period, as noted in the almost flat performance of the Gaussian MS portfolio (gold line).

Lastly, the Gaussian and t-Student MS portfolios had a similar performance in the rest of the simulation, but the Gaussian case led to the more stable performance of both, which suggests that it is preferable to use the Gaussian MS model for active trading in the Mexican stock market.

By following the review of the previous indexes, we found evidence that the better investment timing in the Mexican cases is similar to the Chilean one. That is, the Gaussian MS portfolio is the best performer because, in the previously mentioned 2008 period, the buy trading signal was more accurate. This led to a different performance in the recovering period of 2008–2009. For the specific case of the Mexican index, investing in Gaussian and t-Student MS models led to a more accurate buy signal. This last signal led to a better performance in the 2007–2008 distress time period and to a more accurate buy signal in the March 2001 period. These two accurate buy signals led to better performance and to a higher average investment level in the risky asset or index. As the case of the two previous indexes, the use of MS or MS-GARCH models led to more precise trading signs and to better performance.

Table 9. Performance summary of the Markov-Switching investment strategy applied in the Mexican stock market (from a U.S. dollar-based investor perspective).

Markov-Switching Model Used	Accumulated Return	Mean Return	Return Standard Deviation	Max Drawdown
MS-Gaussian	259.3627 [14.2768]	0.1286	2.9764	−17.4144
MS-t Stud	255.8699 [14.0846]	0.1276	2.8257	−10.1805
MSARCH-Gaussian	75.7788 [4.1713]	0.0567	2.7333	−10.1763
MSARCH-t Stud	95.3431 [5.2482]	0.0673	2.8529	−17.4112
MSGARCH-Gaussian	154.7175 [8.5166]	0.094	2.8376	−10.1853
MSGARCH-t Stud	37.3705 [2.0571]	0.0319	2.8956	−14.0064
Markov-Switching Model Used	**CVaR (95%)**	**CVaR (98%)**	**Sharpe Ratio**	**Mean Risky Exposure**
MS-Gaussian	−6.9409	−8.3678	0.0147	96.76%
MS-t Stud	−6.6954	−7.9479	0.0223	96.36%
MSARCH-Gaussian	−6.6761	−7.8676	−0.0048	94.93%
MSARCH-t Stud	−7.1122	−8.8926	0.0034	97.13%
MSGARCH-Gaussian	−6.6997	−7.8395	0.011	91.72%
MSGARCH-t Stud	−7.182	−8.8326	−0.0054	93.51%

Source: own calculations with data from Thomson Reuters [60].

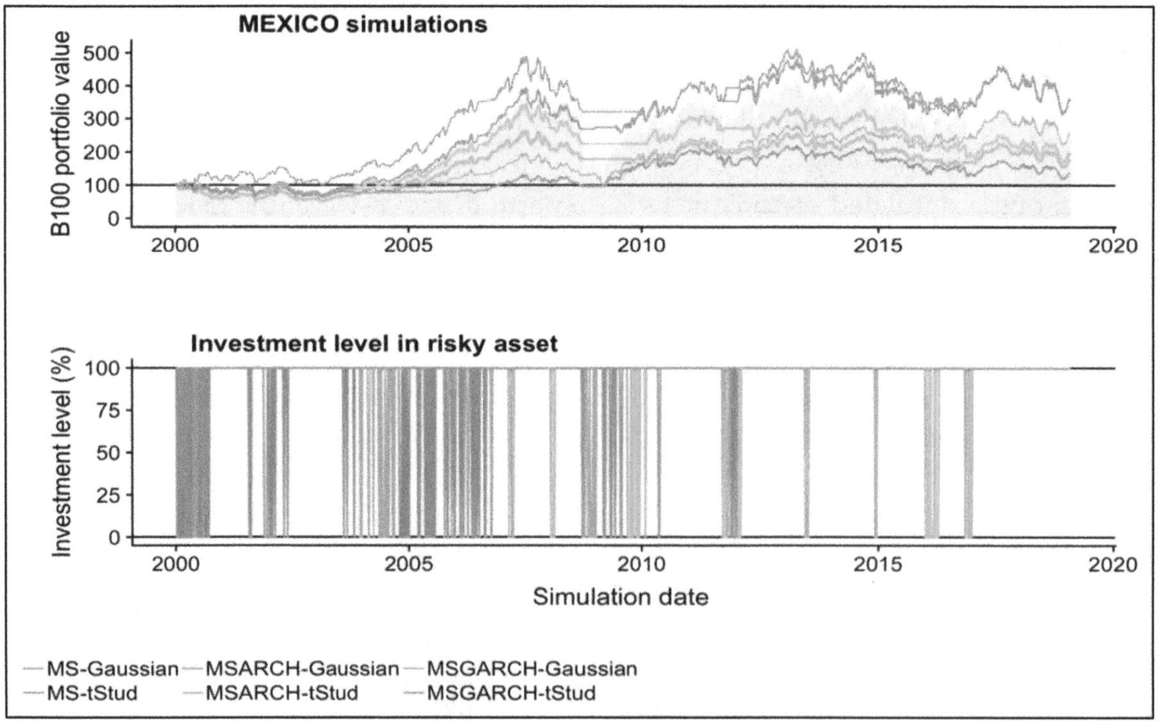

MEXICO simulations

B100 portfolio value: 0, 100, 200, 300, 400, 500; 2000, 2005, 2010, 2015, 2020

Investment level in risky asset

Investment level (%): 0, 25, 50, 75, 100; 2000, 2005, 2010, 2015, 2020

Simulation date

MS-Gaussian MSARCH-Gaussian MSGARCH-Gaussian
MS-tStud MSARCH-tStud MSGARCH-tStud

Figure 3. The historical performance of the six backtested portfolios in the Mexican stock market.

In Table 10, we summarize the results of the full time series, recursive test, and the ones of the active investment strategy analysis. As noted for the Brazilian and Chilean markets, the best model to use for risk measurement is the Gaussian with a constant variance MS model because that is the best (lowest BIC) model in most weeks. Despite this result, the Gaussian MS-GARCH and the t-Student

MS-ARCH models are the best models (respectively) if the investor wants to perform active investment activities in these markets. Lastly, for the Mexican case, there are no conclusive results between the use of the Gaussian MS or the Gaussian MS-GARCH model for risk measurement purposes. We suggest the Gaussian case, given its marginally higher accumulated return against the t-Student one.

Table 10. Summary of the full time series as well as recursive and active investment analysis.

Country	Best Model in Full TS Analysis	Best Model in Recursive Analysis	Most Used Model in the 996 Weeks	Best Model for Active Investment
Brazil	MS-GARCH Gaussian	MS-Gaussian	MS-Gaussian	MS-GARCH Gaussian
Chile	MS-ARCH Gaussian	MS-Gaussian	MS-Gaussian	MS-ARCH t-Student
Mexico	MS-GARCH Gaussian	MS-Gaussian	MS-GARCH Gaussian	MS-Gaussian

Source: own calculations with data from Thomson Reuters [60].

Our results are in line with the previous papers in the literature review by showing proofs that it is better to use MS models for active trading in stocks. More specifically, our contributions show that the use of these models and the use of Gaussian MS-GARCH or t-Student MS-ARCH models are preferable in Brazil and Chile. For the Mexican, the use of a Gaussian time-fixed variance MS model is suggested. Therefore, these results give proof that the use of time-varying variance Markov-switching stochastic processes are preferable for the time series characterization of two of these three Latin-American stock indexes. In addition, our results give proof that the use of either MS or MS-GARCH models is useful for active trading decisions in these three stock markets.

5. Conclusions

Markov-Switching (MS) models, as proposed by Hamilton [14–16], are very useful to determine if a time series' stochastic process has several $s \geq 2$ regimes. This allows us to infer an s number of location (mean) and scale (standard deviation) parameters. Among the most useful applications of these models in the financial industry, we mention the measurement of the risk level in a given investment with two or more possible regimes or states of nature. For the purposes of this paper, the regime $s = 1$ corresponds to a normal or low volatility state of nature at t and $s = 2$ for a distress or high volatility one. This leads to a very useful risk measure, along with the probability $\xi_{s=i,t}$ of being in a given s regime at t or $t+1$ ($\xi_{s=i,t+1}$).

Despite this practical use of the MS models in risk management, little has been written about the use of MS models in stock trading decisions. By knowing the probability $\xi_{s=i,t+1}$ of being in a given regime at $t+1$, an investor can determine whether to invest in a risky (risk-free) asset if the probability of being in a normal (distress) regime is high. This rationale departs from the one proposed by Brooks and Persand [18] and the extensions of Engle, Wahl, and Zagst [21] and De la Torre, Galeana, and Álvarez-García [22]. These previous works test the benefit of investing in a given developed financial market with MS models.

Given the scant literature review of MS models for stock trading, there is no work that test their use in the three most traded Latin American markets from a U.S. dollar-based investor perspective. Additionally, none of the related previous works test the benefit of using a MS model with ARCH (MS-ARCH) or GARCH (MS-GARCH) variances in trading decisions.

Departing from these two issues, the paper tests the use of MS, MS-ARCH, or MS-GARCH models in the Brazilian, Chilean, and Mexican stock markets. With 996 weekly simulations from January 2000 to February 2019 we fitted MS, MS-ARCH, and MS-GARCH models (with Gaussian and t-Student probability density functions) and we ran the next trading rule.

1. To invest in the stock market index if the investor expects to be in the low volatility ($s = 1$) regime at $t + 1$ or
2. To invest in the U.S. risk-free asset if the investor expects a high volatility ($s = 2$) one.

Our results suggest that the Gaussian MS-GARCH and the t-Student MS-GARCH are the best models for investment decisions in the Brazilian and Chilean stock markets, respectively. For the Mexican case, it is preferable to use a Gaussian MS (with constant variance) model.

Despite the beneficial use of these models for investing in these countries, the conclusions in our results change the risk measurement purposes. This means that, for risk management, it is better to use a Gaussian MS (constant variance) model in the Brazilian and Chilean markets. For the Mexican case, it is preferable to use the Gaussian MS-GARCH model.

Our results could be of theoretical use in order to understand the practical use of MS and MS-GARCH models in forecasting during distressful or highly volatile periods. In addition, this can help us to understand and to prove their application in trading decisions. In practical terms, our results contribute to the literature of using MS and MS-GARCH models in trading. This along with the benefits of their use in the three main Latin-American stock markets from a U.S. dollar-based investor perspective.

As extensions to this paper or research guidelines for further research, we suggest testing the use of other probability functions such as the symmetric and asymmetric Gaussian t-Student or GED regime likelihood functions. This test could be made with homogeneous (the same likelihood function in each regime) or homogeneous probabilities in each regime. As another potential review and extension to our results, we propose the use of MS models with asymmetric GARCH processes (also homogeneous and heterogeneous in each regime) along with fractional integration (long memory modelling).

In addition, the use of MS models with more than two regimes in the stochastic process, along with the test in other stock markets or other type of securities, could be of potential interest. We also believe in theoretical interest for Bayesian Statistics to determine what is best to use: either a rolling time window or a recursive (increasing time series) one. This issue is also of practical interest because the potential use of time rolling windows could enhance the informatic efficiency of the MS-GARCH model estimation method. Despite this, this informatic efficiency could come with a cost in terms of accuracy cost in the estimated parameters. Furthermore, the extension of the current MS trading algorithm with a portfolio of these three indexes (or more) is part of the research agenda suggested. In order to manage a portfolio with several stocks or indexes in a regime-switching context, it is needed to make research efforts in the application of MS correlation models [65,66] in trading algorithms such as ours. Lastly, none of the estimations that we made herein used GARCH models that incorporate the influence of external factors (economic or financials). Models like these ones are studied in Billah, Hyndman, and Koehler [67], Conrad and Loch [68], Amendola et al. [69], and Amendola, Candila, and Gallo [70]. We did not consider this type of MS-GARCH models due to the determination of the proper factors that influence the performance of the three stock indexes. Lastly, given these factors, the determination of the proper MS-GARCH model with these could be another extension for our research efforts.

Author Contributions: Conceptualization, Methodology, Formal Analysis, Investigation, Writing-Original Draft Preparation and Writing-Review & Editing, O.V.D.l.T.-T., E.G.-F., J.Á.-G. All authors have read and agreed to the published version of the manuscript.

References

Markowitz, H. Portfolio selection. *J. Financ.* 1952, *7*, 77–91. [CrossRef]

Markowitz, H. *Portfolio Selection. Efficient Diversification of Investments*; Yale University Press: New Haven, CT, USA, 1959.

Alexander, C. Principal component models for generating large GARCH covariance matrices. *Econ. Notes 2002, 31*, 337–359. [CrossRef]

Ang, A.; Bekaert, G. International Asset Allocation With Regime Shifts. *Rev. Financ. Stud. 2002, 15*, 1137–1187. [CrossRef]

Sharpe, W. A simplified model for portfolio analysis. *Manag. Sci. 1963, 9*, 277–293. [CrossRef]

Sharpe, W. Capital asset prices: A theory of market equilibrium under conditions of risk. *J. Financ. 1964, XIX*, 425–442.

Engle, R. Autoregressive Conditional Heteroscedasticity with estimates of the variance of United Kingdom inflation. *Econometrica 1982, 50*, 987–1007. [CrossRef]

Bollerslev, T. A Conditionally Heteroskedastic time series model for speculative prices and rates of return. *Rev. Econ. Stat. 1987, 69*, 542–547. [CrossRef]

Dueker, M. Markov Switching in GARCH Processes and Mean- Reverting Stock-Market Volatility. *J. Bus. Econ. Stat. 1997, 15*, 26–34.

Lamoureux, C.G.; Lastrapes, W.D. Persistence in Variance, Structural Change, and the GARCH Model. *J. Bus. Econ. Stat. 1990, 8*, 225–234. [CrossRef]

Canarella, G.; Pollard, S.K. A switching ARCH (SWARCH) model of stock market volatility: Some evidence from Latin America. *Int. Rev. Econ. 2007, 54*, 445–462. [CrossRef]

Klaassen, F. Improving GARCH volatility forecasts with regime-switching GARCH. In *Advances in Markov-Switching Models*; Physica-Verlag HD: Heidelberg, Germany, 2002; pp. 223–254.

Haas, M.; Mittnik, S.; Paolella, M.S. A New Approach to Markov-Switching GARCH Models. *J. Financ. Econom. 2004, 2*, 493–530. [CrossRef]

Hamilton, J.D. A New Approach to the Economic Analysis of Nonstationary Time Series and the Business Cycle. *Econometrica 1989, 57*, 357–384. [CrossRef]

Hamilton, J.D. Analysis of time series subject to changes in regime. *J. Econom. 1990, 45*, 39–70. [CrossRef]

Hamilton, J.D. *Time Series Analysis*; Princeton University Press: Princeton, NJ, USA, 1994.

Hamilton, D.; Susmel, R. Autorregresive conditional heteroskedasticity and changes in regime. *J. Econom. 1994, 64*, 307–333. [CrossRef]

Brooks, C.; Persand, G. The trading profitability of forecasts of the gilt–equity yield ratio. *Int. J. Forecast.2001, 17*, 11–29. [CrossRef]

Kritzman, M.; Page, S.; Turkington, D. Regime Shifts: Implications for Dynamic Strategies. *Financ. Anal. J.2012, 68*, 22–39. [CrossRef]

Hauptmann, J.; Hoppenkamps, A.; Min, A.; Ramsauer, F.; Zagst, R. Forecasting market turbulence using regime-switching models. *Financ. Mark. Portf. Manag. 2014, 28*, 139–164. [CrossRef]

Engel, J.; Wahl, M.; Zagst, R. Forecasting turbulence in the Asian and European stock market using regime-switching models. *Quant. Financ. Econ. 2018, 2*, 388–406. [CrossRef]

De la Torre, O.; Galeana-Figueroa, E.; Álvarez-García, J. Using Markov-Switching models in Italian, British, U.S. and Mexican equity portfolios: A performance test. *Electron. J. Appl. Stat. Anal. 2018, 11*, 489–505. [CrossRef]

World Federation of Exchanges Statistics—The World Federation of Exchanges. Available online: https: //www.world-exchanges.org/our-work/statistics (accessed on 15 February 2019).

MSCI Inc. MSCI Global Investable Market Indexes Methodology. Available online: http://www.msci.com/ eqb/methodology/meth_docs/MSCI_Jan2015_GIMIMethodology_vf.pdf (accessed on 2 May 2018).

Ang, A.; Bekaert, G. Regime Switches in Interest Rates. *J. Bus. Econ. Stat. 2002, 20*, 163–182. [CrossRef]

Ang, A.; Bekaert, G. Short rate nonlinearities and regime switches. *J. Econ. Dyn. Control 2002, 26*, 1243–Brooks, C.; Persand, G. The trading profitability of forecasts of the gilt–equity yield ratio. *Int. J. Forecast. 2001, 17*, 11–29. [CrossRef]

Klein, A.C. Time-variations in herding behavior: Evidence from a Markov switching SUR model. *J. Int. Financ. Mark. Inst. Money 2013, 26*, 291–304. [CrossRef]

Areal, N.; Cortez, M.C.; Silva, F. The conditional performance of US mutual funds over different market regimes: Do different types of ethical screens matter? *Financ. Mark. Portf. Manag. 2013, 27*, 397–429. [CrossRef]

Zheng, T.; Zuo, H. Reexamining the time-varying volatility spillover effects: A Markov switching causality approach. *N. Am. J. Econ. Financ. 2013, 26*, 643–662. [CrossRef]

Alexander, C.; Kaeck, A. Regime dependent determinants of credit default swap spreads. *J. Bank. Financ.2007*, 1008–1021. [CrossRef]

Castellano, R.; Scaccia, L. Can CDS indexes signal future turmoils in the stock market? A Markov switching perspective. *CEJOR 2014, 22*, 285–305. [CrossRef]

Ma, J.; Deng, X.; Ho, K.-C.; Tsai, S.-B. Regime-Switching Determinants for Spreads of Emerging Markets Sovereign Credit Default Swaps. *Sustainability 2018, 10*, 1–17. [CrossRef]

Piger, J.; Max, J.; Chauvet, M. Smoothed U.S. Recession Probabilities [RECPROUSM156N]. Available online: https://fred.stlouisfed.org/series/RECPROUSM156N (accessed on 22 October 2019).

Chauvet, M. An econometric characterization of business cycle dynamics with factor structure and regime switching. *Int. Econ. Rev. 2000, 10,* 127–142. [CrossRef]

Sottile, P. On the political determinants of sovereign risk: Evidence from a Markov-switching vector autoregressive model for Argentina. *Emerg. Mark. Rev. 2013,* 160–185. [CrossRef]

Riedel, C.; Thuraisamy, K.S.; Wagner, N. Credit cycle dependent spread determinants in emerging sovereign debt markets. *Emerg. Mark. Rev. 2013, 17,* 209–223. [CrossRef]

Zhao, H. Dynamic relationship between exchange rate and stock price: Evidence from China. *Res. Int. Bus. Financ. 2010, 24,* 103–112. [CrossRef]

Walid, C.; Chaker, A.; Masood, O.; Fry, J. Stock market volatility and exchange rates in emerging countries: A Markov-state switching approach. *Emerg. Mark. Rev. 2011, 12,* 272–292. [CrossRef]

Walid, C.; Duc Khuong, D. Exchange rate movements and stock market returns in a regime-switching environment: Evidence for BRICS countries. *Res. Int. Bus. Financ. 2014,* 46–56. [CrossRef]

Rotta, P.N.; Valls Pereira, P.L. Analysis of contagion from the dynamic conditional correlation model with Markov Regime switching. *Appl. Econ. 2016, 48,* 2367–2382. [CrossRef]

Mouratidis, K.; Kenourgios, D.; Samitas, A.; Vougas, D. Evaluating currency crises: A multivariate markov regime switching approach. *Manchester Sch. 2013, 81,* 33–57. [CrossRef]

Miles, W.; Vijverberg, C.-P. Formal targets, central bank independence and inflation dynamics in the UK: A Markov-Switching approach. *J. Macroecon. 2011, 33,* 644–655. [CrossRef]

Lopes, J.M.; Nunes, L.C. A Markov regime switching model of crises and contagion: The case of the Iberian countries in the EMS. *J. Macroecon. 2012, 34,* 1141–1153. [CrossRef]

Kanas, A. Regime linkages between the Mexican currency market and emerging equity markets. *Econ. Model. 2005, 22,* 109–125. [CrossRef]

Alvarez-Plata, P.; Schrooten, M. The Argentinean currency crisis: A Markov-switching model estimation. *Dev. Econ. 2006, 44,* 79–91. [CrossRef]

Parikakis, G.S.; Merika, A. Evaluating volatility dynamics and the forecasting ability of Markov switching models. *J. Forecast. 2009, 28,* 736–744. [CrossRef]

Girdzijauskas, S.; Štreimikiene˙ , D.; Cˇ epinskis, J.; Moskaliova, V.; Jurkonyte˙ , E.; Mackevicˇius, R. Formation of Economic bubles: Cuases and possible interventions. *Technol. Econ. Dev. Econ. 2009, 15,* 267–280. [CrossRef]

Dubinskas, P.; Stunguriene˙ , S. Alterations in the financial markets of the baltic countries and Russia in the period of Economic cownturn. *Technol. Econ. Dev. Econ. 2010, 16,* 502–515. [CrossRef]

Kutty, G. The Relationship Between Exchange Rates and Stock Prices: The Case of Mexico. *N. Am. J. Financ. Bank. Res. 2010, 4,* 1–12.

Ahmed, R.R.; Vveinhardt, J.; Štreimikiene, D.; Ghauri, S.P.; Ashraf, M. Stock returns, volatility and mean reversion in Emerging and Developed financial markets. *Technol. Econ. Dev. Econ. 2018, 24,* 1149–1177. [CrossRef]

Dufrénot, G.; Mignon, V.; Péguin-Feissolle, A. The effects of the subprime crisis on the Latin American financial markets: An empirical assessment. *Econ. Model. 2011, 28,* 2342–2357. [CrossRef]

Sosa, M.; Ortiz, E.; Cabello, A. Dynamic Linkages between Stock Market and Exchange Rate in mila Countries: A Markov Regime Switching Approach (2003-2016). *Análisis Económico 2018, 33,* 57–74. [CrossRef]

Cabrera, G.; Coronado, S.; Rojas, O.; Venegas-Martínez, F. Synchronization and Changes in Volatilities in the Latin American'S Stock Exchange Markets. *Int. J. Pure Appl. Math. 2017, 114.* [CrossRef]

De la Torre, O.; Álvarez-García, J.; Galeana-Figueroa, E. A Comparative Performance Review of the Venezuelan, Latin-American and Emerging Markets Stock Indexes with the North-American Ones Using a Gaussian Two-Regime Markov-Switching Model. *Espacios 2018, 39,* 1–10.

Baum, L.E.; Petrie, T.; Soules, G.; Weiss, N. A maximizaiton thecnique occurring in the Statistical analysis of probabilistic functions of Markov chains. *Ann. Appl. Stat. 1970, 41,* 164–171.

Viterbi, A. Error bounds for convolutional codes and an asymptotically optimum decoding algorithm. *IEEE Trans. Inf. Theory 1967, 13,* 260–269. [CrossRef]

Ardia, D.; Bluteau, K.; Boudt, K.; Trottier, D. Markov–Switching GARCH Models in R: The MSGARCH Package. *J. Stat. Softw. 2019, 91,* 38. [CrossRef]

Kim, C.-J. Dynamic linear models with Markov-switching. *J. Econom. 1994, 60,* 1–22. [CrossRef]

MSCI Inc. End of Day Index Data Search—MSCI. Available online: https://www.msci.com/end-of-day-data- search (accessed on 2 April 2019).

Refinitiv Refinitiv Eikon. Available online: https://eikon.thomsonreuters.com/index.html (accessed on 3 June 2019).

Akaike, H. A new look at the statistical model identification. *IEEE Trans. Automat. Contr. 1974, 19,* 716–723. [CrossRef]

Schwarz, G. Estimating the dimension of a model. *Ann. Stat. 1978*, *6*, 461–464. [CrossRef]

Hannan, E.J.; Quinn, B.G. The Determination of the Order of an Autoregression. *J. R. Stat. Soc. Ser. B 1979*, *41*, 190–195. [CrossRef]

Artzner, P.; Delbaen, F.; Eber, J.M.; Heath, D. Coherent measures of risk. *Math. Financ. 1999*, *9*, 203–228. [CrossRef]

Pelletier, D. Regime Switching for Dynamic Correlations. *J. Econometrics 2006*, *131*, 445–473. [CrossRef]

Haas, M. Covariance forecasts and long-run correlations in a Markov-switching model for dynamic correlations. *Financ. Res. Lett. 2010*, *7*, 86–97. [CrossRef]

Billah, B.; Hyndman, R.J.; Koehler, A.B. Empirical information criteria for time series forecasting model selection. *J. Stat. Comput. Simul. 2005*, *75*, 831–840. [CrossRef]

Conrad, C.; Loch, K. Journal of Applied Econometrics. *J. Appl. Econom. 2015*, *30*, 1090–1114. [CrossRef]

Amendola, A.; Candila, V.; Scognamillo, A. On the influence of US monetary policy on crude oil price volatility more, the out-of-sample forecasting procedure shows that including these additional macroeconomic variables generally improves the forecasting performance. *Empir. Econ. 2017*, *52*, 155–178. [CrossRef]

Amendola, A.; Candila, V.; Gallo, G.M. On the asymmetric impact of macro-variables on volatility. *Econ. Model. 2019*, *76*, 135–152. [CrossRef]

2

A New Continuous-Discrete Fuzzy Model and its Application in Finance

Hoang Viet Long [1,2], Haifa Bin Jebreen [3,*] and Y. Chalco-Cano [4]

1 Division of Computational Mathematics and Engineering, Institute for Computational Science, Ton Duc Thang University, Ho Chi Minh City 70000, Vietnam; hoangvietlong@tdtu.edu.vn
2 Faculty of Mathematics and Statistics, Ton Duc Thang University, Ho Chi Minh City 758307, Vietnam
3 Department of Mathematics, College of Science, King Saud University, P.O. Box 2455, Riyadh 11451, Saudi Arabia
4 Departamento de Matemática, Universidad de Tarapacá, Casilla 7D, Arica 09010, Chile; ychalco@uta.cl
* Correspondence: hjebreen@ksu.edu.sa

Abstract: In this paper, we propose a fuzzy differential-difference equation for modeling of mixed continuous-discrete phenomena. In the special case, we present the general solution of linear fuzzy differential-difference equations. The dynamical process in the intervals is presented by the corresponding fuzzy differential equation and with impulsive jumps in some points. We illustrate the applicability of the model to study the time value of money.

Keywords: fuzzy differential equations; fuzzy difference equations; mixed continuous-discrete model; strongly generalized Hukuhara differentiability; time value of money

---

1. Introduction

Differential and difference equations play a relevant role in modeling problems that arise in physics, engineering, biology, economics, finance, and many other areas. However, in some cases, these equations are restricted in their ability to describe phenomena due to the imprecise or incomplete information about the parameters, variables and initial conditions available. This can result from errors in measurement, observation, or experimental data; application of different operating conditions; or maintenance induced errors [1]. To overcome uncertainties or lack of precision, one can use a fuzzy environment in parameters, variables and initial conditions in place of exact (fixed) ones, by turning general differential and difference equations into fuzzy differential and difference equations, respectively [1–7]. These uncertainties may be modeled by fuzzy set theory when an abundance of data is not available. Accordingly, there is often a need to model, solve and interpret the problems one encounters in the world of uncertainty. The governing differential and difference equations will then become uncertain. Therefore, recently many researchers have studied fuzzy differential equations and fuzzy difference equations in different approaches [8–12].

Fuzzy set theory refers to the uncertainty when we have a lack of knowledge or incomplete information about the variables and parameters. In the financial markets there are elements of uncertainties and lack of precision associates to fluctuation and votality of financial markets. We cannot make forecasts easily, we have incomplete information or some type of uncertainty, about the values of financial factors such as taxes, inflation, interest rate, price change rate among other factors [4,13,14]. In this direction, some problems of the financial field can be approached via fuzzy difference equations [4]. In particular, Papadopoulos at al. in Reference [4] demonstrated the applicability of fuzzy difference equations to the problems of time value of money. The results obtained in the article [4] were motivated by models introduced by Kwapisz in Reference [15], where several difference equations to study the basic problems of finance such as capital deposits and capital investments were presented.

Although small discrete systems are easy to work with, the continuous models are easier to deal with than large discrete systems. Whether or not nature is fundamentally discrete, the most useful models are often continuous because the discreteness can only occur in very small scales. Discrete models are probably useful if nature has genuinely discrete structure. But on larger scales a discrete model would contain some parameters that we cannot measure and might not even be interested in. This is related to the observation that continuous models often work well for large discrete systems. Discreteness is useful to include in the model if it occurs in the situation we are interested in. Therefore, a mixed continuous-discrete model and, in a special case, a differential-difference equation can possess the inherent properties and advantages of both discrete and continuous models which are useful for modeling of real-world phenomena. In this direction, Kwapisz in Reference [15] introduced and studied a general mixed continuous-discrete model describing dynamical processes in some problems that arise in finance. In this model, the dynamical process in each interval is presented by the corresponding differential equation and it displays impulsive jumps, obtained by the corresponding difference equation. Therefore, proposing a mixed continuous-discrete model is a natural way to study a phenomenon which is continuous on some sub-intervals and it has discontinuities in some points.

Motived by the results obatined in Reference [15], on continuous-discrete models, and the recent advances on fuzzy differential equations [2,8], in this article we introduce fuzzy differential-difference equations. We present some results on existence and uniqueness of solutions for this class of models. Finally, we give an example on the time value of money to demonstrate the effectiveness of theoretical results.

2. Preliminaries

We start by recalling some preliminaries about the fuzzy sets defined on $\mathbb{R}$. A fuzzy set on $\mathbb{R}$ is a mapping $u\colon \mathbb{R} \to [0,1]$, where the value $u(x)$ denotes the degree of membership of the element x to the fuzzy set u. For $0 < \alpha \leq 1$, the α-level of u is defined by the set $[u]^\alpha = \{x \in \mathbb{R} \mid u(x) \geq \alpha\}$. For $\alpha = 0$, the support of u is defined as the set $[u]^0 = \text{supp}(u) = \overline{\{x \in \mathbb{R} \mid u(x) > 0\}}$. We denote

$$\mathbb{R}_F = \{u : \mathbb{R} \to [0,1] | u \text{ satisfies } (i) - (iv) \text{ below}\},$$

where

(*i*) u is normal, that is, there exists $x_0 \in \mathbb{R}$ such that $u(x_0) = 1$.
(*ii*) u is fuzzy convex, that is, $u(\lambda x + (1-\lambda)y) \geq \min\{u(x), u(y)\}$, for any $x, y \in \mathbb{R}$ and $0 \leq \lambda \leq 1$.
(*iii*) u is upper semicontinuous.
(*iv*) $[u]^0$ is compact.

If $u \in \mathbb{R}_F$, we say that u is a fuzzy number.

According to Zadeh's Extension Principle [2], operations of addition and scalar multiplication on $\mathbb{R}_F$ are defined as:

$$(u+v)(x) = \sup_{y+z=x} \min\{u(y), v(z)\}, \qquad \text{and} \qquad (\lambda u)(x) = \begin{cases} u(\frac{x}{\lambda}) & \lambda \neq 0, \\ \chi_{\{0\}}(x) & \lambda = 0, \end{cases}$$

where $\chi_{\{0\}}$ is the characteristic function of $\{0\}$. Moreover, the following relations hold:

$$[u+v]^\alpha = [u]^\alpha + [v]^\alpha, \qquad \text{and} \qquad [\lambda u]^\alpha = \lambda [u]^\alpha, \quad \forall u, v \in \mathbb{R}_F, \quad \forall \alpha \in [0,1].$$

Definition 1. *Let $u, v, w \in \mathbb{R}_F$. An element w is called the Hukuhara difference (H-difference, for short) of u and v, if it verifies the equation $u = v + w$. If the H-difference exists, it will be denoted by $u \ominus_H v$. Clearly, $u \ominus_H u = \{0\}$, and if $u \ominus_H v$ exists, it is unique.*

The space $\mathbb{R}_F$ is a complete metric space with the distance $D(u,v)$ given by

$$D(u,v) = \sup_{\alpha\in[0,1]} d([u]^\alpha, [v]^\alpha), \quad \forall u,v \in \mathbb{R}_F,$$

where $d(\cdot,\cdot)$ is the well known Pompeiu-Hausdorff distance on the space $\mathcal{K}_c^n$ of all nonempty, compact and convex subsets of the n-dimensional Euclidean space $\mathbb{R}^n$.

We need the following theorem in this paper.

Theorem 1 ([2]). *(i) For any $u,v,w \in \mathbb{R}_F$, we have*

$$[(u+v)w]^\alpha \subseteq [uw]^\alpha + [vw]^\alpha, \quad \forall \alpha \in [0,1],$$

and, in general, distributivity does not hold.

(ii) For any $u,v,w \in \mathbb{R}_F$ such that none of the supports of u,v,w contain 0, we have

$$u(vw) = (uv)w.$$

In the sequel, we fix $I = (0,T)$, for $T \in \mathbb{R}$. There are several approaches to study fuzzy differential equations [10,12,13,16–19]. In the following, we use the generalized Hukuhara differentiability concept of fuzzy functions [3,8].

Definition 2. *Let $F : I \to \mathbb{R}_F$ and $t_0 \in I$ be fixed. Then, we say that F is differentiable at t_0 if there exists an element $F'(t_0) \in \mathbb{R}_F$ such that either*

(i) For all $h > 0$ sufficiently small, the H-differences $F(t_0+h) \ominus F(t_0), F(t_0) \ominus F(t_0-h)$ exist and the limits (in the metric D)

$$\lim_{h\to 0^+} \frac{F(t_0+h)\ominus F(t_0)}{h} = \lim_{h\to 0^+} \frac{F(t_0)\ominus F(t_0-h)}{h} = F'(t_0),$$

or

(ii) For all $h > 0$ sufficiently small, the H-differences $F(t_0) \ominus F(t_0+h), F(t_0-h) \ominus F(t_0)$ exist and the limits (in the metric D)

$$\lim_{h\to 0^+} \frac{F(t_0)\ominus F(t_0+h)}{-h} = \lim_{h\to 0^+} \frac{F(t_0-h)\ominus F(t_0)}{-h} = F'(t_0).$$

We say that F is (i)-differentiable on I if F is differentiable in the sense (i) of Definition 2. Similarly, we say that F is (ii)-differentiable on I if F is differentiable in the sense (ii) of Definition 2. In this paper, we make use of the following theorem [10].

Theorem 2 ([10]). *Let $F : I \to \mathbb{R}_F$ be a fuzzy function such that $[F(t)]^\alpha = [f_\alpha(t), g_\alpha(t)]$ for each $\alpha \in [0,1]$. Then, we have*

(i) If F is (i)-differentiable, then f_α and g_α are differentiable functions and we have

$$[F'(t)]^\alpha = [f'_\alpha(t), g'_\alpha(t)].$$

(ii) If F is (ii)-differentiable, then f_α and g_α are differentiable functions and we have

$$[F'(t)]^\alpha = [g'_\alpha(t), f'_\alpha(t)].$$

Let us consider the initial value problem to fuzzy differential equation

$$\begin{cases} y'(t) &= f(t, y(t)), \qquad t \in I, \\ y(0) &= y_0, \end{cases} \tag{1}$$

where $f : I \times \mathbb{R}_F \to \mathbb{R}_F$ is a continuous fuzzy mapping and y_0 is a fuzzy number. It is well known from Reference [10] that the sufficient conditions for the existence and uniqueness of the (i)-differentiable solution to the initial value problem (1) are

(a) The fuzzy mapping f is continuous on $I \times \mathbb{R}_F$;
(b) The fuzzy mapping f satisfies Lipschitz condition

$$D(f(t,u), f(t,v) \le LD(u,v), \ L > 0, \ \forall u, v \in \mathbb{R}_F, t \in I.$$

In Reference [2], the sufficient conditions for the unique existence of the (ii)-differentiable solution to the initial value problem (1) are presented.

Now, we consider the first-order fuzzy linear differential equation

$$\begin{cases} y'(t) &= a(t)y(t) + b(t), \qquad t \in I, \\ y(0) &= y_0, \end{cases} \tag{2}$$

where a, $b : I \to \mathbb{R}$ are fuzzy mappings and $y_0 \in \mathbb{R}_F$ is the fuzzy initial condition. The initial value problem (2) was studied in Reference [8] by Bede, Rudas and Bencsik. They have presented the general solution of the problem in some special cases. Later, in Reference [20], the authors have presented the solution of the problem (2) with general conditions.

Theorem 3 ([8]). *Consider the initial value problem (2). Then, we have*

(a) If $a > 0$, then the (i)-differentiable solution to the problem (2) is given by

$$y(t) = e^{\int_0^t a(u)du} \left(y_0 + \int_0^t b(s) e^{-\int_0^s a(u)du} ds \right).$$

(b) If $a < 0$, then the (ii)-differentiable solution to the problem (2) is given by

$$y(t) = e^{\int_0^t a(u)du} \left(y_0 \ominus (-1) \int_0^t b(s) e^{-\int_0^s a(u)du} ds \right),$$

provided the H-difference exists.

Fuzzy difference equation is a difference equation whose parameters or initial data are fuzzy numbers and its solutions are given in the form of fuzzy number sequences. Due to the applicability of fuzzy difference equations in the analysis of phenomena where imprecision is inherent, this class of difference equations is an interesting topic from theoretical point of view. Deeba et al. [9] have studied the first-order fuzzy difference equation $x_{n+1} = wx_n + q$, $n = 0, 1, \ldots$ to investigate the population genetics, where $\{x_n\}$ is a sequence of positive fuzzy numbers and $w, q, x_0 \in \mathbb{R}_F^+$. In Reference [21], Papaschinopoulos et. al. studied the existence and some related properties of the positive solutions of the fuzzy difference equation $x_{n+1} = A + B/x_n, n = 0, 1, \ldots,$ where $\{x_n\}$ is a sequence of positive fuzzy numbers and $A, B \in \mathbb{R}_F^+$.

In the following, we consider the first-order fuzzy difference equation

$$z_{n+1} = \mu_n z_n + \nu_n, \ \ n = 0, 1, 2, \ldots \tag{3}$$

where $\{\mu_n\}$ and $\{\nu_n\}$ are sequences of positive fuzzy numbers and $z_0 \in \mathbb{R}_F^+$. The difference Equation (3) is a generalization of the following fuzzy difference equations, studied in Reference [4]

$$F_{n+1} = F_n + IF_0, \text{ and } F_{n+1} = F_n(I' + 1) + b_n, \quad n = 0, 1, 2, \ldots$$

where $I, I', b_n \in \mathbb{R}_F^+$. Furthermore, the Equation (3) is also a generalization of the fuzzy difference equation $x_{n+1} = wx_n + q$, which was studied in Reference [5,9]. In the following, we study the existence of positive solution to the difference Equation (3).

Since μ_n, ν_n, z_0 are positive fuzzy numbers for each $n = 0, 1, \ldots,$ then the $\alpha-$cuts of z_{n+1} is given by

$$[\underline{z}_{n+1}, \overline{z}_{n+1}]^\alpha = [\underline{\mu}_n \underline{z}_n + \underline{\nu}_n, \overline{\mu}_n \overline{z}_n + \overline{\nu}_n], \qquad n = 0, 1, 2, \ldots$$

Then, we have two classical difference equations

$$\underline{z}_{n+1} = \underline{\mu}_n \underline{z}_n + \underline{\nu}_n, \quad n = 0, 1, 2, \ldots$$

and

$$\overline{z}_{n+1} = \overline{\mu}_n \overline{z}_n + \overline{\nu}_n, \quad n = 0, 1, 2, \ldots$$

Therefore, by using the results of classic difference equations [22], we have

$$\underline{z}_n = \underline{z}_0 \prod_{i=0}^{n-1} \underline{\mu}_i + \sum_{i=0}^{n-1} \underline{\nu}_i \prod_{j=i+1}^{n-1} \underline{\mu}_j, \qquad n = 0, 1, \ldots$$

and

$$\overline{z}_n = \overline{z}_0 \prod_{i=0}^{n-1} \overline{\mu}_i + \sum_{i=0}^{n-1} \overline{\nu}_i \prod_{j=i+1}^{n-1} \overline{\mu}_j, \quad n = 0, 1, \ldots$$

Consequently, we obtain

$$[\underline{z}_n, \overline{z}_n]^\alpha = \left[\underline{z}_0 \prod_{i=0}^{n-1} \underline{\mu}_i + \sum_{i=0}^{n-1} \underline{\nu}_i \prod_{j=i+1}^{n-1} \underline{\mu}_j, \overline{z}_0 \prod_{i=0}^{n-1} \overline{\mu}_i + \sum_{i=0}^{n-1} \overline{\nu}_i \prod_{j=i+1}^{n-1} \overline{\mu}_j \right].$$

Additionally, since μ_n, ν_n, z_0 are positive fuzzy numbers, we have the following result

Theorem 4. *For each $n \in \mathbb{N}$, let $z_0, \nu_n, \mu_n \in \mathbb{R}_F^+$. Then, the positive solution of the first-order fuzzy difference Equation (3) is given by*

$$z_n = z_0 \prod_{i=0}^{n-1} \mu_i + \sum_{i=0}^{n-1} \nu_i \prod_{j=i+1}^{n-1} \mu_j, \qquad n = 0, 1, \ldots$$

There are various methods to compare and arrange fuzzy numbers. In the theoretical point of view, the set of fuzzy numbers can only be partially ordered and hence, it cannot be compared. However, in practical applications such as decision making, scheduling, market analysis or optimization with fuzzy uncertainties, the comparison of fuzzy numbers becomes crucial [23]. In this study, we use the following definition for ordering fuzzy numbers.

Definition 3. *For each $u, v \in \mathbb{R}_F$, we say that the fuzzy number u is greater than the fuzzy number v, denoted by $u \gg v$, if and only if $\underline{u}^\alpha > \underline{v}^\alpha$ and $\overline{u}^\alpha > \overline{v}^\alpha$ for all $\alpha \in [0, 1]$.*

It is well-known that $u \in \mathbb{R}_F^+$ if and only if $u \gg \tilde{0}$, that is, $\overline{u}^\alpha \geq \underline{u}^\alpha > 0$ for all $\alpha \in [0,1]$, where $\tilde{0} = \chi_{\{0\}}$. Similarly, $u \in \mathbb{R}_F^-$ if and only if $\tilde{0} \gg u$, that is, $\underline{u}^\alpha \leq \overline{u}^\alpha < 0$ for all $\alpha \in [0,1]$.

Proposition 1. *Let $u, v, w \in \mathbb{R}_F^+$ and $v \ominus w$ exist such that $v \gg w$. Then, we have*

$$u(v \ominus w) = uv \ominus uw.$$

Proof. Since $v \ominus w$ exists such that $v \gg w$, it implies that

$$\begin{cases} \underline{v} > \underline{w}, \\ \overline{v} > \overline{w} \\ \underline{v} - \underline{w} < \overline{v} - \overline{w}. \end{cases}$$

Thus, it implies that $0 < \underline{v} - \underline{w} < \overline{v} - \overline{w}$ and hence, $v \ominus w \gg \tilde{0}$.

Finally, for each $\alpha \in [0,1]$, we have

$$\begin{aligned} [u(v \ominus w)]^\alpha &= [\underline{u}(\underline{v} - \underline{w}), \overline{u}(\overline{v} - \overline{w})] \\ &= [\underline{uv} - \underline{uw}, \overline{uv} - \overline{uw}] \\ &= [\underline{uv}, \overline{uv}] \ominus [\underline{uw}, \overline{uw}] \\ &= [uv \ominus uw]^\alpha. \end{aligned}$$

□

We have the following lemma.

Lemma 1. *Let $u, v, w \in \mathbb{R}_F$ and the H-differences $u \ominus v$, $(u \ominus v) \ominus w$ exist. Then, the H-difference $u \ominus (v + w)$ exist and we have $(u \ominus v) \ominus w = u \ominus (v + w)$.*

Proof. Let $u \ominus v = \tau_1$ and $(u \ominus v) \ominus w = \tau_2$. Then, $\tau_2 + w = u \ominus v$. So, we have $\tau_2 + w + v = u$. Therefore, $\tau_2 = u \ominus (v + w)$. □

Theorem 5. *Assume that the numbers $\mu_n, \nu_n, z_0 \in \mathbb{R}_F^+$ be such that the H-differences $\mu_n z_n \ominus \nu_n$ exist and $\mu_n z_n \gg \nu_n$ for all $n \geq 0$. Then, the fuzzy solution of the fuzzy difference equation*

$$z_{n+1} = \mu_n z_n \ominus \nu_n, \qquad n = 0, 1, \ldots \tag{4}$$

is given by

$$z_n = z_0 \prod_{i=0}^{n-1} \mu_i \ominus \sum_{i=0}^{n-1} \nu_i \prod_{j=i+1}^{n-1} \mu_j, \qquad n = 0, 1, \ldots$$

Proof. By the assumption that $z_0 \in \mathbb{R}_F^+$, $\mu_n z_n \gg \nu_n$ for each $n \in \mathbb{N}$ and the H-differences $\mu_n z_n \ominus \nu_n$ exist, it implies that $\mu_n z_n \ominus \nu_n \gg \tilde{0}$ and hence, we have $z_{n+1} \in \mathbb{R}_F^+$ for each $n \in \mathbb{N}$. On the other hand, since μ_n and ν_n are positive fuzzy numbers for $n = 0, 1, \ldots$, the α-cuts of z_{n+1} are given by

$$[\underline{z}_{n+1}, \overline{z}_{n+1}]^\alpha = \left[\underline{\mu}_n \underline{z}_n - \underline{\nu}_n, \overline{\mu}_n \overline{z}_n - \overline{\nu}_n\right], \qquad \text{for } n = 0, 1, \ldots \text{ and } \alpha \in [0,1].$$

Then, we have two classical difference equations

$$\underline{z}_{n+1} = \underline{\mu}_n \underline{z}_n - \underline{\nu}_n, \qquad n = 0, 1, \ldots$$

and

$$\overline{z}_{n+1} = \overline{\mu}_n \overline{z}_n - \overline{\nu}_n, \qquad n = 0, 1, \ldots$$

Therefore, by using the results of classic difference equations [22], we have

$$\underline{z}_n = \underline{z}_0 \prod_{i=0}^{n-1} \underline{\mu}_i - \sum_{i=0}^{n-1} \underline{\nu}_i \prod_{j=i+1}^{n-1} \underline{\mu}_j, \quad n = 0, 1, \ldots$$

and

$$\overline{z}_n = \overline{z}_0 \prod_{i=0}^{n-1} \overline{\mu}_i - \sum_{i=0}^{n-1} \overline{\nu}_i \prod_{j=i+1}^{n-1} \overline{\mu}_j, \; n = 0, 1, \ldots$$

Therefore, we obtain

$$z_n = z_0 \prod_{i=0}^{n-1} \mu_i \ominus \sum_{i=0}^{n-1} \nu_i \prod_{j=i+1}^{n-1} \mu_j, \; n = 0, 1, \ldots \tag{5}$$

It is easy to check that the H-difference in (5) exists. Indeed, the corresponding H-differences $z_1 = \mu_0 z_0 \ominus \nu_0$ and $z_2 = \mu_1 z_1 \ominus \nu_1 = \mu_1(\mu_0 z_0 \ominus \nu_0) \ominus \nu_1$ exist. Therefore, by using Lemma 1, the H-difference $\mu_1\mu_0 z_0 \ominus (\mu_1\nu_0 + \nu_1)$ exists and we have

$$z_2 = \mu_1\mu_0 z_0 \ominus (\mu_1\nu_0 + \nu_1).$$

By mathematical induction principle, we can see that the H-difference in (5) exists and hence, the proof is complete. □

3. General Mixed Continuous-Discrete Fuzzy Model

The fuzzy difference equations introduced in Reference [4] are the special cases of the following linear fuzzy difference equation

$$F_{n+1} = a_n F_n + b_n F_{\gamma_n} + f_n, \; n = 0, 1, \ldots \tag{6}$$

where $\{a_n\}, \{b_n\}, \{f_n\}$ are given sequences of fuzzy numbers and $\gamma_n = k\delta_n$ with $\delta_n = [\frac{n}{k}]$ for some integer $k > 0$.

In the following, we consider a positive increasing sequence $\{t_n\}$ satisfying $t_n \to +\infty$, a sequence of fuzzy functions $\{f_n\} \subset C(\overline{J_n} \times \mathbb{R}_F, \mathbb{R}_F)$, $J_n = (t_n, t_{n+1}]$, the fuzzy functions $d_n : \mathbb{R}_F \times \mathbb{R}_F \times \mathbb{R}_F \to \mathbb{R}_F$ for each $n = 0, 1, \ldots$ and the initial value $F_0 \in \mathbb{R}_F$. We introduce a dynamical process as a mixed continuous-discrete fuzzy model by the set of fuzzy differential equations and fuzzy difference equations as follows:

$$\begin{cases} y_n'(t; F_n) = f_n(t, y_n(t; F_n)), & t \in J_n, \\ y_n(t_n^+; F_n) = F_n, & n = 0, 1, \ldots \end{cases} \tag{7}$$

and

$$F_{n+1} = d_n(F_n, F_{\gamma_n}, y_n(t_{n+1}; F_n)). \tag{8}$$

Here, we assume that

$$y_n(t_n^+; F_n) = \lim_{h \to 0^+} y_n(t_n + h; F_n).$$

In order to get the uniqueness of the process, we assume that the sufficient conditions for the existence of (i)-differentiable and (ii)-differentiable solutions are fulfilled (see References [2,10]). Then the unique process in each type of differentiability is defined as

$$y(t; F_0) = y_n(t; F_n), \; t \in (t_n, t_{n+1}], \; n = 0, 1, \ldots$$

Therefore, the dynamical process in interval J_n is presented by the corresponding fuzzy differential equation and it displays impulsive jumps in the points of the sequence $\{t_n\}$. It is easy to see that when the functions d_n do not depend on the third variable or $f_n(t,y) \equiv 0,\ n = 0,1,\ldots$, we have a purely discrete process described by the fuzzy difference equations. For instance, if we assume

$$d_n(x,y,z) = a_n x + b_n y + f_n,$$

then the fuzzy differential-difference Equations (7) and (8) are transformed into (6). On the other hand, if we assume $f_n(t,x) \equiv f(t,x)$ and $d_n(x,y,z) = z$, then we obtain a continuous dynamical process formulated by the fuzzy initial value problem

$$\begin{cases} y'(t) = f(t,y(t)), & t \geq t_0, \\ y(t_0) = F_0. \end{cases}$$

4. Linear Fuzzy Differential-Difference Equations

In this section, we consider the equation of linear form

$$f(t,y) = a(t)y + b(t), \tag{9}$$

where $a : [0,+\infty) \to \mathbb{R}$ and $b : [0,+\infty) \to \mathbb{R}_F$ are continuous functions and

$$d_n(x,y,z) = d_n z + e_n,\ n = 0,1,\ldots \tag{10}$$

is a fuzzy difference equation w.r.t. the sequences $\{d_n\}$ and $\{e_n\}$ of positive fuzzy numbers. Hence, the problem (7)–(8) is transformed into the following fuzzy model

$$\begin{cases} y_n'(t;F_n) = a(t)y_n(t;F_n) + b(t), & t \in (t_n,t_{n+1}], \\ y_n(t_n^+;F_n) = F_n, & n = 0,1,\ldots \end{cases} \tag{11}$$

$$F_{n+1} = d_n y_n(t_{n+1};F_n) + e_n, \qquad n = 0,1,\ldots \tag{12}$$

where the initial value $F_0 \in \mathbb{R}_F^+$. In the following, we will present an explicit formula for the solution $y_n(t;F_n)$ on each interval $J_n = (t_n,t_{n+1}]$. For this aim, we consider three different cases of the real function $a(t)$.

Theorem 6. *Consider the linear mixed continuous-discrete fuzzy model (11)–(12) where $d_n, e_n, F_0 \in \mathbb{R}_F^+$ and $a : [0,+\infty) \to \mathbb{R}^+$, $b : [0,+\infty) \to \mathbb{R}_F^+$ are continuous functions. Then, the (i)-differentiable solution to the model (11)–(12) is given by*

$$\begin{aligned} y_n(t;F_n) = {} & F_0 e^{\int_{t_0}^t a(u)du} \prod_{i=0}^{n-1} d_i + \sum_{i=0}^{n-1} \left[\left(\prod_{j=i}^{n-1} d_j \right) \int_{t_i}^{t_{i+1}} b(s) e^{\int_s^t a(u)du} ds \right] \\ & + \sum_{i=0}^{n-1} \left[e_i e^{\int_{t_{i+1}}^t a(u)du} \prod_{j=i+1}^{n-1} d_j \right] + \int_{t_n}^t b(s) e^{\int_s^t a(u)du} ds, \end{aligned} \tag{13}$$

for each $t \in (t_n,t_{n+1}]$ and each $n = 0,1,\ldots$

Proof. If the function $a(t)$ is positive, then according to Theorem 3, the (i)-differentiable solution of the fuzzy differential-difference Equations (11) and (12) is given by

$$y_n(t;F_n) = e^{\int_{t_n}^t a(u)du} \left(F_n + \int_{t_n}^t b(s) e^{-\int_{t_n}^s a(u)du} ds \right),$$

or equivalently,

$$y_n(t;F_n) = F_n e^{\int_{t_n}^t a(u)du} + \int_{t_n}^t b(s) e^{\int_s^t a(u)du} ds, \tag{14}$$

for all $t_n < t \leq t_{n+1}$ and $n = 0, 1, \ldots$ Therefore, by using the difference Equation (12), we directly have

$$F_{n+1} = d_n \left[F_n e^{\int_{t_n}^{t_{n+1}} a(u)du} + \int_{t_n}^{t_{n+1}} b(s) e^{\int_s^{t_{n+1}} a(u)du} ds \right] + e_n.$$

For each $n \geq 1$, since the terms d_n, $F_n e^{\int_{t_n}^{t_{n+1}} a(u)du}$ and $\int_{t_n}^{t_{n+1}} b(s) e^{\int_s^{t_{n+1}} a(u)du} ds$ are in $\mathbb{R}_F^+$, then Theorem 1 implies that

$$F_{n+1} = F_n d_n e^{\int_{t_n}^{t_{n+1}} a(u)du} + d_n \int_{t_n}^{t_{n+1}} b(s) e^{\int_s^{t_{n+1}} a(u)du} ds + e_n,$$

or equivalently,

$$F_{n+1} = A_n F_n + B_n, \qquad n = 0, 1, \ldots, \tag{15}$$

where

$$A_n = d_n e^{\int_{t_n}^{t_{n+1}} a(u)du},$$
$$B_n = d_n \int_{t_n}^{t_{n+1}} b(s) e^{\int_s^{t_{n+1}} a(u)du} ds + e_n.$$

Therefore, according to Theorem 4, the solution of the fuzzy difference Equation (15) is

$$F_n = F_0 \prod_{i=0}^{n-1} A_i + \sum_{i=0}^{n-1} B_i \prod_{j=i+1}^{n-1} A_j, \qquad n = 0, 1, \ldots$$

and the proof is complete. □

Remark 1. *It is well-known that the Hukuhara differentiable functions have increasing length of support, that is, when the time goes by, the diameter of the fuzzy functions increases, see Reference [10]. Therefore, the solution of the model (11)–(12) has increasing length of support with some impulsive jumps in the points of sequence $\{t_n\}$.*

For $a < 0$, we have the following result.

Theorem 7. *Consider the linear mixed continuous-discrete fuzzy model (11)–(12) where $d_n, e_n, F_0 \in \mathbb{R}_F^+$ and $a : [0, +\infty) \to \mathbb{R}^-$, $b : [0, +\infty) \to \mathbb{R}_F^-$ are continuous functions. Assume that for each $n \geq 0$, the H-differences*

$$F_n e^{\int_{t_n}^{t} a(u)du} \ominus (-1) \int_{t_n}^{t} b(s) e^{\int_s^{t_{n+1}} a(u)du} ds$$

exist and

$$F_n e^{\int_{t_n}^{t} a(u)du} \gg (-1) \int_{t_n}^{t} b(s) e^{\int_s^{t_{n+1}} a(u)du} ds.$$

Then, the (ii)-differentiable solution to the model (11)–(12) is given by

$$\begin{aligned} y_n(t; F_n) = {} & F_0 e^{\int_{t_0}^{t} a(u)du} \prod_{i=0}^{n-1} d_i \ominus \sum_{i=0}^{n-1} \left[\left(\prod_{j=i}^{n-1} d_j \right) \int_{t_i}^{t_{i+1}} b(s) e^{\int_s^{t} a(u)du} ds \right] \\ & + \sum_{i=0}^{n-1} \left[e_i e^{\int_{t_{i+1}}^{t} a(u)du} \prod_{j=i+1}^{n-1} d_j \right] + \int_{t_n}^{t} b(s) e^{\int_s^{t} a(u)du} ds, \end{aligned} \tag{16}$$

for each $t_n < t \leq t_{n+1}$ and each $n = 0, 1, \ldots$

Proof. According to Theorem 3, the (ii)-differentiable solution of (11) is given by

$$y_n(t; F_n) = e^{\int_{t_n}^{t} a(u)du} \left(F_n \ominus (-1) \int_{t_n}^{t} b(s) e^{-\int_{t_n}^{s} a(u)du} ds \right),$$

or equivalently,

$$y_n(t; F_n) = F_n e^{\int_{t_n}^{t} a(u)du} \ominus (-1) \int_{t_n}^{t} b(s) e^{\int_{s}^{t_{n+1}} a(u)du} ds, \tag{17}$$

for each $t_n < t \leq t_{n+1}$, and each $n = 0, 1, \ldots$. Therefore, by using the difference Equation (12), we have

$$F_{n+1} = d_n \left[F_n e^{\int_{t_n}^{t_{n+1}} a(u)du} \ominus (-1) \int_{t_n}^{t_{n+1}} b(s) e^{\int_{s}^{t_{n+1}} a(u)du} ds \right] + e_n.$$

Then, since $d_n, F_n \in \mathbb{R}_F^+$ and $b : [0, \infty) \to \mathbb{R}_F^-$, Proposition 1 follows that

$$F_{n+1} = F_n d_n e^{\int_{t_n}^{t_{n+1}} a(u)du} \ominus (-1) d_n \int_{t_n}^{t_{n+1}} b(s) e^{\int_{s}^{t_{n+1}} a(u)du} ds + e_n,$$

or equivalently

$$F_{n+1} = A_n F_n \ominus B_n, \quad n = 0, 1, \ldots \tag{18}$$

where

$$A_n = d_n e^{\int_{t_n}^{t_{n+1}} a(u)du}, \quad n = 0, 1, \ldots$$
$$B_n = (-1) d_n \int_{t_n}^{t_{n+1}} b(s) e^{\int_{s}^{t_{n+1}} a(u)du} ds + e_n.$$

Thus, by Theorem 5, we obtain the solution as

$$F_n = F_0 \prod_{i=0}^{n-1} A_i \ominus \sum_{i=0}^{n-1} B_i \prod_{j=i+1}^{n-1} A_j, \quad n = 0, 1, \ldots$$

which completes the proof. □

Remark 2. *It is well-known that the (ii)-differentiable functions have non-increasing length of support, that is, when the time goes by, the diameter of the fuzzy functions decrease, see Reference [2]. Therefore, the solution of the model (11)–(12) under the differentiability in type (ii) has non-increasing length of support with some impulsive jumps in the points of sequence $\{t_n\}$.*

In the case $a(t) = 0$, we have the mixed continuous-discrete fuzzy model

$$\begin{cases} y_n'(t; F_n) = b(t), & t \in (t_n, t_{n+1}], \\ y_n(t_n^+; F_n) = F_n, & n = 0, 1, \ldots \end{cases} \tag{19}$$

$$F_{n+1} = d_n y_n(t_{n+1}; F_n) + e_n, \qquad n = 0, 1, \ldots \tag{20}$$

where $F_0 \in \mathbb{R}_F^+$, $b : [0, +\infty) \to \mathbb{R}_F$ is a continuous function and $\{d_n\}$, $\{e_n\}$ are sequences of positive fuzzy numbers. We have the following results for $a(t) = 0$.

Theorem 8. *Consider the mixed continuous-discrete fuzzy model (19)–(20), where the parameters $d_n, e_n, F_0 \in \mathbb{R}_F^+$ and $b : [0, +\infty) \to \mathbb{R}_F$ is a continuous function. Then,*

(i) If the function $b : [0, +\infty) \to \mathbb{R}_F^+$ is continuous, then the (i)-differentiable solution of the fuzzy model (19)–(20) is given by

$$y_n(t;F_n) = F_0 \prod_{i=0}^{n-1} d_i + \sum_{i=0}^{n-1} \left[\prod_{j=i}^{n-1} d_j \int_{t_i}^{t_{i+1}} b(s)ds \right] + \sum_{i=0}^{n-1} \left[e_i \prod_{j=i+1}^{n-1} d_j \right] + \int_{t_n}^{t} b(s)ds,$$

for each $t_n < t \le t_{n+1}$ *and each* $n = 0, 1, \ldots$

(ii) *If* $b : [0, +\infty) \to \mathbb{R}_F^-$ *is a continuous function such that the H-difference* $F_n \ominus (-1) \int_{t_n}^{t} b(s)ds$ *exists and the following term holds*

$$F_n \gg (-1) \int_{t_n}^{t} b(s)ds,$$

then, the (ii)-differentiable solution of the model (19)–(20) is given by

$$y_n(t;F_n) = F_0 \prod_{i=0}^{n-1} d_i \ominus \sum_{i=0}^{n-1} \left[\prod_{j=i}^{n-1} d_j \int_{t_i}^{t_{i+1}} b(s)ds \right] + \sum_{i=0}^{n-1} \left[e_i \prod_{j=i+1}^{n-1} d_j \right] + \int_{t_n}^{t} b(s)ds,$$

for each $t_n < t \le t_{n+1}$ *and each* $n = 0, 1, \ldots$

Proof. By similar arguments as in the case $a(t) > 0$, we obtain the (i)-differentiable solution if $b : [0, +\infty) \to \mathbb{R}_F^+$, while in the case $a(t) < 0$, we receive the (ii)-differentiable solution of the model with $b : [0, +\infty) \to \mathbb{R}_F^-$. □

Example 1. *Consider the following fuzzy differential-difference equation*

$$\begin{cases} y_n'(t, F_n) = y_n(t, F_n) + [\alpha + 1, 3 - \alpha]t, & t \in (t_n, t_{n+1}], \\ y_n(t_n^+, F_n) = F_n, & n = 0, 1, \ldots \end{cases} \tag{21}$$

$$F_{n+1} = 1.1 y_n(t_{n+1}, F_n) + 0.1(n+1)[\alpha + 1, 3 - \alpha], \;\; n = 0, 1, \ldots \tag{22}$$

where $F_0 = [2 + \alpha, 4 - \alpha]$, $t_n = 0.2n$, $n = 0, 1, \ldots$ *Then, the solution of the fuzzy differential-difference Equations (21) and (22) is determined by the formula (13) and its graphical representation is given in Figure 1 for* $\alpha = 0, 1$. *As we see in Figure 1, the length of the support of the solution is increasing. Starting from the triangular fuzzy initial value* $(2, 3, 4)$, *the diameter of the solution increases as time goes by and in the point* $t_1 = 0.2$, *according to Equation (22), we have a jump. Again, starting from the point* t_1 *and using FDE (21), we obtain the solution on* $(0.2, 0.4]$. *We can follow this procedure for* $(0.4, 0.6]$.

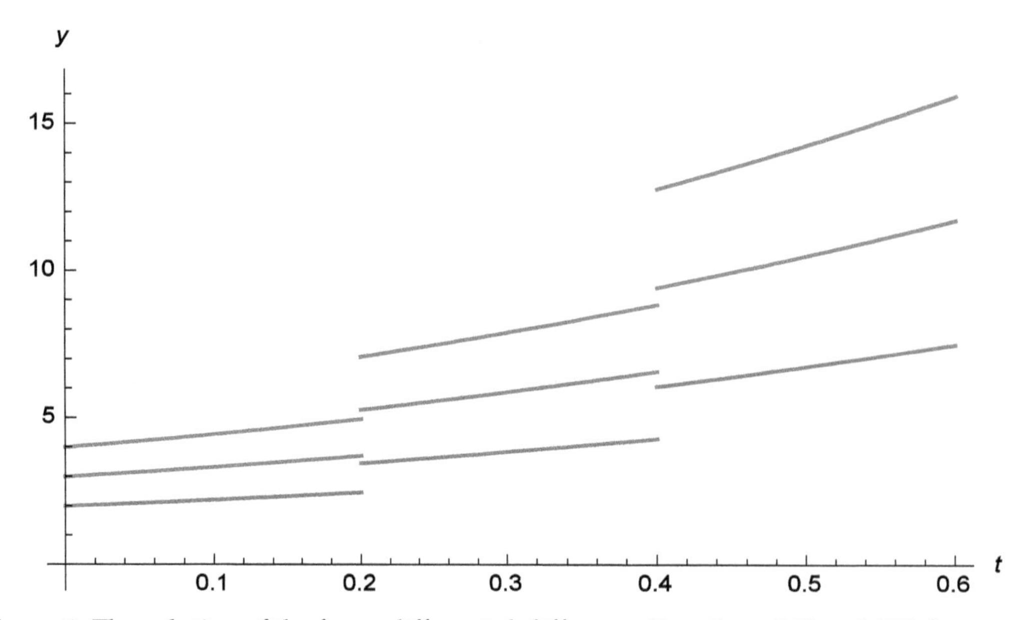

Figure 1. The solution of the fuzzy differential-difference Equations (21) and (22) for $\alpha = 0.1$.

5. Application: Time Value of Money

It is a fact that a fixed amount of money to get after some years is worth less than the same amount today. The main reason is that money due in the future or locked in a fixed term account cannot be spent right away. Meanwhile, prices may rise and the amount will not have the same purchasing power as it would have at present. In addition, there is always a risk that the money will never be received. Therefore, whenever a future payment is uncertain, its value today will be reduced to compensate for the risk. We mention that in this paper, we shall consider situations free from such risk. Bank deposit and bond are generic examples of risk-free assets [24]. A bank deposit is a specific sum of money taken and held on account by a bank, as a service to its customers. Some banks pay the customer through the interest of the funds deposited while others may charge a fee for this service. Therefore a bank deposit is a type of asset. There are many ways that a bank can pay interest on the funds deposited, see for example Reference [24].

Many mathematical methodologies have been developed to study the uncertainty in the estimation of the time value of money. An important effort has been made by Buckley [13] where he has developed fuzzy analogues of the elementary compound interest problem in financial mathematics. Later, in Reference [4], the authors have presented an alternative methodology using fuzzy difference equations. Their method has some advantages such as simplicity and capacity in studying the uncertain factors which cause the change of value of money in different time periods. In this paper, we consider this topic in three following cases:

Case I. Simple Interest: Chrysasif et al. [4] considered a simple capitalization problem. Let us assume that an amount of money is deposited in a bank account to obtain the interest. Then, the future value of this investment consists of the initial value of deposit P, namely the principal, plus all the interest earned during the period of investment. The authors considered the case when the interest is received only by the principal. This motivates the following fuzzy difference equation of simple interest [4]

$$V_{n+1} = V_n + IP, \quad n = 0, 1, \ldots$$

where I is the rate of interest and $V_0 = P$.

Case II. Periodic Compounding: Let us assume that an amount of money P is deposited in a bank account to receive interest at a constant rate I. Here, in contrast to the case of simple interest, we assume that the interest earned will be added to the initial principal periodically. Consequently, the interest will be received not only by the principal, but also by all the interest earned so far. This motivates the following fuzzy difference equation [4]

$$V_{n+1} = V_n(1 + I), \quad V_0 = P, \quad n = 0, 1, \ldots$$

The authors of Reference [4] have studied the compound interest problem considering a new factor e_n, which is added into the equation, denoting the deposits realized during the life of the account

$$V_{n+1} = V_n(1 + I) + e_n, \quad V_0 = P, \quad n = 0, 1, \ldots$$

It is natural to use fuzzy number for the extra deposits because we do not know certainly the number of deposits that the customer will make during the period of investment.

Case III. Continuous Compounding: In this case, the rate of growth of the deposit is proportional to the current wealth. In the periodic compounding, if we consider limit case as $n \to \infty$, we get $V(t) = e^{tI}P$, which is the solution of the following Cauchy problem [24]

$$V'(t) = IV(t), \; V(0) = P.$$

This is known as the continuous compounding, where the corresponding growth factor is e^{tI}.

Remark 3 **([24]).** *For the fixed principal P and interest rate I, the continuous compounding produces the higher future value than periodic compounding with any frequency n.*

Example 2. *In the following, by using the results of Section 4, we introduce a new mixed continuous-discrete fuzzy model to study the future value of money. Consider the following fuzzy differential-difference equation*

$$\begin{cases} V_n'(t;F_n) = IV_n(t;F_n), & t \in (t_n, t_{n+1}], \\ V_n(t_n^+;F_n) = F_n, & n = 0,1,\ldots \end{cases} \tag{23}$$

$$F_{n+1} = V_n(t_{n+1};F_n) + e_n, \quad n = 0,1,\ldots \tag{24}$$

where the initial value $F_0 = P$. *Let us consider triangular fuzzy numbers* $F_0 = \left(\phi, \frac{(\phi+\rho)}{2}, \rho\right)$ *and* $e_n = a_n\left(s, \frac{s+t}{2}, t\right)$, *where their membership functions are given by*

$$F_0(x) = \begin{cases} \frac{-2x+2\phi}{\phi-\rho}, & x \in [\phi, \frac{(\phi+\rho)}{2}), \\ \frac{-2x+2\rho}{\rho-\phi}, & x \in [\frac{(\phi+\rho)}{2}, \rho), \\ 0, & \text{otherwise}, \end{cases}$$

and

$$e_n(x) = \begin{cases} \frac{-2x+2a_n s}{a_n(s-t)}, & x \in [a_n s, \frac{a_n(s+t)}{2}), \\ \frac{-2x+2a_n t}{a_n(t-s)}, & x \in [\frac{a_n(s+t)}{2}, a_n t), \\ 0, & \text{otherwise}. \end{cases}$$

Hence, their level sets are given by

$$[F_0]^\alpha = \left[\frac{\alpha(\rho-\phi)+2\rho}{2}, \frac{\alpha(\phi-\rho)+2\phi}{2}\right],$$
$$[e_n]^\alpha = \frac{a_n}{2}\left[\alpha(t-s)+2s, \alpha(s-t)+2t\right],$$

for all $\alpha \in [0,1]$. *Then, according to the Formula (13), we obtain the solution of the fuzzy differential-difference Equations (23) and (24) is*

$$V_n(t;F_n) = F_0 e^{I(t-t_0)} + \sum_{i=0}^{n-1} e_i e^{I(t-t_{i+1})}. \tag{25}$$

In particular, we consider $I = 3.5$, $t_n = 0.2n, n = 0,1,\ldots$ *and* F_0, e_n *are fuzzy numbers whose level sets are given by*

$$[F_0]^\alpha = 50000 + 5000[-1+\alpha, 1-\alpha],$$
$$[e_n]^\alpha = 200(n+1)[9+\alpha, 11-\alpha].$$

Finally, the solution of fuzzy differential-difference Equations (23) and (24) is determined by the Formula (25) and its graphical representation with $\alpha = 0,1$ *is shown in Figure 2.*

Here, the initial value of the deposit is the triangular fuzzy number $(45,000, 50,000, 55,000)$. *Using the FDE (23), we obtain the solution on* $(0, 0.2]$. *At* $t = 0.2$, *we have an impulsive jump such that we can obtain the value at* t_1^+ *by (24). To obtain the solution on* $(0.2, 0.4]$, *we use the FDE (23) with initial value* F_1. *By following this procedure, we obtain the solution on* $[0, 0.6]$ *in Figure 2.*

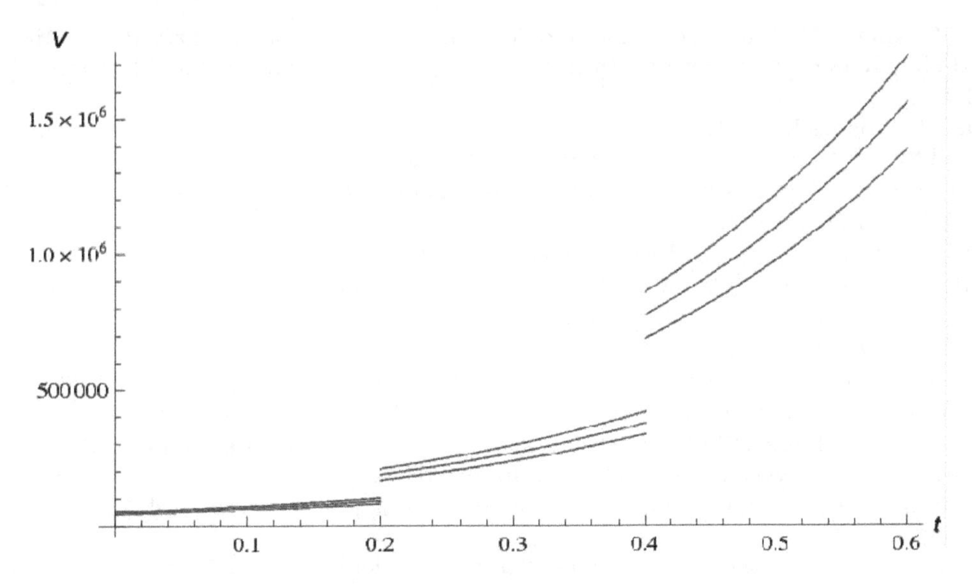

Figure 2. The solution of fuzzy differential-difference Equations (23) and (24) on the interval $[0, 0.6]$ with $\alpha = 0, 1$.

6. Conclusions

In the present paper, a fuzzy differential-difference equation is proposed to model the mixed continuous-discrete phenomena. We presented the dynamical process in the intervals by a fuzzy differential equation and impulsive jumps in some points by the corresponding fuzzy difference equation. This study generalizes the results of Reference [15] to the fuzzy set theory to consider the uncertain factors in differential equations and difference equations. By this approach, we modeled the uncertainty in initial values and parameters of the differential-difference equations. The general solution of linear fuzzy differential-difference equations is presented, too. Finally, the applicability of the model is illustrated by studying the time value of money in finance.

For further research, we propose to extend these results to study the existence of both (i)-differentiable and (ii)-differentiable solutions of the mixed continuous-discrete fuzzy model corresponding to each case of a. The current work opens up many potential results in studying control problems or numerical algorithms for the fuzzy differential-difference equations, that are inspired by pioneer works [25–29].

Author Contributions: All authors contributed equally. All authors have read and agreed to the published version of the manuscript.

References

Chakraverty, S.; Tapaswini, S.; Behera, D. *Fuzzy Differential Equations and Applications for Engineers and Scientists*; Taylor& Francis: Oxfordshire, UK, 2016.

Bede, B. *Mathematics of Fuzzy Sets and Fuzzy Logic*; Springer: London, UK, 2013.

Chalco-Cano, Y.; Román-Flores, H. On new solutions of fuzzy differential equations. *Chaos Solitons Fractals 2008, 38*, 112–119. [CrossRef]

Chrysafis, K.A.; Papadopoulos, B.K.; Papaschinopoulos, G. Papaschinopoulos, On the fuzzy difference equations of finance. *Fuzzy Sets Syst. 2008, 159*, 3259–3270. [CrossRef]

Khastan, A. New solutions for first order linear fuzzy difference equations. *J. Comput. Appl. Math. 2017, 312*, 156–166. [CrossRef]

Papaschinopoulos, G.; Papadopoulos, B.K. On the fuzzy difference equation $x_{n+1} = A + \frac{x_n}{x_{n-m}}$. *Fuzzy Sets Syst. 2002, 129, 73–81. [CrossRef]*

Villamizar-Roa, E.J.; Angulo-Castillo, V.; Chalco-Cano, Y. Existence of solutions to fuzzy differential equations with generalized Hukuhara derivative via contractive-like mapping principles. *Fuzzy Sets Systems 2015, 265*, 24–38. [CrossRef]

Bede, B.; Rudas, I.J.; Bencsik, A.L. First order linear fuzzy differential equations under generalized differentiability. *Inform. Sci. 2007, 177*, 1648–1662. [CrossRef]

Deeba, E.Y.; de Korvin, A. Analysis by fuzzy difference equations of a model of CO_2 level in the blood. *Appl. Math. Lett. 1999, 12*, 33–40. [CrossRef]

Kaleva, O. Fuzzy differential equations. *Fuzzy Sets Syst. 1987, 24*, 301–317. [CrossRef]

Papaschinopoulos, G.; Stefanidou, G. Boundedness and asymptotic behaviour of the solutions of a fuzzy difference equation. *Fuzzy Sets Syst. 2003, 140*, 523–539. [CrossRef]

Rodríguez-López, R. On the existence of solutions to periodic boundary value problems for fuzzy linear differential equations. *Fuzzy Sets Syst. 2013, 219*, 1–26. [CrossRef]

Buckley, J.J. The fuzzy mathematics of finance. *Fuzzy Sets Syst. 1987, 21*, 257–273. [CrossRef]

Córdova, J.D.; Molina, E.C.; López, P.N. Fuzzy logic and financial risk. A proposed classification of financial risk to the cooperative sector. *Contaduría Adm. 2017, 62*, 1687–1703. [CrossRef]

Kwapisz, M. On difference equations arising in mathematics of finance. *Nonlinear Anal. Theory Methods Appl. 1997, 30*, 1207–1218. [CrossRef]

Diamond, P.; Kloeden, P. *Metric Spaces of Fuzzy Sets*; World Scientific: Singapore, 1994.

Gasilov, N.; Amrahov, S.E.; Fatullayev, A.G. Solution of linear differential equations with fuzzy boundary values. *Fuzzy Sets Syst. 2014, 257*, 169–183. [CrossRef]

Nieto, J.J.; Rodríguez-López, R.; Franco, D. Linear first order fuzzy differential equations. *Int. J. Uncertain. Fuzziness Knowl.-Based Syst. 2006, 14*, 687–709. [CrossRef]

Nieto, J.J.; Rodríguez-López, R.; Georgiou, D.N. Fuzzy differential systems under generalized metric spaces approach. *Dyn. Syst. Appl. 2008, 17*, 1–24.

Khastan, A.; Rodríguez-López, R. On the solutions to first order linear fuzzy differential equations. *Fuzzy Sets Syst. 2016, 295*, 114–135. [CrossRef]

Papaschinopoulos, G.; Papadopoulos, B.K. On the fuzzy difference equation $x_{n+1} = A + B$. *Soft Comput. 2002, 6*, 456–461. [CrossRef]

Lakshmikantham, V.; Trigiante, D. *Theory of Difference Equations: Numerical Methods and Applications*; Academic Press: New York, NY, USA, 1988.

Kacprzyk, J.; Fedrizzi, M. *Fuzzy Regression Analysis*; Physica-Verlag: Heidelberg, Germany, 1992.

Capinski, M.; Zastawniak, T. *Mathematics for Finance: An Introduction to Financial Engineering*; Springer: London, UK, 2003.

Dong, N.P.; Long, H.V.; Khastan, A. Optimal control of a fractional order model for granular SEIR epidemic model. *Commun. Nonlinear Sci. Numer. Simulat. 2020, 88*, 105312. [CrossRef]

Mazandarani, M.; Kamyad, A.V. Modified fractional Euler method for solving fuzzy fractional initial value problem. *Commun. Nonlinear Sci. Numer. Simul. 2013, 18*, 12–21. [CrossRef]

Mazandarani, M.; Zhao, Y. Fuzzy Bang-Bang control problem under granular differentiability. *J. Franklin Inst. 2018, 355*, 4931–4951. [CrossRef]

Son, N.T.K.; Dong, N.P.; Son, L.H.; Abdel-Basset, M.; Manogaran, G.; Long, H.V. On the stabilizability for a class of linear time-invariant systems under uncertainty. *Circ. Syst. Signal Process. 2020, 39*, 919–960. [CrossRef]

Son, N.T.K.; Dong, N.P.; Long, H.V.; Son, L.H.; Khastan, A. Linear quadratic regulator problem governed by granular neutrosophic fractional differential equations. *ISA Trans. 2019, 97*, 296–316. [CrossRef] [PubMed]

3

Return and Volatility Transmission between World-Leading and Latin American Stock Markets: Portfolio Implications

Imran Yousaf [1], Shoaib Ali [1] and Wing-Keung Wong [2,3,4,*]

[1] Air University School of Management, Air University, Islamabad 44000, Pakistan; imranyousaf_12@pide.edu.pk (I.Y.); shoaibali_12@pide.edu.pk (S.A.)
[2] Department of Finance, Fintech Center, and Big Data Research Center, Asia University, Taichung 41354, Taiwan
[3] Department of Medical Research, China Medical University Hospital, Taichung 40402, Taiwan
[4] Department of Economics and Finance, The Hang Seng University of Hong Kong, Hong Kong 999077, China
* Correspondence: wong@asia.edu.tw

Abstract: This study uses the BEKK-GARCH model to examine the return-and-volatility spillover between the world-leading markets (USA and China) and four emerging Latin American stock markets over the global financial crisis of 2008 and the crash of the Chinese stock market of 2015. Regarding return spillover, our findings reveal a unidirectional return transmission from Mexico to the US stock market during the global financial crisis. During the crash of the Chinese stock market, the return spillover is found to be unidirectional from the US to the Brazil, Chile, Mexico, and Peru stock markets. Moreover, the results indicate a unidirectional return transmission from China to the Brazil, Chile, Mexico, and Peru stock markets during the global financial crisis and the crash of the Chinese stock market. Regarding volatility spillover, the results show the bidirectional volatility transmission between the US and the stock markets of Chile and Mexico during the global financial crisis. During the Chinese crash, the bidirectional volatility transmission is observed between the US and Mexican stock markets. Furthermore, the volatility spillover is unidirectional from China to the Brazil stock market during the global financial crisis. During the Chinese crash, the volatility spillover is bidirectional between the China and Brazil stock markets. Lastly, a portfolio analysis application has been conducted.

Keywords: return spillover; volatility spillover; optimal weights; hedge ratios; US financial crisis; Chinese stock market crash

JEL Classification: G10; G11; G12; G15

---

1. Introduction

The information transmissions (return and volatility) across equity markets are of greater interest to investors and policymakers with increased financial integration all over the world. For example, if asset volatility is transferred from one market to another during turbulence or crisis, then portfolio managers need to adjust their asset allocation (Bouri 2013; Syriopoulos et al. 2015; Yousaf and Hassan 2019) and financial policymakers need to change their policies to reduce the contagion risk (Yang and Zhou 2017). The linkages between equity markets, especially during a crisis, can also have important implications for asset allocations, portfolio diversification, asset valuation, hedging, and risk management.

In the literature, several studies have examined the linkages between equity markets during the Asian crisis of 1997 (In et al. 2001; Chen et al. 2002; Chancharoenchai and Dibooglu 2006;

Li and Giles 2014; Gulzar et al. 2019) and the global financial crisis (Taşdemir and Yalama 2014; Bekiros 2014; Mensi et al. 2016; Gamba-Santamaria et al. 2017). However, the linkages between equity markets are rarely examined during the crash of the Chinese stock market in 2015. The Chinese stock market crashed in 2015 (Han and Liang 2016; Ahmed and Huo 2019; Yousaf and Hassan 2019). The CSI 300 index had reached up to 5178 points until mid-June in 2015. Then, it took roller-coaster ride and dropped up to 34% in just 20 days, also losing 1000 points within just one week. Around 50% of the Chinese stocks lost more than half of their pre-crash market value. This crash adversely affected the many other financial markets around the globe (Fang and Bessler 2017). Despite the importance of the Chinese crash to international portfolio managers, only Ahmed and Huo (2019) examined the volatility transmission between the Chinese and Asian stock markets during the crash of the Chinese stock market in 2015. The empirical research remains surprisingly limited on the area of linkages between equity markets during the crash of the Chinese stock market.

The US and China are the most significant trading partners of the emerging Latin American economies. From 2000 to 2017, the trade volume of China (US) is increased by 21 (2.5)-fold with emerging Latin American economies. The trade volume of leading economies grew at a different rate with the emerging Latin American (LA) economies; thus, spillover can also be changed between the China-LA and US-LA pairs during the last two decades. Johnson and Soenen (2003) also suggest that trade increases the financial integration between countries' stock markets. Previously, several studies have examined the spillovers between the US and Latin American stock markets (Meric et al. 2001; Arouri et al. 2015; Ben Rejeb and Arfaoui 2016; Cardona et al. 2017; Gamba-Santamaria et al. 2017; Ramirez-Hassan and Pantoja 2018; Yousaf and Ahmed 2018; Fortunato et al. 2019; Coleman et al. 2018). However, the linkages between the China and Latin American stock markets have not yet been explored, especially during the global financial crisis and the crash of the Chinese stock market.

Based on the above-mentioned literature gaps, this study aims to examine the return and volatility spillover between the world-leading (the US and China) and emerging Latin American stock markets during the full sample period, the global financial crisis, and the crash of the Chinese stock market. Additionally, this study estimates the optimal weights and hedge ratios during all the sample periods.

Our study makes the following contributions to the literature. First, regarding return spillover, the findings reveal a unidirectional return transmission from Mexico to the US stock market during the global financial crisis. During the crash of the Chinese stock market, the return spillover is found to be unidirectional from the US to the Brazil, Chile, Mexico, and Peru stock markets. Moreover, the results indicate a unidirectional return transmission from China to the Brazil, Chile, Mexico, and Peru stock markets during the global financial crisis and the crash of the Chinese stock market. Regarding volatility spillover, the results show the bidirectional volatility transmission between the US and the stock markets of Chile and Mexico during the global financial crisis. During the Chinese crash, a bidirectional volatility transmission is observed between the US and Mexican stock markets. Furthermore, the volatility spillover is unidirectional from China to the Brazil stock market during the global financial crisis. During the Chinese crash, the volatility spillover is bidirectional between the China and Brazil stock markets.

The contributions of this study are four-fold. First, this study provides a comprehensive analysis of spillover between the world-leading and emerging LA stock markets during the crash of the Chinese stock market. Second, it contributes to the literature of the China-LA stock markets by examining the spillovers during the global financial crisis. Lastly, the BEKK-GARCH model is applied to estimate the spillovers, optimal weights, and hedge ratios, which provide better statistical properties compared to the many other GARCH models. The rest of the paper is organized as follows: Section 2 provides a review of the literature. The empirical method is described in Section 3. Section 4 consists of the data and the preliminary analysis. The empirical results are reported in Section 5. Finally, Section 5 concludes the discussion.

2. Literature Review

Markowitz's modern portfolio theory can describe the relationship between different stock markets in order to build an optimum portfolio. The rationale behind this concept is to combine risky assets with less risky or risk-free assets in the portfolio (Markovitz 1959). For example, the leading stock market shows a higher volatility during the financial crisis, and as a result the portfolio investors need to diversify their portfolios by investing in weakly integrated emerging stock markets. Therefore, an analysis of risk transmission between different equity markets is essential for portfolio managers to identify opportunities for portfolio diversification across markets and over time.

Over the past decade, there has been a growing body of literature examining the information transmissions (return and volatility) between the US and LA stock markets during the crisis and non-crisis periods. Meric et al. (2001) report significant co-movements between the US and LA (Brazil, Argentina, Chile, and Mexico) stock markets during the period 1984–1995. Fernández-Serrano and Sosvilla-Rivero (2003) report the cointegration across the US and LA equity markets. Sharkasi et al. (2005) investigate the spillover across the US and Brazil stock markets. They provide evidence of co-movements between the US and Brazil stock markets.

Diamandis (2009) investigates the linkages and common trends between the US and four Latin American (Argentina, Brazil, Chile, and Mexico) stock markets. Because the four Latin American countries initiated a phase of financial liberalization in the late 1980s and early 1990s, this study also explores whether the removal of foreign-exchange controls had any effect on the potential linkages. Firstly, this study finds that the US stock market is partially integrated with four LA stock markets. Secondly, the five stock markets have four significant common permanent components/trends which influence their system in the long run. Thirdly, the results indicate significant short-term deviations from standard stochastic patterns during the 1994–1996 Mexican crisis and the 2001 financial crisis.

Beirne et al. (2013) use the tri-variate GARCH-BEKK model to estimate the volatility transmission from mature markets to 41 emerging (including 8 Latin American) stock markets. The volatility transmission is observed to be significant from many mature markets to the emerging stock markets. Additionally, there is evidence of changes in the parameters of volatility spillovers during turbulent or crisis periods. Graham et al. (2012) estimate the integration between the US and 22 emerging equity markets and find evidence of strong co-movements across the US, Brazil, and Mexico equity markets. Hwang (2014) examined the spillover between the US and LA equity markets during the global financial crisis. The study found that the integration between the US and LA equity markets became stronger during the global financial crisis.

Using the VAR-GARCH model, Arouri et al. (2015) estimate the return and volatility transmissions between the US and LA (Brazil, Argentina, Mexico, Chile, and Columbia) stock markets from 1993 through to 2012. The return spillover is seen to be significant from the US to the Argentina, Mexico, and Colombia stock markets. It also provides evidence of a volatility transmission from the US to a few LA stock markets. Syriopoulos et al. (2015) use the VAR-GARCH model and find that the return and volatility spillover is significant between the US and BRICS (Brazil, Russia, India, China, and South Africa) equity markets (at the sectoral level). Mensi et al. (2016) reveal the strong dynamic correlation between US and BRICS equity markets during the global financial crisis of 2008.

Ben Rejeb and Arfaoui (2016) examine the volatility transmission between developed (US and Japan) and emerging (Latin American and Asian) stock markets using standard GARCH models and a quantile regression approach. This study reveals a significant presence of volatility transmission in these markets. The volatility transmission is seen to be closely associated with the crisis period and geographical proximity. A lower and upper quantiles analysis shows that interdependence between markets decreases during a bearish trend, while it increases during bullish markets. Using the GARCH model, Bhuyan et al. (2016) observes return and volatility transmissions from the US to BRICS stock markets.

Al Nasser and Hajilee (2016) provide evidence of short-run integration between developed (US, UK, and Germany) and emerging stock markets (Brazil, Mexico, Russia, China, and Turkey). However, in the long run, the cointegration is only found to be significant between Germany and emerging Asian stock markets. Gamba-Santamaria et al. (2017) examine the directional volatility transmission between the US and the four LA stock markets (Brazil, Chile, Mexico, and Columbia) using the framework of Diebold and Yilmaz (2012). Brazil is found to be the net volatility transmitter for most of the sample period, whereas Columbia, Chile, and Mexico are the net receivers of volatility. Moreover, the US stock market is observed to be the net transmitter of volatility to the four LA stock markets. Besides this, the magnitude of volatility transmission is increased from the US to LA stock markets during the global financial crisis of 2008.[1]

Yousaf and Ahmed (2018) study the influence of the US and Brazil on the Mexico, Argentina, Chile, and Peru stock markets by using GARCH in a mean approach. The study concludes that the return effects are dominantly transmitted from the US to the Mexico, Argentina, Chile, and Peru stock markets. Moreover, the volatility transmission is found to be dominant from Brazil to the Mexico, Argentina, Chile, and Peru stock markets. Cardona et al. (2017) use the MGARCH-BEKK model to estimate the volatility transmission between the US and the six LA stock markets (Brazil, Argentina, Mexico, Chile, Peru, and Colombia). They report the significant volatility transmission from the US to all LA stock markets. Moreover, only Brazil transmits volatility effects to the US stock market.

Ramirez-Hassan and Pantoja (2018) provide evidence of co-movements between the returns of the US and six LA stock markets after the global financial crisis of 2008. Fortunato et al. (2019) provide evidence of return transmission from the US to the Brazil, Chile, Columbia, Mexico, and Peru equity markets. Coleman et al. (2018) find the co-movements between the US and LA (Brazil, Chile, Mexico, Peru, Venezuela, and Argentina) stock markets. Su (2020) reports the dominant risk transmission from the G7 (US, Japan, UK, Germany, France, Italy, and Canada) countries to the BRICS (Brazil, Russia, India, China, and South Africa) stock markets.

However, fewer studies have examined the spillovers between the China and Latin American stock markets during the crisis and non-crisis periods. Garza-García and Vera-Juárez (2010) study the impact of US and Chinese macroeconomic variables on the stock markets of Brazil, Mexico, and Chile. The macroeconomic variables (the US and Chinese) are observed to be integrated with the LA stock markets. Additionally, the US macroeconomic variables Granger affect the Brazilian and Mexican stock markets. On the other hand, the Chinese macroeconomic variables Granger affect the stock markets of Mexico and Chile.

Horvath and Poldauf (2012) find that the Chinese stock market is weakly correlated with the Brazil, Australia, Canada, Germany, Japan, Hong Kong, South Africa, Russia, US, and UK stock markets. Sharma et al. (2013) apply the VAR model to examine the linkages between the BRICS (Brazil, Russia, India, China, and South Africa) stock markets. This study finds a return transmission from Brazil (India) to the Russia, India (Brazil), China, and South Africa equity markets. Moreover, the return transmission is only observed from China to the Russian stock market. Bekiros (2014) looks at the contagion effect between Brazil, Russia, India, and China by using several multivariate GARCH models.

---

[1] Our study is different from the study of Gamba-Santamaria et al. (2017) in the following aspects. Gamba-Santamaria et al. (2017) examine the volatility spillover between the US and four Latin American markets (Brazil, Chile, Mexico, and Columbia) during the US financial crisis, whereas our study is examining the volatility as well as return spillover between the leading (US and China) markets and four Latin American markets (Brazil, Chile, Mexico, and Peru) during the US financial crisis and the crash of the Chinese stock market. More specifically, firstly our study examines the return as well as volatility spillovers, whereas Gamba-Santamaria et al. (2017) examine the directional volatility spillovers. Second, our study is examining the spillovers between two world-leading (the US and China) markets and four LA markets, whereas Gamba-Santamaria et al. (2017) examine the spillovers between US and four LA markets. Third, our study is focusing on the spillovers during the global financial crisis and the crash of the Chinese stock market in 2015, whereas Gamba-Santamaria et al. (2017) examine the spillovers during the US financial crisis. Fourth, our study is using the BEKK-GARCH model, whereas Gamba-Santamaria et al. (2017) employ the approach of Diebold and Yilmaz (2012). Lastly, our full data sample is from January 2001 to May 2020, whereas Gamba-Santamaria et al. (2017) use the sample period from January 2003 to January 2016. Apart from the differences, the study of Gamba-Santamaria et al. (2017) is very beneficial for understanding the linkages among the US and LA stock markets.

This study concludes that there exists a higher integration between Brazil, Russia, India, and China after the global financial crisis.

Ahmad and Sehgal (2015) estimate the volatility of the BRIICKS (Brazil, Russia, India, Indonesia, China, South Korea, and South Africa) stock markets by using the Markov regime-switching (MS) in the mean-variance model. It suggests that investors should allocate investment in the China, Russia, and India emerging stock markets. While investigating the relationship between the Chinese and foreign stock markets (US, Brazil, India, and Germany), Cao et al. (2017) reported a bi-directional causality between the China and foreign stock markets. Previous work does not provide evidence of return and volatility spillover between leading (US and China) and Latin American stock markets during the global financial crisis and the crash of the Chinese stock market. Therefore, this study addresses the above-mentioned literature gaps.

3. Data and Methodology

In this section, we will discuss the data and methodology used in our paper. We first discuss the data.

3.1. Data

This study uses the daily data of benchmark stock indices of the US (S&P 500); China (SSE Composite Index); and four emerging LA stock markets—namely, Brazil (IBOVESPA index), Chile (IPSA index), Mexico (S&P/BMV IPC Index), and Peru (S&P/BVL Peru General TR PEN Index). The data of stock indices are taken from the Data Stream database. The index is assumed to be the same on non-trading days (holidays except weekends) as on the previous trading day, as suggested by Malik and Hammoudeh (2007), and Ali et al. (2020).

This study uses the full sample period from 1 January 2000, to 29 May 2020, and studies the following two sub-samples: the first sub-period from 1 August 2007, to 30 July 2010, presenting the period with the US financial crisis; and the second sub-period from 1 June 2015, to 31 May 2018, presenting the period with the Chinese stock market crash. We note that Yousaf and Hassan (2019) also use similar timeframes for the global financial crisis and the crash of the Chinese stock market. This study follows He (2001) to use three-year data for each crisis for a short-run analysis. Changes in the market correlations take place continuously, not only as a result of the crises but also due to the consequences of many financial, economic, and political events. This study uses the same time for both the crisis periods to make the coefficient comparable. The difference in the opening time of the China and LA stock markets has been adjusted in the estimations.

3.2. Methodology

The econometric specification used in this study has two components. First, a vector autoregression (VAR) with one lag is used to model the returns.[2] This allows for autocorrelations and cross-autocorrelations in the returns. Second, a multivariate BEKK-GARCH model is used to model the time-varying variances and covariances developed by Engle and Kroner (1995).[3] BEKK-GARCH has the attractive characteristics that the conditional covariance matrices are positive definite (Chang et al. 2011). Several studies have used the BEKK-GARCH model to estimate the spillover between different asset classes; see, for example, Chang et al. (2011), Sadorsky (2012), Beirne et al. (2013), Chang et al. (2017), Cardona et al. (2017), and Sarwar et al. (2020). Moreover, we will estimate the optimal weights and hedge ratios using the BEKK-GARCH model.

---

2 The number of lags is selected on the basis of the AIC and SIC criteria.

3 We apply the BEKK-GARCH model on the valuable suggestion of a respected reviewer.

This study aims to examine the return and volatility spillover between the stock markets, and thus we firstly focus on return spillover. For any pair of two series, the following are the specifications for the conditional mean equation:

$$R_t = \mu + \varnothing\, R_{t-1} + e_t \text{ with } e_t = H_t^{1/2}\eta_t. \tag{1}$$

$R_t = (R_t^x, R_t^y)'$ is the vector of returns on the stock market indices x and y at time t, respectively; $\varnothing$ is the 2×2 matrix of parameters, measuring the impacts of own lagged and cross mean transmissions between two series; $e_t = \left(e_t^x, e_t^y\right)'$ is the vector of error terms of the conditional mean equations for the two series at time t; $\eta_t = \left(\eta_t^x, \eta_t^y\right)'$ indicates a sequence of independently and identically distributed random errors; and $H_t = \begin{pmatrix} H_t^x & H_t^{xy} \\ H_t^{xy} & H_t^y \end{pmatrix}$ denotes the conditional variance-covariance matrix of return series of x and y. In addition, $H_t^{1/2}$ is the 2×2 symmetric positive definite matrix.

The full BEKK–GARCH, which imposes positive definiteness restrictions for H_t, is given by:

$$H_t = C'C + A'e_{t-1}e_{t-1}'A + B'H_{t-1}B, \tag{2}$$

where A and B are $(n \times n)$ coefficient matrices and $C'C$ is the decomposition of the intercept matrix. Each element (i,j)th in H_t depends on the corresponding (i,j)th element in $(e_{t-1}e_{t-1}')$ and H_{t-1}. Accordingly, past shocks and volatility are allowed to directly spill over from a market to another, and they are captured by the coefficients of the A and B matrices. More specifically, the BEKK-GARCH matrices can be expanded as follows:

$$h_t^x = C_x + \alpha_x^2\left(e_{t-1}^x\right)^2 + 2\alpha_x\alpha_{yx}e_{t-1}^x e_{t-1}^y + \alpha_{yx}^2\left(e_{t-1}^y\right)^2 + \beta_x^2 h_{t-1}^x + 2\beta_x\beta_{yx}h_{t-1}^{xy} + \beta_{yx}^2 h_{t-1}^y \tag{3}$$

$$h_t^{xy} = C_{xy} + \alpha_x\alpha_{xy}\left(e_{t-1}^x\right)^2 + \left(\alpha_{yx}\alpha_{xy} + \alpha_x\alpha_y\right)e_{t-1}^x e_{t-1}^y + \alpha_{yx}a_y\left(e_{t-1}^y\right)^2 + \beta_s\beta_{xy}h_{t-1}^x + \left(\beta_{yx}\beta_{xy} + \beta_x\beta_y\right)h_{t-1}^{xy} + \beta_{yx}\beta_y h_{t-1}^y \tag{4}$$

$$h_t^y = C_y + \alpha_{xy}^2\left(e_{t-1}^x\right)^2 + 2\alpha_{xy}\alpha_y e_{t-1}^x e_{t-1}^y + \alpha_y^2\left(e_{t-1}^y\right)^2 + \beta_{xy}^2 h_{t-1}^x + 2\beta_{xy}\beta_y h_{t-1}^{xy} + \beta_y^2 h_{t-1}^y \tag{5}$$

The BEKK-GARCH parameters are estimated by the maximum likelihood method using the BFGS algorithm. In addition to the return and volatility spillover, we also compute the optimal weights and hedge ratios for each pair of stocks.

The conditional variance and covariances are used for calculating the optimal portfolio weights and hedge ratios. This study follows Kroner and Ng (1998) in calculating the optimal portfolio weights of different pairs of stock markets:

$$w_t^{xy} = \frac{h_t^y - h_t^{xy}}{h_t^x - 2h_t^{xy} + h_t^y}, \tag{6}$$

$$w_t^{xy} = \begin{cases} 0, If\ w_t^{xy} < 0 \\ w_t^{xy}, If\ 0 \le w_t^{xy} \le 1 \\ 1, If\ w_t^{xy} > 1 \end{cases}$$

where w_t^{xy} is the weight of stock(x) in a \$1 stock(x)-stock(y) portfolio at time t; h_t^{xy} is the conditional covariance between the two stock markets; h_t^x and h_t^y are the conditional variance of stock(x) and stock(y), respectively; and $1 - w_t^{xy}$ is the weight of stock(y) in a \$1 stock(x)-stock(y) portfolio. As suggested by Kroner and Sultan (1993):

$$\beta_t^{xy} = \frac{h_t^{xy}}{h_t^y} \tag{7}$$

where β_t^{xy} represents the hedge ratio. This shows that a short position in the stock (y) market can hedge a long position in stock (x).

4. Empirical Results and Implications

In this section, we will discuss our empirical results and implications. We first discuss our preliminary analysis.

4.1. Descriptive Statistics

Table 1 reports the summary statistics of the daily returns for the US; China; and four emerging LA stock markets—namely, Brazil, Chile, Mexico, and Peru. Among them, Brazil and Peru have the highest mean return, and the US has the smallest mean return during the full sample period. On the other hand, Chile has the smallest standard deviation, while Brazil has the largest standard deviation. Thus, Peru provides the highest mean return, with a relatively smaller risk in the LA stock markets. Overall, the skewness is significantly negative, the kurtosis is significantly higher than three for all stocks, and the Jarque–Bera statistics reject normality hypothesis for all series, inferring that all the returns are negatively skewed and fat-tailed. Moreover, Table 1 also confirms that there are 1% significant autocorrelation and ARCH (autoregressive conditional heteroskedasticity) effects for all returns. We also apply both Augmented Dickey–Fuller (ADF) and Phillip–Perron (PP) tests to examine the stationarity of all the returns and exhibit the results in Table 2. The table indicates that all the series are 1% significant, inferring that all the returns are stationary.

Table 1. Summary statistics.

Markets	Mean	Std. Dev.	Skewness	Kurtosis	J-B Stat	Q-Stat	ARCH
US	0.00016	0.0124	−0.364 ***	14.045 ***	27181.3 ***	56.584 ***	548.40 ***
CHN	0.00040	0.0155	−0.330 ***	8.2116 ***	6121.9 ***	60.119 ***	189.01 ***
BRAZ	0.00047	0.0183	−0.403 ***	9.6439 ***	9937.1 ***	24.957 ***	686.82 ***
CHIL	0.00030	0.0105	−0.878 ***	19.883 ***	37,432.8 ***	148.49 ***	180.34 ***
MEXI	0.00024	0.0128	−0.086 *	8.3698 ***	6403.18 ***	108.33 ***	173.49 ***
PERU	0.00047	0.0133	−0.549 ***	15.441 ***	34605.3 ***	290.64 ***	796.97 ***

Notes: US—United States of America; CHN—China; BRAZ—Brazil; MEXI—Mexico; CHIL—Chile. Q-stat denotes the Ljung–Box Q-statistics. ARCH test refers to the LM-ARCH test. ***, * indicate the statistical significance at 1% and 10%, respectively.

Table 2. Unit root tests

	ADF (*t*-Test)			Phillips-Perron Test		
Markets	None	Constant	Constant and Trend	None	Constant	Constant and Trend
US	−79.73 ***	−79.94 ***	−79.96 ***	−80.00 ***	−80.01 ***	−80.05 ***
CHN	−32.44 ***	−32.48 ***	−32.49 ***	−69.97 ***	−69.89 ***	−69.88 ***
BRAZ	−72.04 ***	−72.08 ***	−72.08 ***	−72.03 ***	−72.08 ***	−72.08 ***
CHIL	−33.59 ***	−33.64 ***	−33.67 ***	−63.11 ***	−63.11 ***	−63.10 ***
MEXI	−50.70 ***	−50.73 ***	−50.73 ***	−63.67 ***	−63.68 ***	−63.68 ***
PERU	−31.06 ***	−31.12 ***	−31.15 ***	−61.82 ***	−61.75 ***	−61.64 ***

Notes: US—United States of America; CHN—China; BRAZ—Brazil; MEXI—Mexico; CHIL—Chile; ADF—Augmented Dickey Fuller. *** indicate the statistical significance at 10%, respectively.

4.2. Return and Volatility Spillover between the US and LA Stock Markets

We turn to apply the BEKK-GARCH model to examine the return and volatility spillovers between the US and LA stock markets in the full sample period, the global financial crisis, and the crash of the Chinese stock market and exhibit the results in Tables 3–5. We note that the 1% significant autocorrelation and ARCH effects for all returns, as shown in Table 1, justify the use of the BEKK-GARCH model in our analysis.

Table 3. Estimates of BEKK-GARCH for the US and Latin American stock markets during the full sample period

	Brazil and US		Chile and US		Mexico and US		Peru and US	
	Coefficient	***p-Value***	**Coefficient**	***p-Value***	**Coefficient**	***p-Value***	**Coefficient**	***p-Value***
				Panel A. Mean Equation				
μ_1	0.075 ***	0.000	0.098 *	0.062	0.058 ***	0.000	0.047 ***	0.000
$\varnothing_{11}$	−0.031 **	0.036	0.007	0.722	0.038 **	0.027	0.137 ***	0.000
$\varnothing_{12}$	0.023 ***	0.005	0.000	0.938	0.021	0.109	−0.019	0.185
μ_2	0.056 ***	0.000	0.062 ***	0.000	0.051 ***	0.000	0.058 ***	0.000
$\varnothing_{21}$	0.055 **	0.028	0.111 ***	0.000	0.050 ***	0.005	0.081 ***	0.000
$\varnothing_{22}$	−0.077 ***	0.000	−0.057 ***	0.001	−0.071 ***	0.000	−0.042 ***	0.006
				Panel B. Variance Equation				
c_{11}	0.219 ***	0.000	2.496 ***	0.000	0.117 ***	0.000	0.161 ***	0.000
c_{21}	0.069 ***	0.004	0.124 ***	0.009	0.050 ***	0.006	0.056 ***	0.003
c_{22}	0.118 ***	0.000	0.042	0.571	0.124 ***	0.000	0.122 ***	0.000
α_{11}	0.225 ***	0.000	0.003	0.562	0.264 ***	0.000	0.309 ***	0.000
α_{12}	−0.014	0.319	0.001	0.611	0.008	0.609	0.004	0.836
α_{21}	0.036	0.421	0.151 **	0.026	−0.026	0.390	0.010	0.541
α_{22}	0.338 ***	0.000	0.341 ***	0.000	0.314 ***	0.000	0.300 ***	0.000
β_{11}	0.966 ***	0.000	0.701 ***	0.000	0.959 ***	0.000	0.942 ***	0.000
β_{12}	0.005	0.158	−0.003	0.400	0.005	0.349	−0.002	0.789
β_{21}	−0.101 **	0.050	0.095	0.203	0.090 **	0.043	−0.004	0.538
β_{22}	0.933 ***	0.000	0.944 ***	0.000	0.938 ***	0.000	0.947 ***	0.000
				Panel C. Diagnostic Tests				
LogL	−16,174.1		−21,397.4		−13,816.1		−14,378.1	
AIC	6.789		8.588		5.970		6.370	
SIC	6.799		8.635		6.017		6.417	
$Q_1[20]$	30.320 *	0.065	1.791	0.720	19.075	0.517	328.759 *	0.031
$Q_2[20]$	18.920	0.527	19.184	0.510	19.493	0.490	17.728	0.605
$Q_1^2[20]$	29.942	0.182	0.004	0.659	34.776 **	0.021	30.071	0.198
$Q_2^2[20]$	27.782	0.115	22.616	0.308	25.161	0.185	33.181	0.146

Notes: US, United States of America; CHN, China; BRAZ, Brazil; CHIL, Chile; MEXI, Mexico. Variable order is the Latin American stock market (1) and China (2). In the mean equations, μ denotes the constant terms, whereas $\varnothing_{12}$ denotes the return spillover from the Latin American stock market to the US stock market. In the variance equation, c denotes the constant terms, α denotes the ARCH terms, and β denotes the GARCH terms. In the variance equation, α_{12} indicates the shock spillover from the Latin American stock market to the US stock market, whereas β_{12} denotes the long−term volatility spillover from the Latin American stock market to the US stock market. Number of lags for VAR is decided using the SIC and AIC criteria. JB, Q(20), and $Q^2(20)$ indicate the empirical statistics of the Jarque–Bera test for normality, Ljung–Box Q statistics of order 20 for autocorrelation applied to the standardized residuals, and squared standardized residuals, respectively. Values in parentheses are the *p-Value*. ***, **, * indicate the statistical significance at 1%, 5%, and 10%, respectively.

Tables 3–5 report the return and volatility spillovers between the US and LA stock markets during the full sample period, the global financial crisis, and the crash of the Chinese stock market, respectively. Referring to coefficients $\varnothing_{11}$ and $\varnothing_{22}$ in Panel A, the results show that the lagged returns significantly influence the current returns in the US and the majority of LA stock markets during the full sample period, the global financial crisis, and the crash of the Chinese stock market. It highlights the possibility of the short-term prediction of current returns through past returns in the US and the majority of the LA stock markets. Our results are consistent with the findings of Syriopoulos et al. (2015) and Arouri et al. (2015), which observe a significant impact of past returns on current returns in the US and LA stock markets.

Table 4. Estimates of BEKK-GARCH for US and Latin American stock markets during the global financial crisis.

	Brazil and US		Chile and US		Mexico and US		Peru and US	
	Coefficient	*p-Value*	Coefficient	*p-Value*	Coefficient	*p-Value*	Coefficient	*p-Value*
Panel A. Mean Equation								
μ_1	0.101 *	0.068	0.113 ***	0.000	0.055	0.246	−0.014	0.744
$\varnothing_{11}$	−0.044	0.450	0.118 ***	0.001	0.023	0.680	0.138 ***	0.000
$\varnothing_{12}$	0.021	0.581	0.019	0.440	0.096 **	0.043	−0.009	0.770
μ_2	0.019	0.665	0.046	0.361	0.064	0.144	0.029	0.518
$\varnothing_{21}$	0.040	0.578	0.020	0.308	0.020	0.707	0.063	0.107
$\varnothing_{22}$	−0.128 ***	0.007	−0.164 ***	0.000	−0.188 ***	0.000	−0.100 ***	0.003
Panel B. Variance Equation								
c_{11}	0.271 **	0.044	0.287 ***	0.000	0.218 **	0.017	0.291 ***	0.000
c_{21}	0.098	0.546	0.040	0.417	−0.035	0.730	0.173 ***	0.000
c_{22}	0.129 **	0.039	0.153 ***	0.000	0.000	0.799	0.109 ***	0.007
α_{11}	0.421 *	0.069	0.483 ***	0.000	0.117	0.220	0.453 ***	0.000
α_{12}	0.139 **	0.022	−0.055	0.333	−0.104 *	0.057	0.087	0.255
α_{21}	−0.237	0.128	−0.020	0.550	0.249 ***	0.000	−0.088	0.581
α_{22}	0.138	0.145	0.292 ***	0.000	0.295 ***	0.000	0.226 ***	0.002
β_{11}	0.902 ***	0.000	0.841 ***	0.000	1.063 ***	0.000	0.896 ***	0.000
β_{12}	-0.041 *	0.082	0.051 **	0.034	0.218 ***	0.001	−0.034	0.197
β_{21}	0.071	0.482	0.024 *	0.083	−0.183 ***	0.000	0.014	0.687
β_{22}	0.990 ***	0.000	0.937 ***	0.000	0.797 ***	0.000	0.969 ***	0.000
Panel C. Diagnostic Tests								
LogL	−2766		−2438.432		−2514.181		−2804.767	
AIC	7.792		7.026		7.132		8.025	
SIC	8.018		7.253		7.359		8.251	
$Q_1[20]$	15.405	0.753	15.396	0.753	15.749	0.732	18.138	0.578
$Q_2[20]$	19.980	0.459	22.469	0.316	25.023	0.201	19.998	0.458
$Q_1^2[20]$	23.671	0.257	17.388	0.628	15.064	0.773	13.734	0.844
$Q_2^2[20]$	37.237 **	0.011	45.203 ***	0.001	33.878 **	0.027	37.570 ***	0.010

Notes: US, United States of America; CHN, China; BRAZ, Brazil; CHIL, Chile; MEXI, Mexico. Variable order is the Latin American stock market (1) and China (2). In the mean equations, μ denotes the constant terms, whereas $\varnothing_{12}$ denotes the return spillover from the Latin American stock market to the US stock market. In the variance equation, c denotes the constant terms, α denotes the ARCH terms, and β denotes the GARCH terms. In the variance equation, α_{12} indicates the shock spillover from the Latin American stock market to the US stock market, whereas β_{12} denotes the long-term volatility spillover from the Latin American stock market to the US stock market. Number of lags for VAR is decided using the SIC and AIC criteria. JB, Q(20), and $Q^2(20)$ indicate the empirical statistics of the Jarque–Bera test for normality, Ljung–Box Q statistics of order 20 for autocorrelation applied to the standardized residuals, and squared standardized residuals, respectively. Values in parentheses are the *p-Value*. ***, **, * indicate the statistical significance at 1%, 5%, and 10%, respectively.

Regarding the interdependence of returns in the mean equation (see coefficients $\varnothing_{12}$ and $\varnothing_{21}$ in Panel A), the results indicate the unidirectional return spillover from the US to the majority of LA stock markets during the full sample period and the crash of the Chinese Stock Market. They imply that the past US returns can be used to predict the current returns of the LA markets during the full sample period and the crash of the Chinese Stock Market. These results are consistent with the previous findings of Arouri et al. (2015), who find the unidirectional return spillover from the US to the LA stock markets. Moreover, the return transmission is also significant from the Brazil to the US stock market during the full sample period. In contrast, the return transmissions are not found to be significant between the US and the majority of the LA stock (except Mexico) markets during the global financial crisis. These results suggest that the US (LA) stock returns are not useful in predicting the returns in the majority of the LA (US) stock markets during the global financial crisis. The results also reveal a unidirectional volatility spillover from Mexico to the US stock market during the global financial crisis.

Based on the variance equation (see coefficients of α_{11} in Panel B), the results show that the conditional volatility of the majority of LA stock markets depends on their past shocks during all the sample periods. In addition, the coefficients of the past own shocks (α_{22}) are highly significant for the US in all the sample periods. Besides this, the sensitivity of past own volatility (β_{11} *and* β_{22}) is significant for the US and LA stock markets during all the sample periods. These results are consistent with the findings of Syriopoulos et al. (2015), which find that the past own volatility is a significant determinant of the future volatility of BRICS countries (including Brazil). Further, the coefficients of past own volatility are higher compared to the coefficients of the past own shocks in the US and LA stock markets, suggesting that the past own volatilities are more critical for the prediction of future volatility than the past own shocks during all the sample periods.

Referring to the coefficient α_{12} and α_{21} in Panel B, the past shocks of the US stock market significantly influence the conditional volatility of just the Chile stock market during the full sample period. During the global financial crisis, the shock transmission is unidirectional from Brazil to the US and bidirectional between the US and Mexican stock markets. Moreover, the conditional volatility of the Mexican stock market is significantly affected by the US during the crash of the Chinese stock market.

Table 5. Estimates of BEKK-GARCH for the US and Latin American stock markets during the crash of the Chinese stock market.

	Brazil and US		Chile and US		Mexico and US		Peru and US	
	Coefficient	*p-Value*	Coefficient	*p-Value*	Coefficient	*p-Value*	Coefficient	*p-Value*
Panel A. Mean Equation								
μ_1	0.090	0.105	0.030	0.210	0.013	0.585	0.088 **	0.029
$\varnothing_{11}$	−0.047	0.172	0.077 *	0.082	−0.032	0.469	0.076 *	0.089
$\varnothing_{12}$	0.016	0.302	−0.035	0.134	0.002	0.937	−0.025	0.550
μ_2	0.064 ***	0.004	0.075 ***	0.001	0.072 ***	0.001	0.061	0.158
$\varnothing_{21}$	0.129 *	0.064	0.120 ***	0.001	0.137 ***	0.000	0.086 *	0.093
$\varnothing_{22}$	−0.066 **	0.040	−0.052	0.112	−0.059 *	0.083	−0.050 *	0.085
Panel B. Variance Equation								
c_{11}	0.268 ***	0.005	0.361 ***	0.004	0.599 ***	0.000	0.159 *	0.066
c_{21}	0.151 **	0.017	0.011	0.720	0.164 ***	0.000	0.117	0.122
c_{22}	0.124 *	0.076	0.186 ***	0.000	0.124	0.150	0.089	0.790
α_{11}	0.196 ***	0.001	0.522 ***	0.001	0.434 ***	0.000	0.278 **	0.011
α_{12}	0.008	0.821	−0.024	0.326	0.033	0.298	0.142	0.331
α_{21}	0.023	0.744	−0.028	0.648	−0.115 *	0.052	0.019	0.850
α_{22}	0.430 ***	0.000	0.421 ***	0.000	0.381 ***	0.000	0.313	0.416
β_{11}	0.958 ***	0.000	0.686 ***	0.001	−0.359*	0.077	0.949 ***	0.000
β_{12}	−0.008	0.571	0.018	0.411	−0.068 ***	0.000	−0.042	0.464
β_{21}	0.013	0.697	0.076	0.227	0.528 ***	0.000	−0.014	0.766
β_{22}	0.880 ***	0.000	0.879 ***	0.000	0.915 ***	0.000	0.917 ***	0.000
Panel C. Diagnostic Tests								
LogL	−2078		−1582.556		−1545.033		−1733.842	
AIC	5.759		4.585		4.429		4.891	
SIC	5.986		4.812		4.655		5.118	
$Q_1[20]$	21.413	0.373	33.001 **	0.034	21.955	0.343	31.804 **	0.045
$Q_2[20]$	24.907	0.205	24.713	0.213	24.601	0.217	25.783	0.173
$Q_1^2[20]$	6.942	0.897	85.117 ***	0.000	29.827 *	0.073	16.276	0.699
$Q_2^2[20]$	8.249	0.890	9.945	0.969	10.383	0.961	8.909	0.984

Notes: US, United States of America; CHN, China; BRAZ, Brazil; CHIL, Chile; MEXI, Mexico. Variable order is the Latin American stock market (1) and China (2). In the mean equations, μ denotes the constant terms, whereas $\varnothing_{12}$ denotes the return spillover from the Latin American stock market to the US stock market. In the variance equation, c denotes the constant terms, α denotes the ARCH terms, and β denotes the GARCH terms. In the variance equation, α_{12} indicates the shock spillover from the Latin American stock market to the US stock market, whereas β_{12} denotes the long-term volatility spillover from the Latin American stock market to the US stock market. Number of lags for VAR is decided using the SIC and AIC criteria. JB, Q(20), and $Q^2(20)$ indicate the empirical statistics of the Jarque–Bera test for normality, Ljung–Box Q statistics of order 20 for autocorrelation applied to the standardized residuals, and squared standardized residuals, respectively. Values in parentheses are the *p-Value*. ***, **, * indicate the statistical significance at 1%, 5%, and 10%, respectively.

Regarding the cross-market volatility spillover (see coefficients β_{12} and β_{21} in Panel B), the results indicate that the volatility transmission is unidirectional from the US to the Brazil and Mexican stock markets during the full sample period. In contrast, the results reveal the bidirectional volatility transmission between the US and two LA stock markets (Chile and Mexico), whereas there was unidirectional volatility transmission from Brazil to the US stock market during the global financial crisis.

These results are in contrast with the findings of Wang et al. (2017), which report an insignificant volatility spillover between the US and Brazil stock markets during the global financial crisis. The considerable trade volumes between the US and two LA stock markets (Brazil and Mexico) explain the volatility linkages between the stock markets of the concerned countries. Johnson and Soenen (2003) also suggest that trade increases the financial contagion effects between the stock markets of concerned countries. From the Latin American region, Mexico is the biggest trading partner of the US; therefore, volatility linkages are also observed between Mexico and the US stock market during the global financial crisis. These findings suggest that portfolio investors can get the maximum benefit of diversification by making a portfolio of US and Peru stocks during the global financial crisis. Lastly, a bidirectional volatility transmission is observed between the US and Mexican stock markets during the crash of the Chinese stock market. It implies that portfolio investors can diversify risk by making a portfolio of the US and LA stock markets (except Mexico) during the crash of the Chinese stock market.

4.3. *Return and Volatility Spillover between China and the LA Stock Markets*

Tables 6–8 represent the return and volatility transmissions between China and the LA stock markets during the full sample period, the global financial crisis, and the crash of the Chinese stock market. The difference in the opening time of the China and LA stock markets has been adjusted where necessary in the estimations. Referring to the coefficient $\varnothing_{11}$ in Panel A, the results indicate that the lagged returns of the majority of LA stock markets (except Brazil) largely determine their current returns during the full sample period and the crash of the Chinese stock market. During the global financial crisis, the past returns significantly affect the current returns of the Chile and Peru stock markets. This implies that the past returns can be used for the short-term prediction of the current LA stock returns. These results confirm the previous findings of Arouri et al. (2015). Referring to the coefficient $\varnothing_{22}$ in Panel A, the lagged returns significantly influence the current returns in the Chinese stock market during the full sample period. In contrast, the current returns of the Chinese stock market are not influenced by their past returns during the global financial crisis and the crash of the Chinese stock market. This implies that the past returns cannot be used for the short-term prediction of the current Chinese stock returns during the crisis period.

Based on the cross-market return spillover (see the coefficients $\varnothing_{12}$ and $\varnothing_{21}$ in Panel A), the results reveal the unidirectional return transmissions from China to the majority of LA stock markets during all the sample periods. These results contradict the previous findings of Aktan et al. (2009) and Sharma et al. (2013), who report the insignificant impact of the Chinese stock returns on the Brazilian stock returns. In addition, the return transmission is also significant from Brazil to China during the crash of the Chinese stock market.

From the variance equation (see coefficients α_{11} and α_{22} Panel B), the findings show that the lagged shocks significantly influence the conditional volatility of the China and LA stock markets during all the sample periods. Referring to the coefficients β_{11} and β_{22}, the results show that the current conditional volatility depends on their past volatility in the China and LA stock markets during the all sample periods. The critical finding is that the coefficients of past own volatility are seen to be higher compared to the past own shocks. This difference suggests that past own volatilities rather than past shocks are more important for the prediction of the current volatility in the China and LA stock markets.

Refer to the coefficients α_{12} and α_{21} in panel B, the shock transmission is unidirectional from Brazil and Peru to the Chinese stock market, whereas bidirectional shock transmission is observed between the China and Mexican stock markets during the full sample period. The results reveal that the past shocks in the Brazil and Mexican stock markets significantly affect the conditional volatility of the Chinese stock market during the global financial crisis. On the other hand, the shock spillover is insignificant between China and the majority of the LA stock markets during the crash of the Chinese stock market.

Table 6. Estimates of BEKK-GARCH for the China and Latin American stock markets during the full sample period.

	Brazil and China		**Chile and China**		**Mexico and China**		**Peru and China**	
	Coefficient	*p-Value*	**Coefficient**	*p-Value*	**Coefficient**	*p-Value*	**Coefficient**	*p-Value*
				Panel A. Mean Equation				
μ_1	0.076 ***	0.001	0.047 ***	0.000	0.038 ***	0.004	0.050 ***	0.000
$\varnothing_{11}$	0.033 **	0.050	0.205 ***	0.000	0.101 ***	0.000	0.234 ***	0.000
$\varnothing_{12}$	0.013	0.208	0.020	0.190	0.014	0.213	0.015	0.287
μ_2	0.044 ***	0.008	0.042 **	0.026	0.050 ***	0.000	0.039 **	0.045
$\varnothing_{21}$	0.132 ***	0.000	0.036 ***	0.000	0.075 ***	0.000	0.053 ***	0.000
$\varnothing_{22}$	0.037 ***	0.005	0.041 **	0.019	0.036 **	0.011	0.042 ***	0.009
				Panel B. Variance Equation				
c_{11}	0.283 ***	0.000	0.186 ***	0.000	0.138 ***	0.000	0.219 ***	0.000
c_{21}	0.009	0.699	−0.001	0.956	0.009	0.721	−0.013	0.500
c_{22}	0.118 ***	0.000	0.121 ***	0.000	0.115 ***	0.000	0.108 ***	0.000
α_{11}	0.273 ***	0.000	0.347 ***	0.000	0.279 ***	0.000	0.394 ***	0.000
α_{12}	−0.023 *	0.059	0.013	0.490	−0.037 ***	0.002	−0.024 *	0.057
α_{21}	0.000	0.899	0.001	0.950	0.021 *	0.087	−0.002	0.920
α_{22}	0.250 ***	0.000	0.243 ***	0.000	0.240 ***	0.000	0.237 ***	0.000
β_{11}	0.948 ***	0.000	0.919 ***	0.000	0.954 ***	0.000	0.904 ***	0.000
β_{12}	0.015 **	0.040	−0.002	0.741	0.009 ***	0.003	0.015 ***	0.007
β_{21}	0.002	0.814	0.002	0.726	−0.004	0.283	0.001	0.840
β_{22}	0.966 ***	0.000	0.968 ***	0.000	0.969 ***	0.000	0.970 ***	0.000
				Panel C. Diagnostic Tests				
LogL	−19,187.432		−15,839.650		−17,037.161		−16,739.334	
AIC	7.720		6.599		7.002		7.045	
SIC	7.767		6.646		7.049		7.092	
$Q_1[20]$	21.935	0.344	19.993	0.458	17.078	0.648	72.725 ***	0.000
$Q_2[20]$	82.861 ***	0.000	78.794 ***	0.000	83.815 ***	0.000	80.555 ***	0.000
$Q_1^2[20]$	26.742	0.133	8.890	0.984	26.056	0.187	18.240	0.572
$Q_2^2[20]$	22.787	0.299	22.412	0.319	25.134	0.196	25.146	0.196

Notes: US, United States of America; CHN, China; BRAZ, Brazil; CHIL, Chile; MEXI, Mexico. Variable order is the Latin American stock market (1) and China (2). In the mean equations, μ denotes the constant terms, whereas $\varnothing_{12}$ denotes the return spillover from the Latin American stock market to the Chinese stock market. In the variance equation, c denotes the constant terms, α denotes the ARCH terms, and β denotes the GARCH terms. In the variance equation, α_{12} indicates the shock spillover from the Latin American stock market to the Chinese stock market, whereas β_{12} denotes the long−term volatility spillover from the Latin American stock market to the Chinese stock market. Number of lags for VAR is decided using the SIC and AIC criteria. JB, Q(20), and $Q^2(20)$ indicate the empirical statistics of the Jarque–Bera test for normality, Ljung–Box Q statistics of order 20 for autocorrelation applied to the standardized residuals, and squared standardized residuals, respectively. Values in parentheses are the *p-Value*. ***, **, * indicate the statistical significance at 1%, 5%, and 10%, respectively.

Based on the cross-market volatility spillover effects (see coefficients β_{12} and β_{21} in Panel B), the results demonstrate that there is unidirectional volatility transmission from Brazil, Mexico, and Peru to China during the full sample period. These volatility transmissions can be explained through the considerable trading volumes between China and two Latin economies (Brazil and Mexico) during the full sample period. During the global financial crisis, the volatility effects are transmitted from the China to Brazil stock markets. Therefore, the majority of LA stock markets provide an opportunity to diversify the risk of Chinese equity portfolios during the global financial crisis. Lastly, the volatility spillover is bidirectional between the China and Brazil stock markets during the crash of the Chinese stock market. Due to the crash of the Chinese stock market, the slowdown of the Chinese economy also affected its major trading partner Brazil and its stock market; therefore, volatility linkages are also observed between China and Brazil. These findings propose that the portfolio investors of Chinese stock markets can get the maximum benefit of diversification by adding Mexico, Chile, and Peru stocks in their portfolios during the crash of the Chinese stock market.

Table 7. Estimates of BEKK-GARCH for the China and Latin American stock markets during the global financial crisis.

	BRAZIL and China		Chile and China		Mexico and China		Peru and China	
	Coefficient	*p-Value*	Coefficient	*p-Value*	Coefficient	*p-Value*	Coefficient	*p-Value*
Panel A. Mean Equation								
μ_1	0.106 *	0.081	0.116 ***	0.001	0.035	0.476	0.014	0.773
$\varnothing_{11}$	0.029	0.465	0.195 ***	0.000	0.034	0.340	0.211 ***	0.000
$\varnothing_{12}$	−0.003	0.938	−0.073	0.189	−0.034	0.424	−0.023	0.554
μ_2	0.044	0.590	0.001	0.987	0.090	0.285	0.030	0.731
$\varnothing_{21}$	0.186 ***	0.000	0.030 *	0.083	0.101 ***	0.000	0.103 ***	0.000
$\varnothing_{22}$	0.032	0.422	0.022	0.572	0.033	0.431	0.027	0.488
Panel B. Variance Equation								
c_{11}	−0.201	0.105	0.142 **	0.025	0.127 ***	0.008	0.344 ***	0.000
c_{21}	0.134	0.577	1.932 ***	0.000	0.134	0.479	0.081	0.250
c_{22}	0.143	0.385	0.000	0.920	0.195 **	0.037	0.163 *	0.098
α_{11}	0.297 ***	0.000	0.408 ***	0.000	0.253 ***	0.000	0.477 ***	0.000
α_{12}	−0.041	0.409	−0.108	0.124	−0.041	0.392	−0.005	0.902
α_{21}	−0.069 *	0.089	0.007	0.686	0.060 **	0.020	−0.031	0.215
α_{22}	0.188 ***	0.001	0.316 ***	0.000	0.209 ***	0.000	0.181 ***	0.000
β_{11}	0.947 ***	0.000	0.893 ***	0.000	0.960 ***	0.000	0.870 ***	0.000
β_{12}	0.006	0.580	0.021	0.130	0.003	0.820	0.005	0.777
β_{21}	0.028 *	0.085	0.047	0.260	−0.009	0.292	0.005	0.515
β_{22}	0.977 ***	0.000	−0.259	0.143	0.972 ***	0.000	0.979 ***	0.000
Panel C. Diagnostic Tests								
LogL	−3318.981		−2879.180		−3090.113		−3166.538	
AIC	8.973		7.826		8.388		8.689	
SIC	9.200		8.052		8.614		8.915	
$Q_1[20]$	14.730	0.792	11.805	0.923	17.935	0.592	17.407	0.626
$Q_2[20]$	30.922 *	0.056	32.099 **	0.042	30.614 *	0.060	31.335 *	0.051
$Q_1^2[20]$	22.021	0.339	12.016	0.916	14.074	0.827	9.990	0.968
$Q_2^2[20]$	28.559	0.127	37.567	0.161	26.806	0.141	28.213	0.104

Notes: US, United States of America; CHN, China; BRAZ, Brazil; CHIL, Chile; MEXI, Mexico. Variable order is the Latin American stock market (1) and China (2). In the mean equations and μ denotes the constant terms, whereas $\varnothing_{12}$ denotes the return spillover from the Latin American stock market to the Chinese stock market. In the variance equation, c denotes the constant terms, α denotes the ARCH terms, and β denotes the GARCH terms. In the variance equation, α_{12} indicates the shock spillover from the Latin American stock market to the Chinese stock market, whereas β_{12} denotes the long-term volatility spillover from the Latin American stock market to the Chinese stock market. Number of lags for VAR is decided using the SIC and AIC criteria. JB, Q(20), and $Q^2(20)$ indicate the empirical statistics of the Jarque–Bera test for normality, Ljung–Box Q statistics of order 20 for autocorrelation applied to the standardized residuals, and squared standardized residuals, respectively. Values in parentheses are the *p-Value*. ***, **, * indicate the statistical significance at 1%, 5%, and 10%, respectively.

Table 8. Estimates of BEKK-GARCH for the China and Latin American stock markets during the crash of the Chinese stock market.

	BRAZIL and China		Chile and China		Mexico and China		Peru and China	
	Coefficient	*p-Value*	Coefficient	*p-Value*	Coefficient	*p-Value*	Coefficient	*p-Value*
Panel A. Mean Equation								
μ_1	0.055	0.282	0.039 *	0.063	0.008	0.765	0.059 *	0.087
$\varnothing_{11}$	0.036	0.292	0.181 ***	0.000	0.077 **	0.017	0.229 ***	0.000
$\varnothing_{12}$	0.043 **	0.048	0.026	0.350	0.012	0.791	0.033	0.333
μ_2	0.022	0.495	0.020	0.474	0.020	0.420	0.022	0.410
$\varnothing_{21}$	0.105 ***	0.001	0.057 ***	0.001	0.074 ***	0.000	0.062 ***	0.003
$\varnothing_{22}$	0.032	0.360	0.040	0.228	0.039	0.255	0.026	0.470

Table 8. *Cont.*

	BRAZIL and China		Chile and China		Mexico and China		Peru and China	
	Coefficient	*p-Value*	Coefficient	*p-Value*	Coefficient	*p-Value*	Coefficient	*p-Value*
				Panel A. Mean Equation				
μ_1	0.055	0.282	0.039 *	0.063	0.008	0.765	0.059 *	0.087
$\varnothing_{11}$	0.036	0.292	0.181 ***	0.000	0.077 **	0.017	0.229 ***	0.000
$\varnothing_{12}$	0.043 **	0.048	0.026	0.350	0.012	0.791	0.033	0.333
μ_2	0.022	0.495	0.020	0.474	0.020	0.420	0.022	0.410
$\varnothing_{21}$	0.105 ***	0.001	0.057 ***	0.001	0.074 ***	0.000	0.062 ***	0.003
$\varnothing_{22}$	0.032	0.360	0.040	0.228	0.039	0.255	0.026	0.470
				Panel B. Variance Equation				
c_{11}	0.915 ***	0.000	0.232 *	0.057	0.390 ***	0.000	0.259 ***	0.009
c_{21}	0.126 ***	0.001	0.049	0.132	0.033	0.553	0.089 **	0.045
c_{22}	0.000	0.865	0.000	0.799	0.058	0.237	0.051	0.320
α_{11}	0.334 ***	0.000	0.452 ***	0.004	0.405 ***	0.000	0.345 ***	0.004
α_{12}	−0.043	0.159	0.096 **	0.021	−0.017	0.778	0.006	0.934
α_{21}	0.051	0.336	0.005	0.845	−0.069	0.394	0.054	0.182
α_{22}	0.236 ***	0.000	0.208 ***	0.000	0.229 ***	0.000	0.256 ***	0.000
β_{11}	0.691 ***	0.000	0.844 ***	0.000	0.751 ***	0.000	0.890 ***	0.000
β_{12}	−0.069 *	0.071	−0.065	0.109	−0.032	0.662	−0.032	0.410
β_{21}	−0.072 *	0.056	0.005	0.525	0.031	0.260	−0.012	0.404
β_{22}	0.968 ***	0.000	0.978 ***	0.000	0.974 ***	0.000	0.965 ***	0.000
				Panel C. Diagnostic Tests				
LogL	−2506.043		−1947.155		−2014.255		−2107.485	
AIC	7.227		5.934		5.989		6.295	
SIC	7.454		6.160		6.216		6.521	
$Q_1[20]$	21.462	0.370	29.255 *	0.083	24.444	0.224	23.882	0.248
$Q_2[20]$	23.562	0.262	27.467	0.123	24.664	0.215	26.564	0.148
$Q_1^2[20]$	10.480	0.959	61.006 ***	0.000	15.298	0.759	14.951	0.779
$Q_2^2[20]$	21.376	0.375	27.907	0.112	25.661	0.177	20.186	0.446

Notes: US, United States of America; CHN, China; BRAZ, Brazil; CHIL, Chile; MEXI, Mexico. Variable order is the Latin American stock market (1) and China (2). In the mean equations and μ denotes the constant terms, whereas $\varnothing_{12}$ denotes the return spillover from the Latin American stock market to the Chinese stock market. In the variance equation, c denotes the constant terms, α denotes the ARCH terms, and β denotes the GARCH terms. In the variance equation, α_{12} indicates the shock spillover from the Latin American stock market to the Chinese stock market, whereas β_{12} denotes the long-term volatility spillover from the Latin American stock market to the Chinese stock market. Number of lags for VAR is decided using the SIC and AIC criteria. JB, Q(20), and $Q^2(20)$ indicate the empirical statistics of the Jarque–Bera test for normality, Ljung–Box Q statistics of order 20 for autocorrelation applied to the standardized residuals, and squared standardized residuals, respectively. Values in parentheses are the *p-Value*. ***, **, * indicate the statistical significance at 1%, 5%, and 10%, respectively.

4.4. Optimal Weights and Hedge Ratio Portfolio Implications

In the above-mentioned results, volatility transmission is observed between the several pairs of stock markets during the different sample periods. Thus, investment in these pairs of stock markets reduces the benefit of diversification. Therefore, the risk transmission across stock markets push investors to adjust their asset allocation and to hedge their portfolio risk over time. For this reason, this study estimates the optimal weights and hedge ratios.

Tables 9 and 10 report the optimal weights for the pairs of LA-US and LA-China during all the sample periods. The findings reveal that the optimal weight is 0.11 for BRAZ/US during the full sample period, revealing that for a $1 portfolio in Brazil-US, 11 cents should be invested in the Brazil stock market and the remaining 89 cents in the US stock market. The interpretations of all the optimal weights are not interpreted here for the sake of brevity. For the LA-US portfolio (see Table 9), the results show that the average optimal weights are seen to be higher in the global financial crisis and the crash of the Chinese stock market as compared to the full sample period. For the LA-US portfolio, the investors are suggested to allocate a higher proportion of investment in LA stocks during the global financial crisis and the crash of the Chinese stock market. For the pair of LA-China (see Table 10), the results show that the optimal weights are higher during the global financial crisis and the crash of the Chinese stock market compared to the full sample period. For the LA-China portfolio, investors should increase their investment in LA stocks during the global financial crisis and the crash of the Chinese stock market.

Table 9. Optimal weights and hedge ratios for Latin America (LA)/US

	BRAZ/US	CHIL/US	MEXI/US	Peru/US
		Full Sample Period		
w_t^{LU}	0.11	0.51	0.29	0.27
β_t^{LU}	0.93	0.63	0.25	0.28
		US Financial Crisis		
w_t^{LU}	0.17	0.77	0.49	0.41
β_t^{LU}	0.94	0.42	0.77	0.56
		Chinese Stock Market Crash		
w_t^{LU}	0.09	0.54	0.46	0.39
β_t^{LU}	0.98	0.34	0.59	0.43

Note: w_t^{LU} and β_t^{LU} represent the optimal weight and hedge ratio for the LA-US pair. L and U in superscripts denote the Latin American and US stock markets, respectively.

Table 10. Optimal weights and hedge ratios for LA/China

Header	BRAZ/CHN	CHIL/CHN	MEXI/CHN	Peru/CHN
		Full Sample Period		
w_t^{LC}	0.41	0.70	0.61	0.63
β_t^{LC}	0.14	0.07	0.07	0.08
		US Financial Crisis		
w_t^{LC}	0.53	0.81	0.68	0.63
β_t^{LC}	0.21	0.09	0.15	0.13
		Chinese Stock Market Crash		
w_t^{LC}	0.43	0.68	0.64	0.64
β_t^{LC}	0.23	0.11	0.07	0.11

Note: w_t^{LC} and β_t^{LC} represent the optimal weight and hedge ratio for LA-China pair. L and C in superscripts denote the Latin American and Chinese stock markets, respectively.

It is also essential to estimate the risk-minimizing optimal hedge ratios for portfolios of different stocks. Referring to Table 9, the optimal hedge ratio range is 0.93 for BRAZ/US during the full sample period, showing that a $1-long position in Brazil stocks can be hedged for 93 cents with a short position in the US stocks. The interpretations of all the optimal hedge ratios are not interpreted here for the sake of brevity. For the LA-US portfolio (see Table 9), the average optimal hedge ratios are found to be higher for most of the pairs during the global financial crisis and the crash of the Chinese stock market compared to the full sample period. It implies that less LA stocks are needed to minimize the risk of US stock during crisis periods compared to the full sample period. For the LA-China portfolio (see Table 10), the optimal hedge ratios are also higher during both crises, which implies that the lesser LA stocks are required to minimize the risk of the Chinese stock market during both crises compared to the full sample period.

5. Conclusions

This study examines the return and volatility spillover between the world-leading (the US and China) and emerging Latin American (Brazil, Chile, Mexico, and Peru) stock markets during the full sample period, the global financial crisis, and the crash of the Chinese stock market. Moreover, this study also estimates the optimal weights and hedge ratios during all the sample periods. The BEKK-GARCH model is applied to estimate the return and volatility spillover between the stock markets.

Regarding return spillover, the results reveal a unidirectional return spillover from the US to the majority of the LA stock markets during the full sample period and the Chinese crash. This implies that the US stock market prices play an important role in predicting the prices of the majority of LA stock markets during the full sample period and the Chinese crash. During the global financial crisis, the return transmissions are not significant between the US and the majority of Latin American stock markets. This implies that the prices of the US (LA) stock markets do not contribute to the role of price discovery in the LA (US) stock markets during the global financial crisis. For the China-LA nexus, the results reveal a unidirectional return transmission from China to Brazil, Chile, Mexico, and Peru stock markets during all the sample periods. Thus, the Chinese stock returns can be useful in predicting the returns of the LA stock markets.

Regarding the volatility spillover between the US and LA stock markets, the results reveal the bidirectional volatility transmission between the US and two stock markets of Chile and Mexico, as well as the unidirectional volatility transmission from Brazil to the US stock market during the global financial crisis. During the Chinese crash, a bidirectional volatility transmission is observed between the US and Mexican stock markets. This implies that portfolio investors can diversify risk by making a portfolio of the US and LA stock markets (except Mexico) during the crash of the Chinese stock market.

Regarding the volatility spillover between the China and LA stock markets, the volatility spillover is unidirectional from the China to Brazil stock markets during the global financial crisis. Therefore, the majority of the LA stock markets provide an opportunity to diversify the risk of Chinese equity portfolios during the global financial crisis. During the Chinese crash, the volatility spillover is bidirectional between the China and Brazil stock markets. These findings propose that the portfolio investors of the Chinese stock markets can get the maximum benefit of diversification by adding Mexico, Chile, and Peru stocks to their portfolios during the crash of the Chinese stock market. These findings are also important because understanding the stock market volatility behavior can play a vital role during the valuation of derivatives and for hedging purposes. Moreover, policymakers should consider the "prices and volatilities of the world-leading stock market" as one of the critical factors while devising the policies to stabilize their emerging financial markets.

Based on optimal weights, investors are suggested to allocate a higher proportion of investment to the LA stocks in the LA-US portfolio during the global financial crisis and the crash of the Chinese stock market. For the LA-China portfolio, investors should increase their investment in the LA stocks during the global financial crisis and the crash of the Chinese stock market. Based on hedge ratios, less LA stocks are needed to minimize the risk of the US and Chinese stocks during the periods of both crises compared to the full sample period. Overall, these findings provide useful information for policymakers and portfolio managers regarding optimal asset allocation, diversification, hedging, forecasting, and risk management.

This study employs the BEKK-GARCH model to examine the linkages between the world-leading countries and the emerging Latin American stock markets. Extensions could include other models to examine the return and volatility spillover—for example, cointegration and causality (Lv et al. 2019; Demirer et al. 2019), Copulas (Ly et al. 2019a, 2019b; Yuan et al. 2020), Stochastic Dominance (Chiang et al. 2008; Abid et al. 2014; Guo et al. 2017; Wong et al. 2018), and many others. See, for example, Chang et al. (2018), Woo et al. (2020), and the references therein for more information.

Author Contributions: Conceptulization, estimations, formal analysis, original draft preparation (I.Y.); Data collection, methodology writing, and review of draft (S.A.); review, editing, and funding (W.-K.W.). All authors have read and agreed to the published version of the manuscript.

Acknowledgments: The first author gratefully acknowledge Arshad Hassan (Professor/Dean, department of Management and Social Sciences, Capital University of Science and Technology, Islamabad) for their valuable suggestions. The third author would like to thank Robert B. Miller and Howard Thompson for their continuous guidance and encouragement. All the errors remain with the authors.

References

Abid, Fathi, Pui Leung, Mourad Mroua, and Wing Wong. 2014. International Diversification Versus Domestic Diversification: Mean-Variance Portfolio Optimization and Stochastic Dominance Approaches. *Journal of Risk and Financial Management* 7: 45–66. [CrossRef]

Ahmad, Wasim, and Sanjay Sehgal. 2015. Regime Shifts and Volatility in BRIICKS Stock Markets: An Asset Allocation Perspective. *International Journal of Emerging Markets* 10: 383–408. [CrossRef]

Ahmed, Abdullahi D., and Rui Huo. 2019. Impacts of China's Crash on Asia-Pacific Financial Integration: Volatility Interdependence, Information Transmission and Market Co-Movement. *Economic Modelling* 79: 28–46. [CrossRef]

Aktan, Bora, Pinar Evrim Mandaci, Baris Serkan Kopurlu, and Bulent Ersener. 2009. Behaviour of Emerging Stock Markets in the Global Financial Meltdown: Evidence from Bric-A. *African Journal of Business Management* 3: 396–404.

Al Nasser, Omar M., and Massomeh Hajilee. 2016. Integration of Emerging Stock Markets with Global Stock Markets. *Research in International Business and Finance* 36: 1–12. [CrossRef]

Ali, Shoaib, Imran Yousaf, and Muhammad Naveed. 2020. Role of credit rating in determining capital structure: Evidence from non-financial sector of Pakistan. *Studies of Applied Economics* 38. [CrossRef]

Arouri, Mohamed El Hédi, Amine Lahiani, and Duc Khuong Nguyen. 2015. Cross-Market Dynamics and Optimal Portfolio Strategies in Latin American Equity Markets. *European Business Review* 27: 161–81. [CrossRef]

Beirne, John, Guglielmo Maria Caporale, Marianne Schulze-Ghattas, and Nicola Spagnolo. 2013. Volatility Spillovers and Contagion from Mature to Emerging Stock Markets. *Review of International Economics* 21: 1060–75. [CrossRef]

Bekiros, Stelios D. 2014. Contagion, decoupling and the spillover effects of the US financial crisis: Evidence from the BRIC markets. *International Review of Financial Analysis* 33: 58–69. [CrossRef]

Bhuyan, Rafiqul, Mohammad G. Robbani, Bakhtear Talukdar, and Ajeet Jain. 2016. Information Transmission and Dynamics of Stock Price Movements: An Empirical Analysis of BRICS and US Stock Markets. *International Review of Economics & Finance* 46: 180–95.

Bouri, Elie I. 2013. Correlation and Volatility of the MENA Equity Markets in Turbulent Periods, and Portfolio Implications. *Economics Bulletin* 33: 1575–93.

Cao, Guangxi, Yan Han, Qingchen Li, and Wei Xu. 2017. Asymmetric MF-DCCA Method Based on Risk Conduction and Its Application in the Chinese and Foreign Stock Markets. *Physica A: Statistical Mechanics and Its Applications* 468: 119–30. [CrossRef]

Cardona, Laura, Marcela Gutiérrez, and Diego A. Agudelo. 2017. Volatility Transmission between US and Latin American Stock Markets: Testing the Decoupling Hypothesis. *Research in International Business and Finance* 39: 115–27. [CrossRef]

Chancharoenchai, Kanokwan, and Sel Dibooglu. 2006. Volatility Spillovers and Contagion During the Asian Crisis: Evidence from Six Southeast Asian Stock Markets. *Emerging Markets Finance and Trade* 42: 4–17. [CrossRef]

Chang, Chia-Lin, Michael McAleer, and Roengchai Tansuchat. 2011. Crude Oil Hedging Strategies Using Dynamic Multivariate GARCH. *Energy Economics* 33: 912–23. [CrossRef]

Chang, Chia-Lin, Michael McAleer, and Guangdong Zuo. 2017. Volatility Spillovers and Causality of Carbon Emissions, Oil and Coal Spot and Futures for the EU and USA. *Sustainability* 9: 1789. [CrossRef]

Chang, Chia-Lin, Michael McAleer, and Wing-Keung Wong. 2018. Big Data, Computational Science, Economics, Finance, Marketing, Management, and Psychology: Connections. *Journal of Risk and Financial Management* 11: 15. [CrossRef]

Chen, Gong-Meng, Michael Firth, and Oliver Meng Rui. 2002. Stock Market Linkages: Evidence from Latin America. *Journal of Banking & Finance* 26: 1113–41.

Chiang, Thomas, Hooi Hooi Lean, and Wing-Keung Wong. 2008. Do REITs Outperform Stocks and Fixed-Income Assets? New Evidence from Mean-Variance and Stochastic Dominance Approaches. *Journal of Risk and Financial Management* 1: 1–40. [CrossRef]

Coleman, Simeon, Vitor Leone, and Otavio R. de Medeiros. 2018. Latin American Stock Market Dynamics and Comovement. *International Journal of Finance & Economics* 24: 1109–29.

Demirer, Riza, Rangan Gupta, Zhihui Lv, and Wing-Keung Wong. 2019. Equity Return Dispersion and Stock Market Volatility: Evidence from Multivariate Linear and Nonlinear Causality Tests. *Sustainability* 11: 351. [CrossRef]

Diamandis, Panayiotis F. 2009. International Stock Market Linkages: Evidence from Latin America. *Global Finance Journal* 20: 13–30. [CrossRef]

Diebold, Francis X., and Kamil Yilmaz. 2012. Better to Give than to Receive: Predictive Directional Measurement of Volatility Spillovers. *International Journal of Forecasting* 28: 57–66. [CrossRef]

Engle, Robert F., and Kenneth F. Kroner. 1995. Multivariate Simultaneous Generalized ARCH. *Econometric Theory* 11: 122–50. [CrossRef]

Fang, Lu, and David A. Bessler. 2017. Is It China That Leads the Asian Stock Market Contagion in 2015? *Applied Economics Letters* 25: 752–57. [CrossRef]

Fernández-Serrano, José L., and Simón Sosvilla-Rivero. 2003. Modelling the Linkages between US and Latin American Stock Markets. *Applied Economics* 35: 1423–34. [CrossRef]

Fortunato, Graziela, Nathalia Martins, and Carlos de Lamare Bastian-Pinto. 2019. Global Economic Factors and the Latin American Stock Markets. *Latin American Business Review* 21: 61–91. [CrossRef]

Gamba-Santamaria, Santiago, Jose Eduardo Gomez-Gonzalez, Jorge Luis Hurtado-Guarin, and Luis Fernando Melo-Velandia. 2017. Stock Market Volatility Spillovers: Evidence for Latin America. *Finance Research Letters* 20: 207–16. [CrossRef]

Garza-García, Jesus Gustavo, and Maria Eugenia Vera-Juárez. 2010. Who influences Latin American stock market returns? China versus USA. *International Research Journal of Finance and Economics* 55: 22–35.

Graham, Michael, Jarno Kiviaho, and Jussi Nikkinen. 2012. Integration of 22 Emerging Stock Markets: A Three-Dimensional Analysis. *Global Finance Journal* 23: 34–47. [CrossRef]

Gulzar, Saqib, Ghulam Mujtaba Kayani, Hui Xiaofeng, Usman Ayub, and Amir Rafique. 2019. Financial Cointegration and Spillover Effect of Global Financial Crisis: A Study of Emerging Asian Financial Markets. *Economic Research-Ekonomska Istraživanja* 32: 187–218. [CrossRef]

Guo, Xu, Xuejun Jiang, and Wing-Keung Wong. 2017. Stochastic Dominance and Omega Ratio: Measures to Examine Market Efficiency, Arbitrage Opportunity, and Anomaly. *Economies* 5: 38. [CrossRef]

Han, Qian, and Jufang Liang. 2016. Index Futures Trading Restrictions and Spot Market Quality: Evidence from the Recent Chinese Stock Market Crash. *Journal of Futures Markets* 37: 411–28. [CrossRef]

He, Ling T. 2001. Time Variation Paths of International Transmission of Stock Volatility—US vs. Hong Kong and South Korea. *Global Finance Journal* 12: 79–93. [CrossRef]

Horvath, Roman, and Petr Poldauf. 2012. International Stock Market Comovements: What Happened during the Financial Crisis? *Global Economy Journal* 12: 1850252. [CrossRef]

Hwang, Jae-Kwang. 2014. Spillover Effects of the 2008 Financial Crisis in Latin America Stock Markets. *International Advances in Economic Research* 20: 311–24. [CrossRef]

In, Francis, Sangbae Kim, Jai Hyung Yoon, and Christopher Viney. 2001. Dynamic Interdependence and Volatility Transmission of Asian Stock Markets. *International Review of Financial Analysis* 10: 87–96. [CrossRef]

Johnson, Robert, and Luc Soenen. 2003. Economic Integration and Stock Market Comovement in the Americas. *Journal of Multinational Financial Management* 13: 85–100. [CrossRef]

Kroner, Kenneth F., and Victor K. Ng. 1998. Modeling Asymmetric Comovements of Asset Returns. *Review of Financial Studies* 11: 817–44. [CrossRef]

Kroner, Kenneth F., and Jahangir Sultan. 1993. Time-Varying Distributions and Dynamic Hedging with Foreign Currency Futures. *The Journal of Financial and Quantitative Analysis* 28: 535. [CrossRef]

Li, Yanan, and David E. Giles. 2014. Modelling Volatility Spillover Effects Between Developed Stock Markets and Asian Emerging Stock Markets. *International Journal of Finance & Economics* 20: 155–77.

Lv, Zhihui, Amanda M. Y. Chu, Michael McAleer, and Wing-Keung Wong. 2019. Modelling Economic Growth, Carbon Emissions, and Fossil Fuel Consumption in China: Cointegration and Multivariate Causality. *International Journal of Environmental Research and Public Health* 16: 4176. [CrossRef] [PubMed]

Ly, Sel, Kim-Hung Pho, Sal Ly, and Wing-Keung Wong. 2019a. Determining Distribution for the Product of Random Variables by Using Copulas. *Risks* 7: 23. [CrossRef]

Ly, Sel, Kim-Hung Pho, Sal Ly, and Wing-Keung Wong. 2019b. Determining Distribution for the Quotients of Dependent and Independent Random Variables by Using Copulas. *Journal of Risk and Financial Management* 12: 42. [CrossRef]

Malik, Farooq, and Shawkat Hammoudeh. 2007. Shock and Volatility Transmission in the Oil, US and Gulf Equity Markets. *International Review of Economics & Finance* 16: 357–68.

Markovitz, Harry M. 1959. *Portfolio Selection: Efficient Diversification of Investments.* Cowles Foundation Monograph 16. London: Yale University Press.

Mensi, Walid, Shawkat Hammoudeh, Duc Khuong Nguyen, and Sang Hoon Kang. 2016. Global Financial Crisis and Spillover Effects among the U.S. and BRICS Stock Markets. *International Review of Economics & Finance* 42: 257–76.

Meric, Gulser, Ricardo P. C. Leal, Mitchell Ratner, and Ilhan Meric. 2001. Co-Movements of U.S. and Latin American Equity Markets before and after the 1987 Crash. *International Review of Financial Analysis* 10: 219–235. [CrossRef]

Ramirez-Hassan, Andres, and Javier Orlando Pantoja. 2018. Co-Movements between Latin American and US Stock Markets: Convergence After the Financial Crisis? *Latin American Business Review* 19: 157–72. [CrossRef]

Ben Rejeb, Aymen, and Mongi Arfaoui. 2016. Financial Market Interdependencies: A Quantile Regression Analysis of Volatility Spillover. *Research in International Business and Finance* 36: 140–57. [CrossRef]

Sadorsky, Perry. 2012. Correlations and Volatility Spillovers between Oil Prices and the Stock Prices of Clean Energy and Technology Companies. *Energy Economics* 34: 248–55. [CrossRef]

Sarwar, Suleman, Aviral Kumar Tiwari, and Cao Tingqiu. 2020. Analyzing Volatility Spillovers between Oil Market and Asian Stock Markets. *Resources Policy* 66: 101608. [CrossRef]

Sharkasi, Adel, Heather J. Ruskin, and Martin Crane. 2005. Interrelationships among international stock market indices: Europe, Asia and the Americas. *International Journal of Theoretical and Applied Finance* 8: 603–22. [CrossRef]

Sharma, Gagan Deep, Mandeep Mahendru, and Sanjeet Singh. 2013. Are the Stock Exchanges of Emerging Economies Inter-Linked: Evidence from BRICS. *Indian Journal of Finance* 7: 26–37.

Su, Xianfang. 2020. Measuring extreme risk spillovers across international stock markets: A quantile variance decomposition analysis. *The North American Journal of Economics and Finance* 51: 101098. [CrossRef]

Syriopoulos, Theodore, Beljid Makram, and Adel Boubaker. 2015. Stock Market Volatility Spillovers and Portfolio Hedging: BRICS and the Financial Crisis. *International Review of Financial Analysis* 39: 7–18. [CrossRef]

Taşdemir, Murat, and Abdullah Yalama. 2014. Volatility spillover effects in interregional equity markets: Empirical evidence from Brazil and Turkey. *Emerging Markets Finance and Trade* 50: 190–202. [CrossRef]

Wang, Gang-Jin, Chi Xie, Min Lin, and H. Eugene Stanley. 2017. Stock Market Contagion during the Global Financial Crisis: A Multiscale Approach. *Finance Research Letters* 22: 163–68. [CrossRef]

Wong, Wing-Keung, Hooi Lean, Michael McAleer, and Feng-Tse Tsai. 2018. Why Are Warrant Markets Sustained in Taiwan but Not in China? *Sustainability* 10: 3748. [CrossRef]

Woo, Kai-Yin, Chulin Mai, Michael McAleer, and Wing-Keung Wong. 2020. Review on Efficiency and Anomalies in Stock Markets. *Economies* 8: 20. [CrossRef]

Yang, Zihui, and Yinggang Zhou. 2017. Quantitative Easing and Volatility Spillovers Across Countries and Asset Classes. *Management Science* 63: 333–54. [CrossRef]

Yousaf, Imran, and Junaid Ahmed. 2018. Mean and Volatility Spillover of the Latin American Stock Markets. *Journal of Business & Economics* 10: 51–63.

Yousaf, Imran, and Arshad Hassan. 2019. Linkages between Crude Oil and Emerging Asian Stock Markets: New Evidence from the Chinese Stock Market Crash. *Finance Research Letters* 31: 207–17. [CrossRef]

Yuan, Xinyu, Jiechen Tang, Wing-Keung Wong, and Songsak Sriboonchitta. 2020. Modeling Co-Movement among Different Agricultural Commodity Markets: A Copula-GARCH Approach. *Sustainability* 12: 393. [CrossRef]

4

CVaR Regression based on the Relation between CVaR and Mixed-Quantile Quadrangles

Alex Golodnikov [1], Viktor Kuzmenko [1] and Stan Uryasev [2,*]

[1] V.M. Glushkov Institute of Cybernetics, 40, pr. Akademika Glushkova, 03187 Kyiv, Ukraine
[2] Applied Mathematics & Statistics, Stony Brook University, B-148 Math Tower, Stony Brook, NY 11794, USA
* Correspondence: Stanislav.Uryasev@stonybrook.edu

Abstract: A popular risk measure, conditional value-at-risk (CVaR), is called expected shortfall (ES) in financial applications. The research presented involved developing algorithms for the implementation of linear regression for estimating CVaR as a function of some factors. Such regression is called CVaR (superquantile) regression. The main statement of this paper is: CVaR linear regression can be reduced to minimizing the Rockafellar error function with linear programming. The theoretical basis for the analysis is established with the quadrangle theory of risk functions. We derived relationships between elements of CVaR quadrangle and mixed-quantile quadrangle for discrete distributions with equally probable atoms. The deviation in the CVaR quadrangle is an integral. We present two equivalent variants of discretization of this integral, which resulted in two sets of parameters for the mixed-quantile quadrangle. For the first set of parameters, the minimization of error from the CVaR quadrangle is equivalent to the minimization of the Rockafellar error from the mixed-quantile quadrangle. Alternatively, a two-stage procedure based on the decomposition theorem can be used for CVaR linear regression with both sets of parameters. This procedure is valid because the deviation in the mixed-quantile quadrangle (called mixed CVaR deviation) coincides with the deviation in the CVaR quadrangle for both sets of parameters. We illustrated theoretical results with a case study demonstrating the numerical efficiency of the suggested approach. The case study codes, data, and results are posted on the website. The case study was done with the Portfolio Safeguard (PSG) optimization package, which has precoded risk, deviation, and error functions for the considered quadrangles.

Keywords: quantile; VaR; quadrangle; CVaR; conditional value-at-risk; expected shortfall; ES; superquantile; deviation; risk; error; regret; minimization; CVaR estimation; regression; linear regression; linear programming; portfolio safeguard; PSG

---

1. Introduction

We start the introduction with a quick outline of the main result of this paper. The conditional value-at-risk (CVaR) is a popular risk measure. It is called expected shortfall (ES) in financial applications and it is included in financial regulations. This paper provides algorithms for the estimation of CVaR with linear regression as a function of factors. This task is of critical importance in practical applications involving low probability events.

By definition, CVaR is an integral of the value-at-risk (VaR) in the tail of a distribution. VaR can be estimated with the quantile regression by minimizing the Koenker–Bassett error function. This paper shows that CVaR can be estimated by minimizing a mixture of the Koenker–Bassett errors with an additional constraint. This mixture is called the Rockafellar error and it has been earlier used for CVaR estimation without a rigorous mathematical justification. One more equivalent variant of CVaR regression can be done by minimizing a mixture of CVaR deviations for finding all coefficients, except the intercept. In this case, the intercept is calculated using an analytical expression, which is the CVaR of the optimal residual without an intercept. The new mathematical result links quantile and CVaR regressions and shows that convex and linear programming methods can be straightforwardly used for CVaR estimation. Mathematical justification of the results involves a risk quadrangle concept combining regret, error, risk, deviation, and statistic notions.

Quantiles evaluating different parts of a distribution of a random value are quite popular in various applications. In particular, quantiles are used to estimate tail of a distribution (e.g., 90%, 95%, and 99% quantiles). This paper is motivated by finance applications, where a quantile is called VaR. Risk measure VaR is included in finance regulations for the estimation of market risk. VaR has several attractive properties, such as the simplicity of calculation, stability of estimation, and availability of quantile regression, for the estimation of VaR as a function of explanatory factors. The quantile regression (see Koenker and Bassett (1978), Koenker (2005)) is an important factor supporting the popularity of VaR. For instance, a quantile regression was used by Adrian and Brunnermeier (2016) to estimate institution's contribution to systemic risk.

However, VaR also has some undesirable properties:

- Lack of convexity: portfolio diversification may increase VaR.
- VaR is not sensitive to outcomes exceeding VaR, which allows for stretching of the distribution without an increasing of the risk measured by VaR.
- VaR has poor mathematical properties, such as discontinuity with respect to (w.r.t.) portfolio positions for discrete distributions based on historical data.

Shortcomings of VaR led financial regulators to use an alternative measure of risk, which is called conditional value-at-risk (CVaR) in this paper. This risk measure was introduced in Rockafellar and Uryasev (2000) and further studied in Rockafellar and Uryasev (2002) and many other papers. CVaR for continuous distributions equals the conditional expectation of losses exceeding VaR. An important mathematical fact is that CVaR is a coherent risk measure (see Acerbi and Tasche (2002), Rockafellar and Uryasev (2002)). Ziegel (2014) shows that CVaR is elicitable in a week sense. Fissler and Ziegel (2015) proved that (VaR, CVaR) is jointly elicitable, meaning elic(CVaR) ≤ 2, and more generally, that spectral risk measures have a low elicitation complexity. These results clarify the regression procedure of Rockafellar et al. (2014); their algorithm implicitly tracks the quantiles suggested by elicitation complexity.

Rockafellar and Uryasev (2000, 2002) have shown that CVaR of a convex function of variables is also a convex function. Due to this property, CVaR optimization problems can be reduced to convex and linear optimization problems.

This paper is based on risk quadrangle theory, which defines quadrangles (i.e., groups) of stochastic functionals Rockafellar and Uryasev (2013). Every quadrangle contains risk, deviation, error, and regret (negative utility). These elements of the quadrangle are linked by the statistic function.

The relation of quantile regression and CVaR optimization was explained using a quantile quadrangle (see Rockafellar and Uryasev (2013)). It was shown that the Koenker–Bassett error function and CVaR belong to the same quantile quadrangle. By minimizing the Koenker–Bassett error function with respect to one parameter, we obtain the CVaR deviation (which is the CVaR for the centered random value). The optimal value of the parameter, which is called statistic, equals VaR. Therefore, the linear regression with the Koenker–Bassett error estimates VaR as a function of factors. The fact that statistic equals VaR and is also used for building the optimization approach for CVaR (see Rockafellar and Uryasev (2000, 2002)).

Another important contribution that takes advantage of quadrangle theory is the regression decomposition theorem proved in Rockafellar et al. (2008). With this decomposition theorem, the regression problem is decomposed in two steps: (1) minimization of deviation from the corresponding quadrangle, and (2) calculation of the intercept by using statistic from this quadrangle. For instance, by applying the decomposition theorem to the quantile quadrangle, we can do quantile regression by minimizing CVaR deviation for finding all regression coefficients, except the intercept. Then, the intercept is calculated by using VaR statistic.

CVaR can be approximated using the weighted average of VaRs with different confidence levels, which is called the mixed VaR method. Rockafellar and Uryasev (2013) demonstrated that mixed VaR is a statistic in the mixed-quantile quadrangle. The error function, corresponding to this quadrangle (called the Rockafellar error) can be minimized for the estimation of the mixed VaR with linear regression. The Rockafellar error is a solution of a minimization problem with one linear constraint. Linear regression for estimating mixed VaR can be done by minimizing the Rockafellar error with convex and linear programming (Appendix A contains these formulations). Alternatively, this regression can be done in two steps with the decomposition theorem. The deviation in the mixed-quantile quadrangle is the mixed CVaR deviation, therefore all regression coefficients, except the intercept, can be found by minimizing this deviation. Further, the intercept can be found by using statistic, which is the mixed VaR.

Rockafellar et al. (2014) developed the CVaR quadrangle with the statistic equal to CVaR. Risk envelopes and identifiers for this quadrangle were calculated in Rockafellar and Royset (2018). This CVaR quadrangle is a theoretical basis for constructing the regression for estimating CVaR. Rockafellar et al. (2014) called the linear regression for estimation of CVaR using superquantile (CVaR) regression. The superquantile is an equivalent term for CVaR. Here we use the term "CVaR regression". The CVaR regression plays a major role in various engineering areas, especially in financial applications. For instance, Huang and Uryasev (2018) used CVaR regression for the estimation of risk contributions of financial institutions and Beraldi et al. (2019) used CVaR for solving portfolio optimization problems with transaction costs.

This paper considers only discrete random values with a finite number of equally probable atoms. This special case is considered because it is needed for the implementation of the linear regression for the CVaR estimation. We have explained with an example how parameters of the optimization problems are calculated.

The equal probabilities property was used for calculating parameters of optimization problems. It is possible to calculate parameters with non-equal probabilities of atoms, but this is beyond the scope of the paper, which is focused on the linear regression.

We suggested two sets (Sets 1 and 2) of parameters for the mixed-quantile quadrangle. Set 1 corresponds to the two-step implementation of the CVaR regression in Rockafellar et al. (2014), and Set 2 is a new set of parameters. We proved that with Set 1, the statistic, risk, and deviation of the mixed-quantile and CVaR quadrangles coincide. Therefore, CVaR regression can be done by minimizing the Rockafellar error with convex and linear programming. For Set 2, the mixed-quantile and CVaR quadrangle share risk and deviation parameters. Also, the statistic of this mixed-quantile quadrangle (which may not be unique) includes statistic of the CVaR quadrangle. Therefore, minimizing the Rockafellar error correctly calculates all regression coefficients, but may provide an incorrect intercept. This is actually not a big concern because we know that the intercept is equal to the CVaR of an optimal residual without intercept.

Also, we demonstrated that the CVaR regression can be done in two steps with the decomposition theorem by using parameters from Sets 1 and 2 in the mixed-quantile deviation. A similar two-step procedure was used for CVaR regression in Rockafellar et al. (2014). Here we justify this two-step procedure through the equivalence of deviations in CVaR and mixed-quantile quadrangles with parameters from Sets 1 and 2.

This paper is organized as follows. Section 2 provides general results about quadrangles. In particular, we considered quantile, mixed-quantile, and CVaR quadrangles. Sections 3 and 4 introduced and investigated the parameters from Sets 1 and 2, respectively. Section 5 provided optimization problem statements based on CVaR and mixed-quantile quadrangles and described the linear regression for CVaR estimation. Section 6 presented a case study and applied CVaR regression to the financial style classification problem. The case study is posted on the web with codes, data, and solutions. Appendix A provides convex and linear programming problems for minimization of the Rockafellar error; Appendix B provides Portfolio Safeguard (PSG) codes implementing regression optimization problems.

2. Quantile, Mixed-Quantile, and CVaR Quadrangles

Rockafellar and Uryasev (2013) developed a new paradigm called the risk quadrangle, which linked risk management, reliability, statistics, and stochastic optimization theories. The risk quadrangle methodology united risk functions for a random value X in groups (quadrangles) consisting of five elements:

- Risk $\mathcal{R}(X)$, which provides a numerical surrogate for the overall hazard in X.
- Deviation $\mathcal{D}(X)$, which measures the "nonconstancy" in X as its uncertainty.
- Error $\mathcal{E}(X)$, which measures the "nonzeroness" in X.
- Regret $\mathcal{V}(X)$, which measures the "regret" in facing the mix of outcomes of X.
- Statistic $\mathcal{S}(X)$ associated with X through $\mathcal{E}$ and $\mathcal{V}$.

These elements of a risk quadrangle are related as follows:

$$\mathcal{V}(X) = \mathcal{E}(X) + E(X)$$

$$\mathcal{R}(X) = \mathcal{D}(X) + E(X)$$

$$\mathcal{R}(X) = \min_C\{C + \mathcal{V}(X - C)\}$$

$$\mathcal{D}(X) = \min_C\{\mathcal{E}(X - C)\}$$

$$\underset{C}{\operatorname{argmin}}\{C + \mathcal{V}(X - C)\} = \mathcal{S}(X) = \underset{C}{\operatorname{argmin}}\{\mathcal{E}(X - C)\}$$

where $E(X)$ denotes the mean of X and the statistic, $\mathcal{S}(X)$, can be a set, if the minimum is achieved for multiple points.

Further, we use the following notations. The cumulative distribution function is denoted by $F_X(x) = prob\{X \le x\}$. The positive and negative part of a number are denoted using:

$$[t]^+ = \begin{cases} t, \text{ for } t > 0 \\ 0, \text{ for } t \le 0 \end{cases} \quad \text{and} \quad [t]^- = \begin{cases} -t, \text{ for } t < 0 \\ 0, \text{ for } t \ge 0 \end{cases}$$

The lower and upper VaR (quantile) are defined as follows:

lower VaR:

$$VaR_\alpha^-(X) = \begin{cases} sup\{x, F_X(x) < \alpha\} for\ 0 < \alpha \le 1 \\ inf\{x, F_X(x) \ge \alpha\}\ for\ \alpha = 0 \end{cases}$$

upper VaR:

$$VaR_\alpha^+(X) = \begin{cases} inf\{x, F_X(x) > \alpha\}\ for\ 0 \le \alpha < 1 \\ sup\{x, F_X(x) \le \alpha\}\ for\ \alpha = 1 \end{cases}$$

VaR (quantile) is a set if the lower and upper quantiles do not coincide:

$$VaR_\alpha(X) = \left[VaR_\alpha^-(X), VaR_\alpha^+(X)\right]$$

otherwise VaR is a singleton $VaR_\alpha(X) = VaR_\alpha^-(X) = VaR_\alpha^+(X)$.

Conditional value-at-risk (CVaR) with the confidence level $\alpha \in (0,1)$ can be defined in many ways. We prefer the following constructive definition:

$$CVaR_\alpha(X) = \min_C\left\{C + \frac{1}{1-\alpha}E[X - C]^+\right\}$$

In financial applications, however, the most popular definition of CVaR is

$$CVaR_\alpha(X) = \frac{1}{1-\alpha}\int_\alpha^1 VaR_\beta^-(X)d\beta.$$

For $\alpha = 0$, $CVaR_0(X)$ is defined as $CVaR_0(X) = \lim_{\varepsilon\to 0} CVaR_\varepsilon(X) = E(X)$.

For $\alpha = 1$, $CVaR_1(X)$ is defined as $CVaR_1(X) = VaR_1^-(X)$ if a finite value of $VaR_1^-(X)$ exists.

Quadrangles are named after statistic functions. The most famous quadrangle is the quantile quadrangle (see Rockafellar and Uryasev (2013)), named after the VaR (quantile) statistic. This quadrangle establishes relations between the CVaR optimization technique described in Rockafellar and Uryasev (2000, 2002) and quantile regression (see Koenker and Bassett (1978), Koenker (2005)). In particular, it was shown that CVaR minimization and the quantile regression are similar procedures based on the VaR statistic in the regret and error representation of risk and deviation.

Here is the definition of the quantile quadrangle for $\alpha \in (0,1)$:

Quantile Quadrangle (Rockafellar and Uryasev (2013))

Statistic: $\mathcal{S}_\alpha(X) = VaR_\alpha(X) =$ VaR (quantile) statistic.
Risk: $\mathcal{R}_\alpha(X) = CVaR_\alpha(X) = \min_C\{C + \mathcal{V}_\alpha(X - C)\} =$ CVaR risk.
Deviation: $\mathcal{D}_\alpha(X) = CVaR_\alpha(X) - E[X] = \min_C\{\mathcal{E}_\alpha(X - C)\} =$ CVaR deviation.
Regret: $\mathcal{V}_\alpha(X) = \frac{1}{1-\alpha}E[X]^+ =$ average absolute loss, scaled.
Error: $\mathcal{E}_\alpha(X) = E\left[\frac{\alpha}{1-\alpha}[X]^+ + [X]^-\right] = \mathcal{V}_\alpha(X) - E[X] =$ normalized Koenker–Bassett error.

The quantile quadrangle sets an example for development of more advances quadrangles. The following mixed-quantile quadrangle includes statistic, which is equal to the weighted average of VaRs (quantiles) with specified positive weights. Therefore, the error in this quadrangle can be used to build a regression for the weighted average of VaRs (quantiles). Since CVaR can be approximated by a weighted average of VaRs, the error function in this quadrangle can be used to build linear regression for the estimation of CVaR.

Mixed-Quantile Quadrangle (Rockafellar and Uryasev (2013)).

Confidence levels $\alpha_k \in (0,1)$, $k = 1,\ldots,r$, and weights $\lambda_k > 0$, $\sum_{k=1}^r \lambda_k = 1$. The error in this quadrangle is called the Rockafellar Error.

Statistic: $\mathcal{S}(X) = \sum_{k=1}^r \lambda_k VaR_{\alpha_k}(X) =$ mixed VaR (quantile).
Risk: $\mathcal{R}(X) = \sum_{k=1}^r \lambda_k CVaR_{\alpha_k}(X) =$ mixed CVaR.
Deviation: $\mathcal{D}(X) = \sum_{k=1}^r \lambda_k CVaR_{\alpha_k}(X - E[X]) =$ mixed CVaR deviation.
Regret: $\mathcal{V}(X) = \min_{B_1,\ldots,B_r}\left\{\sum_{k=1}^r \lambda_k \mathcal{V}_{\alpha_k}(X - B_k) \middle| \sum_{k=1}^r \lambda_k B_k = 0\right\} =$ the minimal weighted average of regrets $\mathcal{V}_{\alpha_k}(X - B_k) = \frac{1}{1-\alpha_k}E[X - B_k]^+$ satisfying the linear constraint on $B_1,\ldots,B_r$.
Error: $\mathcal{E}(X) = \min_{B_1,\ldots,B_r}\left\{\sum_{k=1}^r \lambda_k \mathcal{E}_{\alpha_k}(X - B_k) \middle| \sum_{k=1}^r \lambda_k B_k = 0\right\} =$ Rockafellar error $=$ the minimal weighted average of errors $\mathcal{E}_{\alpha_k}(X - B_k) = E\left[\frac{\alpha_k}{1-\alpha_k}[X - B_k]^+ + [X - B_k]^-\right]$ satisfying the linear constraint on $B_1,\ldots,B_r$.

The following CVaR quadrangle can be considered as the limiting case of the mixed-quantile quadrangle when the number of terms in this quadrangle tends to infinity. The statistic in this quadrangle is CVaR; therefore, the error in this quadrangle can be used for the estimation of CVaR with linear regression.

CVaR Quadrangle (Rockafellar et al. (2014)) for $\alpha \in (0,1)$.

Statistic: $\overline{\mathcal{S}}_\alpha(X) = CVaR_\alpha(X) =$ CVaR.
Risk: $\overline{\mathcal{R}}_\alpha(X) = \frac{1}{1-\alpha}\int_\alpha^1 CVaR_\beta(X)d\beta =$ CVaR2 risk.
Deviation: $\overline{\mathcal{D}}_\alpha(X) = \overline{\mathcal{R}}_\alpha(X) - E[X] = \frac{1}{1-\alpha}\int_\alpha^1 CVaR_\beta(X)d\beta - E[X] =$ CVaR2 deviation.
Regret: $\overline{\mathcal{V}}_\alpha(X) = \frac{1}{1-\alpha}\int_0^1 \left[CVaR_\beta(X)\right]^+ d\beta =$ CVaR2 regret.
Error: $\overline{\mathcal{E}}_\alpha(X) = \overline{\mathcal{V}}_\alpha(X) - E[X] =$ CVaR2 error.

The following section proves that for a discretely distributed random value with equally probable atoms, the CVaR quadrangle is "equivalent" to a mixed-quantile quadrangle with some parameters in the sense that statistic, risk, and deviation in these quadrangles coincide. This fact was proved for a set of random values with equal probabilities and variable locations of atoms.

The set of parameters considered in the following section is used in two-step CVaR regression in Rockafellar et al. (2014).

3. Set 1 of Parameters for Mixed-Quantile Quadrangle

Set 1 of parameters for the mixed-quantile quadrangle for a discrete uniformly distributed random value X consists of confidence levels $\alpha_k \in (0,1), k = 1,\ldots,r$ and weights $\lambda_k > 0$ such that $\sum_{k=1}^r \lambda_k = 1$. Parameter r depends only on the number of atoms in X and the confidence level α of the CVaR quadrangle. We proved that statistic, risk, and deviation of the mixed-quantile quadrangle with the Set 1 of parameters coincide with the statistic, risk, and deviation of the CVaR quadrangle.

Let X be a discrete random value with support x^i and $\text{Prob}\big(X = x^i\big) = 1/\nu$ for $i = 1,2,\ldots,\nu$, where ν is the number of atoms. Denote $x^{max} = \max\limits_{i=1,\ldots,\nu} x^i$. For this random value, $CVaR_1(X) = VaR_1^-(X) = x^{max}$.

Set 1 of parameters:

- partition of the interval $[\alpha,1]$: $\beta_{\nu_\alpha - 1} = \alpha$, and $\beta_i = i\delta$, for $i = \nu_\alpha,\ \nu_\alpha + 1,\ldots,\nu$, where $\delta = 1/\nu$, $\nu_\alpha = \lfloor \nu\alpha \rfloor + 1$, with $\lfloor z \rfloor$ being the largest integer less than or equal to z; $\delta_\alpha = \beta_{\nu_\alpha} - \alpha$.
- weights: $p_{\nu_\alpha} = \frac{\delta_\alpha}{1-\alpha},\ p_i = \frac{\delta}{1-\alpha},\ \ i = \nu_\alpha + 1,\ldots,\nu.$
- confidence levels: $\gamma_i = 1 - \frac{\beta_i - \beta_{i-1}}{\ln\left(\frac{1-\beta_{i-1}}{1-\beta_i}\right)}, i = \nu_\alpha,\ldots,\nu - 1;\ \gamma_\nu = 1.$

Lemma 1. *Let X be a discrete random value with ν equally probable atoms. Then, statistic, risk, and deviation of the CVaR quadrangle for X are given by the following expressions with parameters specified by Set 1:*

1. *CVaR statistic:*

$$\overline{\mathcal{S}}_\alpha(X) = CVaR_\alpha(X) = \sum\nolimits_{i=\nu_\alpha}^{\nu} p_i VaR_{\gamma_i}(X) \tag{1}$$

2. *CVaR2 risk:*

$$\overline{\mathcal{R}}_\alpha(X) = \frac{1}{1-\alpha}\int_\alpha^1 CVaR_\beta(X)d\beta = \sum\nolimits_{i=\nu_\alpha}^{\nu} p_i CVaR_{\gamma_i}(X) \tag{2}$$

3. *CVaR2 deviation:*

$$\overline{\mathcal{D}}_\alpha(X) = \frac{1}{1-\alpha}\int_\alpha^1 CVaR_\beta(X)d\beta - E[X] = \sum\nolimits_{i=\nu_\alpha}^{\nu} p_i CVaR_{\gamma_i}(X) - E[X] \tag{3}$$

Proof. Appendix C contains proof of the lemma. □

Note. Expression (1) is valid for arbitrary $\gamma_i \in (\beta_{i-1}, \beta_i)$, $i = \nu_\alpha, \ldots, \nu$. Equations (2) and (3) are valid for arbitrary $\gamma_\nu \in [\beta_{\nu-1}, 1]$.

We want to emphasize that the statement of Lemma 1 is valid for any discrete random value with equally probable atoms. The statement does not depend upon atom locations.

Corollary 1. *For the random value X defined in Lemma 1, statistic, risk, and deviation of the CVaR quadrangle coincide with statistic, risk, and deviation of the mixed-quantile quadrangle with* $r = \nu - \nu_\alpha + 1$, $\lambda_k = p_{\nu_\alpha - 1 + k}$, $\alpha_k = \gamma_{\nu_\alpha - 1 + k}$, $k = 1, \ldots, r$.

Proof. Right hand sides in Equations (1)–(3) define statistic, risk, and deviation of the mixed-quantile quadrangle because $p_i > 0$, $i = \nu_\alpha, \ldots, \nu$, $\sum_{i=\nu_\alpha}^{\nu} p_i = 1$ and $VaR_{\gamma_i}(X)$, $i = \nu_\alpha, \ldots, \nu$, are singletons. □

Example 1. *Let X be a discrete random value with five atoms (−40; −10; 20; 60 100) and equal probabilities = 0.2.*

Figure 1 explains how to calculate statistic in the CVaR quadrangle and mixed-quantile quadrangle with Set 1 of parameters for $\alpha = 0.5$. Bold lines show $VaR_\alpha^-(X)$ as a function of α. $CVaR_{0.5}(X)$ equals the dark area under the $VaR_\alpha(X)$ divided by $1 - \alpha$. CVaR can be calculated as integral of $VaR_\alpha(X)$ or as the sum of areas of rectangles. Figure 2 explains how to calculate risk in the CVaR quadrangle and mixed-quantile quadrangle with Set 1 of parameters for $\alpha = 0.5$. The bold continuous curve shows $CVaR_\alpha(X)$ as a function of α. Risk $\overline{\mathcal{R}}_{0.5}(X)$ is equal to the area under the CVaR curve divided by $1 - \alpha$. This area can be calculated as the integral of CVaR or as the sum of areas of rectangles. The area of every rectangle is equal to the area under CVaR in the appropriate range of α. The equality of areas defines values of γ_i. Parameters p_i, γ_i do not depend on the values of atoms.

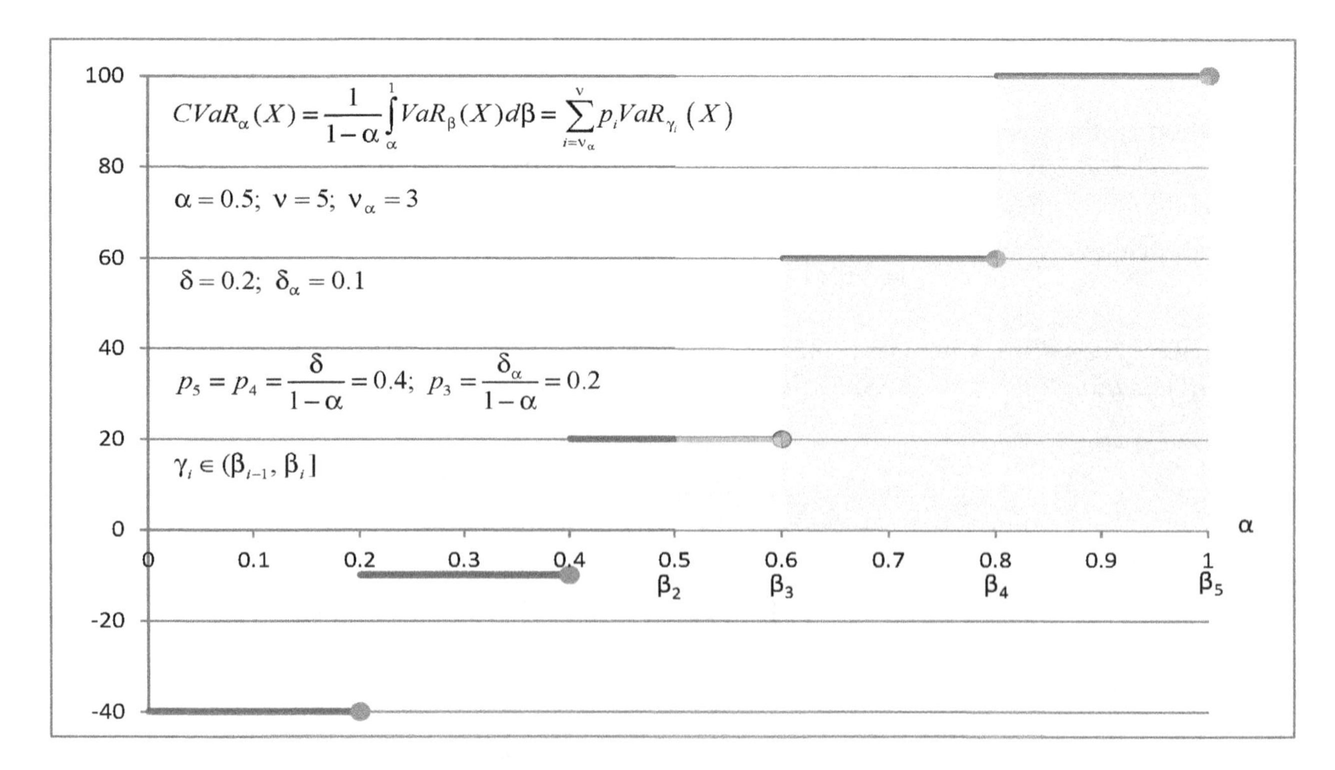

Figure 1. Five equally probable atoms. VaR and CVaR with Set 1 of parameters for $\alpha = 0.5$.

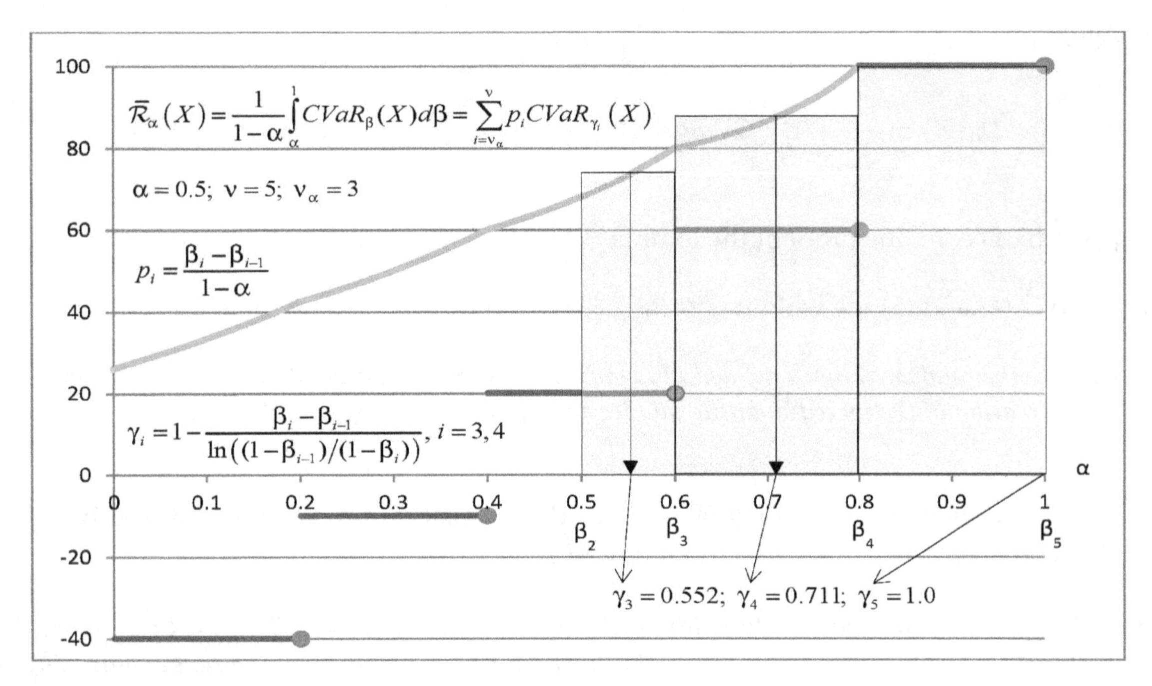

Figure 2. Five equally probable atoms. Risk in the CVaR quadrangle and mixed-quantile quadrangle with Set 1 of parameters for $\alpha = 0.5$.

4. Set 2 of Parameters for the Mixed-Quantile Quadrangle

This section gives an alternative expression for the risk $\overline{\mathcal{R}}_\alpha(X)$ and deviation $\overline{\mathcal{D}}_\alpha(X)$ in the CVaR quadrangle for a discrete uniformly distributed random variable. This expression is based on the following Set 2 of parameters. This set of parameters has the same number of parameters as Set 1 but different values of weights and confidence levels. Similar to Section 3, let X be a discrete random value with support, x^i, $i = 1,2,\ldots,\nu$, and $\text{Prob}\big(X = x^i\big) = 1/\nu$ for $i = 1,2,\ldots,\nu$. Denote $x^{max} = \max_{i=1,\ldots,\nu} x^i$. For this random value $CVaR_1(X) = VaR_1^-(X) = x^{max}$.

Set 2 of parameters:

- partition of the interval $[\beta_{\nu_\alpha-1}, 1]$: $\delta = 1/\nu$, $\beta_i = i\delta$, for $i = \nu_\alpha - 1,\ \nu_\alpha, \ldots, \nu$, where $\nu_\alpha = \lfloor \nu\alpha \rfloor + 1$, with $\lfloor z \rfloor$ being the largest integer less than or equal to z; $\delta_\alpha = \beta_{\nu_\alpha} - \alpha$.
- confidence levels: β_i, $i = \nu_\alpha - 1,\ \nu_\alpha, \ldots, \nu$.
- weights:
  $q_\nu = 0;$
  $q_{\nu-1} = \frac{\delta}{1-\alpha} \times [2\ln(2)] \approx \frac{\delta}{1-\alpha} \times 1.386294361$, (if $\nu - 1 > \nu_\alpha$)
  $q_{\nu-2} = \frac{\delta}{1-\alpha} \times 2\big[3\ln\big(\frac{3}{2}\big) + \ln\big(\frac{1}{2}\big)\big] \approx \frac{\delta}{1-\alpha} \times 1.046496288$, (if $\nu - 2 > \nu_\alpha$)
  $q_{\nu-j} = \frac{\delta}{1-\alpha} \times j\big[(j+1)\ln\big(\frac{j+1}{j}\big) + (j-1)\ln\big(\frac{j-1}{j}\big)\big]$, (if $j > 2, \nu - j > \nu_\alpha$)
  $q_{\nu_\alpha} = \frac{\delta}{1-\alpha} \times j\big[\delta - \delta_\alpha + (j+1)\ln\big(\frac{1-\alpha}{\delta j}\big) + (j-1)\ln\big(\frac{j-1}{j}\big)\big]$, (if $\nu_\alpha < \nu - 1,\ j = \nu - \nu_\alpha$)
  $q_{\nu_\alpha-1} = \frac{\delta}{1-\alpha} \times j\Big[\delta_\alpha + (j-1)\ln\Big(\frac{\delta(j-1)}{1-\alpha}\Big)\Big]$, (if $\nu_\alpha - 1 < \nu - 1,\ j = \nu - \nu_\alpha + 1$)
  if $\nu_\alpha = \nu - 1$, then $q_{\nu_\alpha} = \frac{\delta}{1-\alpha} \times 2\big[1 + \ln\big(\frac{1-\alpha}{\delta}\big)\big] - 1$, $q_{\nu_\alpha-1} = \frac{\delta}{1-\alpha} \times 2\big[\ln\big(\frac{\delta}{1-\alpha}\big) - 1\big] + 2$
  if $\nu_\alpha = \nu$, then $q_{\nu_\alpha-1} = 1$.

Lemma 2. *Let X be a discrete random value with ν equally probable atoms. Then, risk and deviation of the CVaR quadrangle for X are given by the following expressions with parameters from Set 2.*

1. *CVaR2 Risk:*

$$\overline{\mathcal{R}}_\alpha(X) = \frac{1}{1-\alpha}\int_\alpha^1 CVaR_\beta(X)d\beta = \sum\nolimits_{i=\nu_\alpha-1}^{\nu-1} q_i CVaR_{\beta_i}(X) \tag{4}$$

2. *CVaR2 Deviation:*

$$\overline{\mathcal{D}}_\alpha(X) = \frac{1}{1-\alpha}\int_\alpha^1 CVaR_\beta(X)d\beta - E[X] = \sum_{i=\nu_\alpha-1}^{\nu-1} q_i CVaR_{\beta_i}(X) - E[X] \quad (5)$$

Proof. Appendix D contains proof of the lemma. □

Note. Equations (4) and (5) are valid if $CVaR_{\beta_{\nu-1}}$ is replaced by $CVaR_\gamma$ with an arbitrary $\gamma \in [\beta_{\nu-1}, 1]$.

Corollary 2. *For the random value X defined in Lemma 2, risk and deviation of the CVaR quadrangle coincide with risk and deviation of the mixed-quantile quadrangle with* $r = \nu - \nu_\alpha + 1$, $\lambda_k = q_{\nu_\alpha-2+k}$, $\alpha_k = \beta_{\nu_\alpha-2+k}$, $k = 1,\ldots,r$.

Proof. Right hand sides in Equations (4) and (5) define risk and deviation of the mixed-quantile quadrangle because $q_i > 0$, $i = \nu_\alpha - 1,\ldots,\nu-1$, and $\sum_{i=\nu_\alpha-1}^{\nu-1} q_i = 1$. □

Lemma 3. *Let X be a discrete random value with equally probable atoms* x^i, $i = 1,2,\ldots,\nu$, $Prob\left(X = x^i\right) = 1/\nu$. *Then, statistic of the mixed-quantile quadrangle defined by the Set 2 of parameters is a range containing statistic of the CVaR quadrangle.*

Proof. Appendix E contains proof of the lemma. □

5. On the Estimation of CVaR with Mixed-Quantile Linear Regression

This section formulates regression problems using the CVaR quadrangle and mixed-quantile quadrangle. For discrete final distributions with equally probable atoms, we prove some equivalence statements for the CVaR and mixed-quantile quadrangles. Further, we demonstrate how to estimate CVaR by using the linear regression with error and deviation from the mixed-quantile quadrangle.

We want to estimate variable V using a linear function $f(\boldsymbol{Y}) = C_0 + \boldsymbol{C}^T\boldsymbol{Y}$ of the explanatory factors $\boldsymbol{Y} = (Y_1,\ldots,Y_n)$. Let $\widetilde{\mathcal{E}}$ be an error from some quadrangle (further we consider, mixed-quantile and CVaR quadrangles), and $\widetilde{\mathcal{D}}$ and $\widetilde{\mathcal{S}}$ be a deviation and a statistic, respectively, corresponding to this quadrangle. Below we consider optimization statements for solving regression problems.

General Optimization Problem 1

Minimize error $\widetilde{\mathcal{E}}$ and find optimal C_0^*, $\boldsymbol{C}^*$

$$\min_{C_0\in\mathbb{R},\ \boldsymbol{C}\in\mathbb{R}^m} \widetilde{\mathcal{E}}\left(Z(C_0,\boldsymbol{C})\right)$$

where $Z(C_0,\boldsymbol{C}) = V - C_0 - \boldsymbol{C}^T\boldsymbol{Y}$.

General Optimization Problem 2

- *Step 1. Find an optimal vector* $\boldsymbol{C}^*$ *by minimizing deviation:*

$$\min_{\boldsymbol{C}\in\mathbb{R}^m} \widetilde{\mathcal{D}}(Z_0(\boldsymbol{C}))$$

where $Z_0(\boldsymbol{C}) = V - \boldsymbol{C}^T\boldsymbol{Y}$.

- *Step 2. Assign* C_0^*:

$$C_0^* \in \widetilde{\mathcal{S}}(Z_0(\boldsymbol{C}^*))$$

Error $\widetilde{\mathcal{E}}(X)$ is called nondegenerate if:

$$\inf_{X:EX=D} \widetilde{\mathcal{E}}(X) > 0 \text{ for constants } D \neq 0.$$

Rockafellar et al. (2008), p. 722, proved the following decomposition theorem.

Theorem 1. *(Error-Shaping Decomposition of Regression). Let $\widetilde{\mathcal{E}}$ be a nondegenerate error and $\widetilde{\mathcal{D}} = \min_{C} \{\widetilde{\mathcal{E}}(X - C)\}$ be the corresponding deviation, and let $\widetilde{\mathcal{S}}$ be the associated statistic. Point $(C_0^*, \boldsymbol{C}^*)$ is a solution of the General Optimization Problem 1 if and only if $\boldsymbol{C}^*$ is a solution of the General Optimization Problem 2, Step 1 and $C_0^* \in \widetilde{\mathcal{S}}(Z_0(\boldsymbol{C}^*))$ with Step 2.*

According to the decomposition theorem, when $\widetilde{\mathcal{E}}$, $\widetilde{\mathcal{D}}$, and $\widetilde{\mathcal{S}}$ are elements of the CVaR quadrangle, the following Optimization Problems 1 and 2 are equivalent.

Optimization Problem 1

Minimize error from the CVaR quadrangle:

$$\min_{C_0 \in \mathbb{R},\ \boldsymbol{C} \in \mathbb{R}^m} \overline{\mathcal{E}}_\alpha(Z(C_0, \boldsymbol{C}))$$

where $Z(C_0, \boldsymbol{C}) = V - C_0 - \boldsymbol{C}^T \boldsymbol{Y}$.

Optimization Problem 2

- *Step 1. Find an optimal vector $\boldsymbol{C}^*$ by minimizing deviation from the CVaR quadrangle:*

$$\min_{\boldsymbol{C} \in \mathbb{R}^m} \overline{\mathcal{D}}_\alpha(Z_0(\boldsymbol{C}))$$

where $Z_0(\boldsymbol{C}) = V - \boldsymbol{C}^T \boldsymbol{Y}$.

- *Step 2. Calculate:*

$$C_0^* = CVaR_\alpha(Z_0(\boldsymbol{C}^*))$$

In Optimization Problem 2 in Step 2 the statistic equals $\overline{S}(Z_0(\boldsymbol{C}^*)) = CVaR_\alpha(Z_0(\boldsymbol{C}^*))$, which is the specification of the inclusion operation in the Optimization Problem General 2 in Step 2.

Optimization Problem 2 is used in Rockafellar et al. (2014) for the construction of the linear regression algorithms for estimating CVaR.

According to the decomposition theorem, when $\widetilde{\mathcal{E}}$, $\widetilde{\mathcal{D}}$, and $\widetilde{\mathcal{S}}$ are elements of a mixed-quantile quadrangle, the following Optimization Problems 3 and 4 are equivalent.

Optimization Problem 3

Minimize error from the mixed-quantile quadrangle:

$$\min_{C_0 \in \mathbb{R},\ \boldsymbol{C} \in \mathbb{R}^m} \mathcal{E}(Z(C_0, \boldsymbol{C}))$$

Optimization Problem 4

- *Step 1. Find an optimal vector $\boldsymbol{C}^*$ by minimizing deviation from the mixed-quantile quadrangle:*

$$\min_{\boldsymbol{C} \in \mathbb{R}^m} \mathcal{D}(Z_0(\boldsymbol{C}))$$

- *Step 2. Assign:*

$$C_0^* \in \mathcal{S}(Z_0(\boldsymbol{C}^*))$$

Corollaries 1 and 2 can be used for constructing the linear regression for estimating CVaR. Let $\boldsymbol{Y}$ be a random vector of factors for estimating the random value V. We consider that the linear regression function $f(\boldsymbol{Y}) = C_0 + \boldsymbol{C}^T\boldsymbol{Y}$ approximates CVaR of V, where $C_0 \in \mathbb{R}$, $\boldsymbol{C} \in \mathbb{R}^m$ are variables in the linear regression. The residual is denoted by $Z(C_0, \boldsymbol{C}) = V - \left(C_0 + \boldsymbol{C}^T\boldsymbol{Y}\right)$ and $Z_0(\boldsymbol{C}) = V - \boldsymbol{C}^T\boldsymbol{Y}$.

Further we provide a lemma about linear regression problems based on Corollary 1 with the Set 1 of parameters. The main statement here is that the Optimization Problems 3 and 4 for the mixed-quantile quadrangle can be used to solve linear regression problems for estimating CVaR. This is the case because CVaR and mixed-quantile quadrangles have the same Statistic and Deviation.

Lemma 4. *Let the residual random value $Z(C_0, \boldsymbol{C}) = V - \left(C_0 + \boldsymbol{C}^T\boldsymbol{Y}\right)$ be discretely distributed with ν equally probable atoms. Let us consider the CVaR quadrangle with error $\overline{\mathcal{E}}_\alpha(Z(C_0, \boldsymbol{C}))$, deviation $\overline{\mathcal{D}}_\alpha(Z_0(\boldsymbol{C}))$, and statistic $\overline{S}_\alpha(Z_0(\boldsymbol{C}^*))$. Let us also consider the mixed-quantile quadrangle with the error $\mathcal{E}(Z(C_0, \boldsymbol{C}))$, deviation $\mathcal{D}(Z_0(\boldsymbol{C}))$, and statistic $\mathcal{S}(Z_0(\boldsymbol{C}^*))$ with parameters defined by Set 1 and $r = \nu - \nu_\alpha + 1$, $\lambda_k = p_{\nu_\alpha - 1 + k}$, $\alpha_k = \gamma_{\nu_\alpha - 1 + k}$, $k = 1, \ldots, r$. Then, the Optimization Problems 1–4 are equivalent, i.e., the sets of optimal vectors of these optimization problems coincide. Moreover, let $\left(C_0^*, \boldsymbol{C}^*\right)$ be a solution vector of the equivalent Optimization Problems 1–4. Then:*

$$C_0^* = CVaR_\alpha(Z_0(\boldsymbol{C}^*))$$

$$\overline{\mathcal{E}}_\alpha\left(C_0^*, \boldsymbol{C}^*\right) = \mathcal{E}\left(C_0^*, \boldsymbol{C}^*\right) = \mathcal{D}(Z_0(\boldsymbol{C}^*)) = \overline{\mathcal{D}}_\alpha(Z_0(\boldsymbol{C}^*))$$

Proof. This lemma is a direct corollary of the decomposition Theorem 1 and Corollary 1 of Lemma 1. Indeed, Corollary 1 implies that the Optimization Problems 2 and 4 are equivalent. Further, the decomposition theorem implies that Optimization Problems 1 and 2 and the Optimization Problems 3 and 4 are equivalent. □

Further we provide a lemma about linear regression problems based on Corollary 2 with the Set 2 of parameters. The main statement is that the Optimization Problem 4, Step 1 for the Mixed-Quantile Quadrangle can be used to solve linear regression problem for estimating CVaR. This is the case because CVaR and Mixed-Quantile Quadrangles have the same Deviation. After obtaining vector of coefficients $\boldsymbol{C}^*$, intercept is calculated, $C_0^* = CVaR_\alpha(Z_0(\boldsymbol{C}^*))$.

Lemma 5. *Let the residual random value $Z(C_0, \boldsymbol{C}) = V - \left(C_0 + \boldsymbol{C}^T\boldsymbol{Y}\right)$ be discretely distributed with ν equally probable atoms. Let the mixed-quantile quadrangle with deviation $\mathcal{D}(Z_0(\boldsymbol{C}))$ be defined by parameters of Set 2 and $r = \nu - \nu_\alpha + 1$, $\lambda_k = q_{\nu_\alpha - 2 + k}$, $\alpha_k = \beta_{\nu_\alpha - 2 + k}$, $k = 1, \ldots, r$. Then, $\left(C_0^*, \boldsymbol{C}^*\right)$ is a solution of Optimization Problem 1, if and only if, $\left(C_0^*, \boldsymbol{C}^*\right)$ is a solution of the following two-step procedure:*

- *Step 1. Find an optimal vector $\boldsymbol{C}^*$ by minimizing deviation from the mixed-quantile quadrangle:*

$$\min_{\boldsymbol{C} \in \mathbb{R}^m} \mathcal{D}(Z_0(\boldsymbol{C}))$$

- *Step 2. Calculate $C_0^* = CVaR_\alpha(Z_0(\boldsymbol{C}^*))$.*

Proof. This lemma is a direct corollary of the decomposition Theorem 1 and Corollary 2 or Lemma 2. Indeed, the Corollary 2 implies that the Optimization Problems 2 and 4 are equivalent. Further, since deviations of the CVaR and mixed-quantile quadrangles coincide, we can use Step 1 to calculate optimal coefficients $\boldsymbol{C}^*$. Further, the intercept is calculated with $C_0^* = CVaR_\alpha(Z_0(\boldsymbol{C}^*))$ because CVaR is the statistic in the CVaR quadrangle. □

For the Set 2 of parameters, the deviations in the CVaR and mixed-quintile quadrangles coincide. The two-step procedure in Optimization Problem 4 can be used to solve linear regression problems with Set 2 parameters for the mixed-quantile deviation. Also, the minimization of the Rockafellar error with the Set 2 of parameters may result in a correct C^*. The statistic of CVaR belongs to the statistic of the mixed-quantile quadrangle. Therefore, the optimization of the Rockafellar error with the Set 2 of parameters may lead to a wrong value of intercept C_0^*. This potential incorrectness can be fixed by assigning $C_0^* = CVaR_\alpha(Z_0(C^*))$.

6. Case Study: Estimation of CVaR with Linear Regression and Style Classification of Funds

The case study described in this section is posted online (see Case Study (2016)). The codes and data are available for downloading and verification. Every optimization problem is presented in three formats: Text, MATLAB, and R. Calculations were done with a PC with a 3.14 GHz processor.

We have applied CVaR regression to the return-based style classification of a mutual fund. We regress a fund return by several indices as explanatory factors. The estimated coefficients represent the fund's style with respect to each of the indices.

A similar problem with a standard regression based on the mean squared error was considered by Carhart (1997) and Sharpe (1992). They estimated the conditional expectation of a fund return distribution (under the condition that a realization of explanatory factors is observed). Basset and Chen (2001) extended this approach and conducted the style analyses of quantiles of the return distribution. This extension is based on the quantile regression suggested by Koenker and Bassett (1978). The Case Study (2014), "Style Classification with Quantile Regression" implemented this approach and applied quantile regression to the return-based style classification of a mutual fund.

For the numerical implementation of CVaR linear regression, we used the Portfolio Safeguard (2018) package. Portfolio Safeguard (PSG) can solve nonlinear and mixed-integer nonlinear optimization problems. A special feature of PSG is that it includes precoded nonlinear functions: CVaR2 error (`cvar2_err`) and CVaR2 deviation (`cvar2_dev`) from the CVaR quadrangle, Rockafellar error (`ro_err`) from the mixed-quantile quadrangle, and CVaR deviation (`cvar_dev`) from the quantile quadrangle.

We implemented the following equivalent variants of CVaR regression:

- Minimization of the CVaR2 error (PSG function `cvar2_err`).
- Two-step procedure with CVaR2 deviation (PSG function `cvar2_dev`).
- Minimization of the Rockafellar error (PSG function `ro_err`) with the Set 1 of parameters.
- The two-step procedure using mixed CVaR deviation with the Set 1 and Set 2 of parameters. This is calculated as a weighted sum of CVaR deviations (PSG function `cvar_dev`) from the quantile quadrangle.

PSG automatically converts the analytic problem formulations to the mathematical programming codes and solves them. We included in Appendix A convex and linear programming problems for the minimization of the Rockafellar error with the Set 1 of parameters. These formulations are provided for verification purposes. They can be implemented with standard commercial software. For instance, the linear programming formulation can be implemented with the Gurobi optimization package. If Gurobi is installed in the computer, PSG can use Gurobi code as a subsolver. With the CARGRB solver in PSG, by setting the linearize option to 1, it is possible to solve the linear programming problem with Gurobi. However, this conversion will deteriorate the performance, compared to the default PSG solver VAN. For small problems it will not be noticeable. However, for problems with a large number of scenarios (e.g., with 10^8 observations), the standard PSG solver VAN will dramatically outperform the Gurobi linear programming implementation. In this case, Gurobi may not even start on a small PC because of a shortage of memory. Nevertheless, if the number of observations is small (e.g., 10^3) and the number of factors is very large (e.g., 10^7), it is recommended that the linear programming formulation is used.

We regressed the CVaR of the return distribution of the Fidelity Magellan Fund on the explanatory variables: Russell 1000 Growth Index (RLG), Russell 1000 Value Index (RLV), Russell Value Index (RUJ), and Russell 2000 Growth Index (RUO). The dataset includes 1264 historical daily returns of the Magellan Fund and the indices, which were downloaded from the Yahoo Finance website. The data (design matrix for the regression) is posted on the Case Study (2016) website.

The CVaR regression was done with the confidence levels $\alpha = 0.75$ and $\alpha = 0.9$. Calculation results are in Tables 1 and 2, respectively. Here is the description of the columns of the tables:

- Optimization Problem #: Optimization Problem number, as denoted in Section 5; also it is the problem number in the case study posted online, see, Case Study (2016).
- Set #: Set of parameters for the mixed-quantile quadrangle.
- Objective: Optimal value of the objective function.
- RLG: coefficient for the Russell Value Index.
- RLV: coefficient for the Russell 1000 Value Index.
- RUJ: coefficient for the Russell Value Index.
- RUO: coefficient for the Russell 2000 Growth Index.
- Intercept: regression intercept.
- Solving Time: solver optimization time.

Table 1. Optimization outputs: estimating CVaR with the linear regression, $\alpha = 0.75$.

Optimization Problem #	Set #	Objective	RLG	RLV	RUJ	RUO	Intercept	Solving Time (s)
1	N/A	0.01248	0.486	0.581	−0.0753	-6.22×10^{-3}	6.98×10^{-3}	0.02
2	N/A	0.01248	0.486	0.582	−0.0753	-6.22×10^{-3}	6.98×10^{-3}	0.02
3	Set 1	0.01247	0.486	0.582	−0.0753	-6.22×10^{-3}	6.96×10^{-3}	0.03
4	Set 1	0.01251	0.486	0.582	−0.0752	-6.22×10^{-3}	6.98×10^{-3}	0.11
4	Set 2	0.01248	0.486	0.582	−0.0753	-6.23×10^{-3}	6.98×10^{-3}	0.03

Table 2. Optimization outputs: estimating CVaR with the linear regression, $\alpha = 0.9$.

Optimization Problem #	Set #	Objective	RLG	RLV	RUJ	RUO	Intercept	Solving Time (s)
1	N/A	0.016656	0.472	0.606	−0.078	-7.052×10^{-3}	1.05×10^{-2}	0.02
2	N/A	0.016656	0.472	0.606	−0.078	-7.052×10^{-3}	1.05×10^{-2}	0.02
3	Set 1	0.016656	0.472	0.606	−0.078	-7.052×10^{-3}	1.05×10^{-2}	0.02
4	Set 1	0.016656	0.472	0.606	−0.078	-7.052×10^{-3}	1.05×10^{-2}	0.11
4	Set 2	0.016656	0.472	0.606	−0.078	-7.052×10^{-3}	1.05×10^{-2}	0.04

Tables 1 and 2 show calculation results for the considered equivalent problems. We observe that regression coefficients coincide for all problems in Tables 1 and 2. This confirms the correctness of theoretical results and the numerical implementation. Also, we want to point out that the regression coefficients are quite similar for $\alpha = 0.75$ (Table 1) and $\alpha = 0.9$ (Table 2).

The calculation time in majority of cases was around 0.02–0.04 s, except for the case with the mixed CVaR deviation for Set 1, which took 0.11 s. The PSG calculation times were quite low because the solver "knows" analytical expressions for the functions and can take advantage of this knowledge.

7. Conclusions

The quadrangle risk theory Rockafellar and Uryasev (2013) and the decomposition theorem Rockafellar et al. (2008) provided a framework for building a regression with relevant deviations. Solution of a regression problem is split in two steps: (1) minimization of deviation from the corresponding quadrangle, and (2) determining of intercept by using statistic from this quadrangle. For CVaR regression, Rockafellar et al. (2014) reduced the optimization problem at Step 1 to a high-dimension linear programming problem. We suggested two sets of parameters for the mixed-quantile quadrangle and investigated its relationship with the CVaR quadrangle. The Set 1 of parameters corresponds to CVaR regression in Rockafellar et al. (2014), where the Set 2 is a new set of parameters.

For the Set 1 of parameters, the minimization of error from CVaR Quadrangle was reduced to the minimization of the Rockafellar error from the mixed-quantile quadrangle. For both sets of parameters, the minimization of deviation in CVaR quadrangle is equivalent to the minimization of deviation in mixed-quantile quadrangle.

We presented optimization problem statements for CVaR regression problems using CVaR and mixed-quantile quadrangles. Linear regression problem for estimating CVaR were efficiently implemented in Portfolio Safeguard (2018) with convex and linear programming. We have done a case study for the return-based style classification of a mutual fund with CVaR regression. We regressed the fund return by several indices as explanatory factors. Numerical results validating the theoretical statements are placed to the web (see Case Study (2016)).

Author Contributions: Conceptualization, S.U.; Formal analysis, V.K.; Investigation, A.G.; Methodology, S.U.; Software, V.K.; Supervision, S.U.; Writing—original draft, A.G.

Appendix A CVaR Regression with Rockafellar Error: Convex and Linear Programming

The value of the Rockafellar error with given set of parameters λ_k, α_k ($\alpha_k \in (0,1), k = 1,\ldots r$, $\sum_{k=1}^{r} \lambda_k = 1$) for a random value X is a minimum w.r.t. a set of variables $B_1,\ldots,B_r$ of a mixture of Koenker–Bassett Error functions with one linear constraint on these variables:

$$Rockafellar\_Error\ (X)_{\lambda_1,\alpha_1,\ldots,\lambda_r,\alpha_r} = \min_{B_1,\ldots,B_r} \left\{ \sum_{k=1}^{r} \lambda_k \mathcal{E}_{\alpha_k}(X - B_k) \middle| \sum_{k=1}^{r} \lambda_k B_k = 0 \right\}$$

where $\mathcal{E}_{\alpha_k}(X - B_k) = E\left[\frac{\alpha_k}{1-\alpha_k}[X - B_k]^+ + [X - B_k]^-\right]$ is the normalized Koenker–Bassett error.

By using regret from the mixed-quantile quadrangle, we express the Rockafellar error as follows:

$$Rockafellar\_Error\ (X)_{\lambda_1,\alpha_1,\ldots,\lambda_r,\alpha_r} = \min_{B_1,\ldots,B_r} \left\{ \sum_{k=1}^{r} \frac{\lambda_k}{1-\alpha_k} E[X - B_k]^+ \middle| \sum_{k=1}^{r} \lambda_k B_k = 0 \right\} - E[X].$$

For the linear regression problem, the random variable X is defined by a set of differences between observed values V_i and linear functions $C_0 + \boldsymbol{C}^T \boldsymbol{Y}_i$, where $\boldsymbol{Y}_i$ is a vector of explanatory factors, $i = 1, 2, \ldots, \nu$. Vectors $\boldsymbol{C}$ and $\boldsymbol{Y}$ have m components, $\boldsymbol{C} = (C_1, \ldots, C_m)$, $\boldsymbol{Y} = (Y_1, \ldots, Y_m)$, and C_0 is a scalar. Residuals $X_i = V_i - C_0 - \boldsymbol{C}^T \boldsymbol{Y}_i$ are values (scenarios) of atoms of the random value X. We consider that all atoms have equal probabilities. The estimation of V with factors $\boldsymbol{Y}$ is done by minimizing the error w.r.t. variables $\boldsymbol{C}$, C_0. Further we use the Set 1 of parameters. Let us denote:

$$E[V] = \frac{1}{\nu} \sum_{i=1}^{\nu} V_i,\ E[\boldsymbol{Y}] = \frac{1}{\nu} \sum_{i=1}^{\nu} \boldsymbol{Y}_i.$$

Appendix A.1 Convex Programming Formulation for CVaR Regression

Minimize the Rockafellar error:

$$\min_{B_1,\ldots,B_r,C_0,C_1,\ldots,\,C_m,}\left\{\sum_{k=1}^{r}\frac{\lambda_k}{(1-\alpha_k)v}\sum_{i=1}^{v}\left[V_i-C_0-\boldsymbol{C}^T\boldsymbol{Y}_i-B_k\right]^+-E[V]+C_0+\boldsymbol{C}^TE[\boldsymbol{Y}]\right\} \tag{A1}$$

subject to the constraint:

$$\sum_{k=1}^{r}\lambda_k B_k=0. \tag{A2}$$

This optimization problem has a convex objective and one linear constraint.

Appendix A.2 Linear Programming Formulation for CVaR Regression

Equations (A1) and (A2) are reduced to linear programming with additional variables and constraints:

$$\min_{\substack{A_{11},\ldots,A_{rv}\\ B_1,\ldots,B_r,C_0,C_1,\ldots,\,C_m,}}\left\{\sum_{k=1}^{r}\frac{\lambda_k}{(1-\alpha_k)v}\sum_{i=1}^{v}A_{ki}-E[V]+C_0+\boldsymbol{C}^TE[\boldsymbol{Y}]\right\} \tag{A3}$$

subject to constraints:

$$\sum_{k=1}^{r}\lambda_k B_k=0 \tag{A4}$$

$$A_{ki}\geq V_i-C_0-\boldsymbol{C}^T\boldsymbol{Y}_i-B_k,\ k=1,\ldots,r,\ i=1,\ldots,v \tag{A5}$$

$$A_{ki}\geq 0,\ k=1,\ldots,r,\ i=1,\ldots,v \tag{A6}$$

The linear function $C_0^*+\boldsymbol{C}^{*T}\boldsymbol{Y}$ estimates $CVaR_\alpha(V)$ as a function of explanatory factors $\boldsymbol{Y}$, where C_0^* and $\boldsymbol{C}^*$ are optimal values of variables for Equations (A1) and (A2) or (A3)–(A6).

Appendix B Codes Implementing Regression Optimization Problems

This appendix contains codes implementing Optimization Problems 1–4 described in Section 5. Codes and solution results are posted at internet link: Case Study (2016). Codes are written in Portfolio Safeguard (PSG) Text, MATLAB, and R environments. Here are codes in the Text environment.

Optimization Problem 1

Code in PSG Text format:

```
minimize
cvar2_err(0.75,matrix_s)
```

The keyword "`minimize`" indicates that the objective function is minimized. The objective function `cvar2_err(0.75,matrix_s)` calculates error $\overline{\mathcal{E}}_\alpha(Z(C_0,\boldsymbol{C}))$ in CVaR quadrangle with confidence level $\alpha=0.75$. The `matrix_s` contains scenarios of residual of the regression $Z(C_0,\boldsymbol{C})=V-C_0-\boldsymbol{C}^T\boldsymbol{Y}$.

Optimization Problem 2

Code in PSG Text format:

```
minimize
cvar2_dev(0.75,matrix_s)
value:
cvar_risk(0.75,matrix_s)
```

The code includes two parts. The first part begins with the keyword "`minimize`" indicating that the objective function is minimized. It implements Step 1 of the Optimization Problem 2, which minimizes deviation from the CVaR quadrangle (for determining the optimal vector C^* of regression coefficients without an intercept). The PSG function `cvar2_dev(0.75,matrix_s)` calculates deviation $\overline{\mathcal{D}}_\alpha(Z_0(C))$ with $\alpha = 0.75$. The `matrix_s` contains scenarios of the residual of the regression without an intercept $Z_0(C) = V - C^T Y$.

The second part of the code begins with the keyword "`value`." This part implements Step 2 of the Optimization Problem 2 for calculating the optimal value of intercept C_0^*. The PSG function `cvar_risk(0.75,matrix_s)` calculates $CVaR_\alpha(Z_0(C^*))$ with $\alpha = 0.75$ at the optimal point C^*.

Optimization Problem 3

Code in PSG Text format with parameters from Set 1.

```
minimize
ro_err(matrix_s, matrix_coeff)
```

The keyword "`minimize`" indicates that the objective function is minimized. The objective function `ro_err(matrix_s, matrix_coeff)` calculates the Rockafellar error $\mathcal{E}(Z(C_0, C))$ in the mixed-quantile quadrangle. The `matrix_s` contains scenarios of the residual of the regression $Z(C_0, C) = V - C_0 - C^T Y$. The `matrix_coeff` includes vectors of weights and confidence levels for Set 1 with $\alpha = 0.75$.

Optimization Problem 4

Code in PSG Text format with parameters from Set 1 for $\alpha = 0.75$.

```
minimize
vector_c*cvar_dev(vector_a, matrix_s)
value:
cvar_risk(0.75,matrix_s)
```

The code includes two parts. The first part begins with the keyword "`minimize`" indicating that the objective function is minimized. It implements Step 1 of the Optimization Problem 4, which minimizes deviation from the mixed-quantile quadrangle (for determining optimal vector C^* of regression coefficients without intercept). The inner product `vector_c*cvar_dev(vector_a, matrix_s)` calculates the mixed CVaR deviation $\mathcal{D}(Z_0(C))$. The function `cvar_dev` corresponds to the CVaR deviation from the quantile quadrangle. Vector `vector_c` contains weights for CVaR Deviation mix corresponding to Set 1. Vector `vector_a` contains confidence levels defined by Set 1. The `matrix_s` contains scenarios of the residual of the regression $Z(C_0, C) = V - C_0 - C^T Y$.

The second part of the code begins with the keyword "`value`." This part implements Step 2 of the Optimization Problem 4, calculating the optimal value of intercept C_0^*. The PSG function `cvar_risk(0.75,matrix_s)` calculates $CVaR_\alpha(Z_0(C^*))$ with $\alpha = 0.75$ at the optimal point C^*.

Appendix C Proof of the Lemma 1

Proof. According to the definition, $\beta_v = 1$. However, while proving this lemma, we consider β_v a bit smaller than 1, i.e., $\beta_v = 1 - \varepsilon > \beta_{v-1}, \ \varepsilon > 0$, to avoid division by 0. Then, we will consider limit $\varepsilon \to 0$ to finish the proof. Note that for $\alpha < 1$ and for the considered partition, $\beta_{i-1} < \beta_i$ for all $i = \nu_\alpha, \ldots, \nu$.

First, let us prove that $\gamma_i \in (\beta_{i-1}, \beta_i)$ for $\beta_i < 1$. Consider three functions of σ, δ for $\sigma > 0, \ \delta \geq 0$:

$$f_1(\delta) = \frac{\delta}{\sigma + \delta}, \ f_2(\delta) = \ln\left(1 + \frac{\delta}{\sigma}\right), \ f_3(\delta) = \frac{\delta}{\sigma}$$

When $\delta = 0$, all functions equal to 0 and have equal derivatives. When $\delta > 0$, it is true for derivatives that:

$$f'_1(\delta) < f'_2(\delta) < f'_3(\delta)$$

Hence, when > 0, similar inequalities are valid for functions:

$$\frac{\delta}{\sigma+\delta} < \ln\left(1+\frac{\delta}{\sigma}\right) < \frac{\delta}{\sigma}.$$

There exist some δ_γ such that $0 < \delta_\gamma < \delta$ and $\frac{\delta}{\sigma+\delta_\gamma} = \ln\left(1+\frac{\delta}{\sigma}\right)$. If $\delta = \beta_i - \beta_{i-1}$, $\sigma = 1-\beta_i$, $\gamma = 1-\sigma-\delta_\gamma$, then $\gamma = \gamma_i$ and $\beta_{i-1} < \gamma < \beta_i$. Therefore, $\gamma_i \in (\beta_{i-1}, \beta_i)$.

Further, we show how to calculate the integral $\int_\alpha^{1-\varepsilon} CVaR_\beta(X)d\beta$ as a sum of integrals over the partition:

$$\overline{\mathcal{R}}_\alpha(X) = \lim_{\varepsilon\to 0}\frac{1}{1-\alpha}\int_\alpha^{1-\varepsilon} CVaR_\beta(X)d\beta = \lim_{\beta_v\to 1}\frac{1}{1-\alpha}\sum_{i=\nu_\alpha}^{\nu}\int_{\beta_{i-1}}^{\beta_i} CVaR_\beta(X)d\beta.$$

Let us denote $C_{\beta_i} = CVaR_{\beta_i}(X)$ and $V_i = VaR_{\gamma_i}(X)$. Note that $VaR_\gamma(X)$ is a singleton for every $\gamma \in (\beta_{i-1}, \beta_i)$ and it equals V_i, because $\gamma_i \in (\beta_{i-1}, \beta_i)$.

Below, we use value V_i for the closed interval $[\beta_{i-1}, \beta_i]$ while calculating the integral over this interval because the value of the integral does not depend on the finite values of $VaR_\gamma(X)$ at the boundary points β_{i-1}, β_i.

Using the definition of CVaR (Rockafellar and Uryasev (2002), Proposition 8 CVaR for scenario models) we write:

$$\begin{aligned}\int_{\beta_{i-1}}^{\beta_i} CVaR_\beta(X)d\beta &= \int_{\beta_{i-1}}^{\beta_i}\tfrac{1}{1-\beta}\left(C_{\beta_i}(1-\beta_i)+V_i(\beta_i-\beta)\right)d\beta \\ &= C_{\beta_i}(1-\beta_i)\int_{\beta_{i-1}}^{\beta_i}\tfrac{1}{1-\beta}d\beta + V_i\int_{\beta_{i-1}}^{\beta_i}\tfrac{\beta_i-\beta}{1-\beta}d\beta \\ &= C_{\beta_i}(1-\beta_i)\ln\left(\tfrac{1-\beta_{i-1}}{1-\beta_i}\right) + V_i\left[(\beta_i-\beta_{i-1})-(1-\beta_i)\ln\left(\tfrac{1-\beta_{i-1}}{1-\beta_i}\right)\right]\end{aligned} \tag{A7}$$

Let us make the transformation of Equation (A7) using the expression for γ_i in the Set 1 definition:

$$\begin{aligned}\int_{\beta_{i-1}}^{\beta_i} CVaR_\beta(X)d\beta &= (\beta_i-\beta_{i-1})\left[V_i + \tfrac{1}{\beta_i-\beta_{i-1}}\ln\left(\tfrac{1-\beta_{i-1}}{1-\beta_i}\right)\left(C_{\beta_i}-V_i\right)(1-\beta_i)\right] \\ &= (\beta_i-\beta_{i-1})\left[V_i+\tfrac{1}{1-\gamma_i}\left(C_{\beta_i}-V_i\right)(1-\beta_i)\right] = (\beta_i-\beta_{i-1})\tfrac{1}{1-\gamma_i}\left[C_{\beta_i}(1-\beta_i)+V_i(\beta_i-\gamma_i)\right] \\ &= (\beta_i-\beta_{i-1})CVaR_{\gamma_i}(X)\end{aligned}$$

The last equality is valid because $\gamma_i \in (\beta_{i-1}, \beta_i)$.

Taking into account that $\lim_{\varepsilon\to 0}\gamma_\nu = 1$, $CVaR_{1-\varepsilon}(X) = CVaR_1(X)$, and $VaR_{1-\varepsilon}(X) = VaR_1(X)$, we obtain:

$$\overline{\mathcal{R}}_\alpha(X) = \lim_{\varepsilon\to 0}\sum_{i=\nu_\alpha}^{\nu}\frac{\beta_i-\beta_{i-1}}{1-\alpha}CVaR_{\gamma_i}(X) = \sum_{i=\nu_\alpha}^{\nu}p_i CVaR_{\gamma_i}(X)$$

Deviation is calculated as follows:

$$\overline{\mathcal{D}}_\alpha(X) = \sum_{i=\nu_\alpha}^{\nu}p_i CVaR_{\gamma_i}(X) - E[X]$$

By the definition of CVaR (Rockafellar and Uryasev (2002)):

$$CVaR_\alpha(X) = \sum_{i=\nu_\alpha}^{\nu}p_i VaR_{\gamma_i}(X) = \overline{\mathcal{S}}_\alpha(X)$$

Lemma 1 is proved. □

Appendix D Proof of Lemma 2

Proof. Similar to proof of Lemma 1, we consider β_v as a bit smaller than 1, i.e., $\beta_v = 1 - \varepsilon > \beta_{v-1}$, $\varepsilon > 0$, to avoid division by 0. Then, we consider limit $\varepsilon \to 0$ to finish the proof. We denote $C_{\beta_i} = CVaR_{\beta_i}(X)$ and $V_i = VaR_\gamma(X)$ for any $\gamma \in (\beta_{i-1}, \beta_i)$ because $VaR_\gamma(X)$ does not change on this interval.

Additionally we denote $\delta_i = \beta_i - \beta_{i-1}$, $i = \nu_\alpha, \ldots, \nu$, and $\sigma_i = 1 - \beta_i$, $i = \nu_\alpha - 1, \ldots, \nu$. Note that all $\delta_i > 0$, all $\sigma_i > 0$, and $\delta_i = \delta_j = \delta$ for all $i, j < \nu$.

$$\overline{\mathcal{R}}_\alpha(X) = \lim_{\varepsilon\to 0}\frac{1}{1-\alpha}\int_\alpha^{1-\varepsilon} CVaR_\beta(X)d\beta = \lim_{\beta_v\to 1}\frac{1}{1-\alpha}\sum_{i=\nu_\alpha}^{\nu}\int_{\beta_{i-1}}^{\beta_i} CVaR_\beta(X)d\beta. \tag{A8}$$

Equation (A7) in Lemma 1 is valid for any interval $[\theta, \beta_i]$ such that $\beta_{i-1} \le \theta \le \beta_i$, therefore:

$$\int_\theta^{\beta_i} CVaR_\beta(X)d\beta = C_{\beta_i}(1-\beta_i)\ln\left(\frac{1-\theta}{1-\beta_i}\right) + V_i\left[(\beta_i - \theta) - (1-\beta_i)\ln\left(\frac{1-\theta}{1-\beta_i}\right)\right] \tag{A9}$$

Let us express V_i from $CVaR_{\beta_{i-1}}(X)$ using the definition of CVaR from Rockafellar and Uryasev (2002) (Equation (25)), then insert this expression into Equation (A9):

$$C_{\beta_{i-1}} = \tfrac{1}{1-\beta_{i-1}}\big(C_{\beta_i}(1-\beta_i) + V_i(\beta_i - \beta_{i-1})\big) = \tfrac{1}{\sigma_{i-1}}\big(C_{\beta_i}\sigma_i + V_i\delta_i\big),$$
$$V_i = \tfrac{1}{\delta_i}\big(C_{\beta_{i-1}}\sigma_{i-1} - C_{\beta_i}\sigma_i\big)$$

By substituting V_i into Equation (A9), we obtain:

$$\int_\theta^{\beta_i} CVaR_\beta(X)d\beta = C_{\beta_i}\sigma_i\ln\left(\frac{1-\theta}{\sigma_i}\right) + \frac{1}{\delta_i}\big(C_{\beta_{i-1}}\sigma_{i-1} - C_{\beta_i}\sigma_i\big)\left[(\beta_i - \theta) - \sigma_i\ln\left(\frac{1-\theta}{\sigma_i}\right)\right] \tag{A10}$$

The last equation contains two CVaRs, C_{β_i} and $C_{\beta_{i-1}}$. Let us express the coefficients of these CVaRs. C_{β_i} in Equation (A10) has the following coefficient:

$$q_{1i} = \sigma_i\left[\ln\left(\frac{1-\theta}{\sigma_i}\right)\left(1 + \frac{\sigma_i}{\delta_i}\right) - \frac{\beta_i - \theta}{\delta_i}\right] \tag{A11}$$

and $C_{\beta_{i-1}}$ has the coefficient:

$$q_{2i-1} = \frac{\sigma_{i-1}}{\delta_i}\left[(\beta_i - \theta) - \sigma_i\ln\left(\frac{1-\theta}{\sigma_i}\right)\right] \tag{A12}$$

Coefficient $q_{2i-1} \ge 0$ because the value in square brackets is the same as in Equations (A7) and (A9), and is a result of the integration of a non-negative function over the interval $[\theta, \beta_i]$. The coefficient $q_{1i} \ge 0$ because it is an integral of the non-negative function $\frac{\sigma_i}{\delta_i}\left[\frac{\delta_i - \beta_i + \beta}{1-\beta}\right]$ over the same interval.

When summing up integrals to obtain $\frac{1}{1-\alpha}\int_\alpha^{1-\varepsilon} CVaR_\beta(X)d\beta$, every C_{β_i}, aside from $C_{\beta_{\nu_\alpha}-1}$ and C_{β_ν}, enters the sum two times with coefficients depending on $\nu, \alpha, \beta_{i-1}, \beta_i$, and β_{i+1}. Once in Equation (A11) for i, and the second time in Equation (A12) for $i+1$. All coefficients are non-negative.

Let us explain this in more detail.

If i is such that $\nu_\alpha < i < \nu$, then $\beta_{\nu_\alpha} < \beta_i < \beta_\nu$ and $\theta = \beta_{i-1}$ for i in Equation (A11) and $\theta = \beta_i$ for $i+1$ in Equation (A12). Then, the coefficient for C_{β_i} in Equation (A8) equals:

$$q_i = \tfrac{1}{1-\alpha}(q_{1i} + q_{2i}) = \tfrac{1}{1-\alpha}\Big(\sigma_i\Big[\ln\Big(\tfrac{\sigma_{i-1}}{\sigma_i}\Big)\Big(\tfrac{\sigma_{i-1}}{\delta_i}\Big) - 1\Big] + \tfrac{\sigma_i}{\delta_{i+1}}\Big[\delta_{i+1} - \sigma_{i+1}\ln\Big(\tfrac{\sigma_i}{\sigma_{i+1}}\Big)\Big]\Big)$$
$$= \tfrac{\sigma_i}{1-\alpha}\Big[\tfrac{\sigma_{i-1}}{\delta_i}\ln\Big(\tfrac{\sigma_{i-1}}{\sigma_i}\Big) - \tfrac{\sigma_{i+1}}{\delta_{i+1}}\ln\Big(\tfrac{\sigma_i}{\sigma_{i+1}}\Big)\Big]$$

If $i = \nu_\alpha < \nu$, then $\theta = \alpha$ in Equation (A11) and $\theta = \beta_i$ in Equation (A12). Then, the coefficient for C_{β_i} in Equation (A8) equals:

$$q_i = \tfrac{1}{1-\alpha}(q_{1i} + q_{2i}) = \tfrac{1}{1-\alpha}\Big(\sigma_i\Big[\ln\Big(\tfrac{1-\alpha}{\sigma_i}\Big)\Big(\tfrac{\sigma_{i-1}}{\delta_i}\Big) - \tfrac{\delta_\alpha}{\delta_i}\Big] + \tfrac{\sigma_i}{\delta_{i+1}}\Big[\delta_{i+1} - \sigma_{i+1}\ln\Big(\tfrac{\sigma_i}{\sigma_{i+1}}\Big)\Big]\Big)$$
$$= \tfrac{\sigma_i}{1-\alpha}\Big[1 - \tfrac{\delta_\alpha}{\delta_i} + \tfrac{\sigma_{i-1}}{\delta_i}\ln\Big(\tfrac{1-\alpha}{\sigma_i}\Big) - \tfrac{\sigma_{i+1}}{\delta_{i+1}}\ln\Big(\tfrac{\sigma_i}{\sigma_{i+1}}\Big)\Big]$$

If $i = \nu_\alpha - 1$, then C_{β_i} enters in the sum only in Equation (A12) with $\theta = \alpha$. Then:

$$q_i = \frac{1}{1-\alpha}q_{2i} = \frac{\sigma_i}{1-\alpha}\left[\frac{\delta_\alpha}{\delta_{i+1}} - \frac{\sigma_{i+1}}{\delta_{i+1}}\ln\left(\frac{1-\alpha}{\sigma_{i+1}}\right)\right]$$

If $i = \nu > \nu_\alpha$, then C_{β_i} enters in the sum only in Equation (A11) with $\theta = \beta_{i-1}$. Then:

$$q_i = \frac{1}{1-\alpha}q_{1i} = \frac{\sigma_i}{1-\alpha}\left[\ln\left(\frac{\sigma_{i-1}}{\sigma_i}\right)\left(\frac{\sigma_{i-1}}{\delta_i}\right) - 1\right]$$

Also, in the case when $i = \nu = \nu_\alpha$, C_{β_i} enters in the sum only in Equation (A11) with $\theta = \alpha$. Then:

$$q_i = \frac{1}{1-\alpha}q_{1i} = \frac{\sigma_i}{1-\alpha}\left[\ln\left(\frac{1-\alpha}{\sigma_i}\right)\left(1 + \frac{\sigma_i}{\delta_i}\right) - \frac{\delta_\alpha}{\delta_i}\right]$$

Then, the risk in the CVaR quadrangle equals:

$$\overline{\mathcal{R}}_\alpha(X) = \lim_{\varepsilon\to 0}\sum\nolimits_{i=\nu_\alpha-1}^{\nu} q_i CVaR_{\beta_i}(X).$$

Deviation is calculated as follows: $\overline{\mathcal{D}}_\alpha(X) = \overline{\mathcal{R}}_\alpha(X) - E[X]$.

Taking into account that all β_i are fixed for $i < \nu$, the limit operation affects only coefficients for $i = \nu$. Namely, $\lim\limits_{\varepsilon\to 0}\beta_\nu = 1$, $\lim\limits_{\varepsilon\to 0}\sigma_\nu = 0$, $\lim\limits_{\varepsilon\to 0}\sigma_\nu\ln(\sigma_\nu) = 0$, $\lim\limits_{\varepsilon\to 0}\delta_\nu = \delta$. Then, the limit values of coefficients q_i are equal to:

$q_\nu = 0$ in both cases, when $\nu > \nu_\alpha$ and $\nu = \nu_\alpha$.
$q_i = \frac{1}{1-\alpha}\times\frac{\sigma_i}{\delta}\Big[\sigma_{i-1}\ln\Big(\frac{\sigma_{i-1}}{\sigma_i}\Big)\Big]$ for i such that $\nu_\alpha < i = \nu - 1$.
$q_i = \frac{1}{1-\alpha}\times\frac{\sigma_i}{\delta}\Big[\sigma_{i-1}\ln\Big(\frac{\sigma_{i-1}}{\sigma_i}\Big) + \sigma_{i+1}\ln\Big(\frac{\sigma_{i+1}}{\sigma_i}\Big)\Big]$ for i such that $\nu_\alpha < i < \nu - 1$.
$q_i = \frac{1}{1-\alpha}\times\frac{\sigma_i}{\delta}\Big[\delta - \delta_\alpha + \sigma_{i-1}\ln\Big(\frac{1-\alpha}{\sigma_i}\Big) + \sigma_{i+1}\ln\Big(\frac{\sigma_{i+1}}{\sigma_i}\Big)\Big]$ for $i = \nu_\alpha < \nu - 1$.
$q_i = \frac{1}{1-\alpha}\times\frac{\sigma_i}{\delta}\Big[\delta_\alpha + \sigma_{i+1}\ln\Big(\frac{\sigma_{i+1}}{1-\alpha}\Big)\Big]$ for $i = \nu_\alpha - 1 < \nu - 1$.

If the number of atoms used in calculating $\overline{\mathcal{R}}_\alpha(X)$ is 3 or 2, then the coefficients have values:

$q_i = \frac{1}{1-\alpha}\times\frac{\sigma_i}{\delta}\Big[\delta - \delta_\alpha + \sigma_{i-1}\ln\Big(\frac{1-\alpha}{\sigma_i}\Big)\Big]$ for $i = \nu_\alpha = \nu - 1$.
$q_i = \frac{1}{1-\alpha}\times\frac{\sigma_i}{\delta}[\delta_\alpha] = 1$ for $i = \nu_\alpha - 1 = \nu - 1$.

It can be shown that $\sum_{i=\nu_\alpha-1}^{\nu} q_i = 1$ by sequentially summing up coefficients.

By recalling that $\sigma_i = \delta(v - i)$, we can rewrite the equations for coefficients q_i using δ and $j = v - i$.

Lemma 2 is proved. □

Appendix E Proof of the Lemma 3

Proof. Distribution of random value X defines the partition of the interval $[0, 1]$: $\delta = 1/\nu$, $\beta_i = i\delta$, for $i = 0, 1, \ldots, \nu$.

CDF $F_X(x) = prob\{X \leq x\}$ is a non-decreasing and right-continuous function and it is constant for every right-open interval $[\beta_{i-1}, \beta_i)$, $i = 1, \ldots, \nu$. Therefore, from the definitions of $VaR_\gamma^-(X)$ and $VaR_\gamma^+(X)$, we have:

$$VaR_\gamma^-(X) = VaR_\gamma^+(X) = VaR_\gamma(X) = const \text{ for } \gamma \in (\beta_{i-1}, \beta_i).$$

$$VaR_{\beta_i}^+(X) = VaR_\gamma(X) \text{ for } \gamma \in (\beta_i, \beta_{i+1}) \text{ and } i = 0, \ldots, \nu - 1.$$

$$VaR_{\beta_i}^-(X) = VaR_\gamma(X) \text{ for } \gamma \in (\beta_{i-1}, \beta_i) \text{ and } i = 1, \ldots, \nu.$$

For $i = 0$ we have $\beta_i = 0$ and $VaR_0^-(X) = -\infty$.

Thus, according to the definition of statistic for the mixed-quantile quadrangle, for Set 2:

$$\mathcal{S}_\alpha^{II}(X) = \sum_{i=\nu_\alpha-1}^{\nu-1} q_i VaR_{\beta_i}(X)$$

where $VaR_{\beta_i}(X)$ are intervals (that may have zero lengths for some i), and q_i and ν_α are defined above for the Set 2. Therefore, $\mathcal{S}_\alpha^{II}(X)$ is also an interval and it has a non-zero length if not all $VaR_{\beta_i}^-(X)$, $i = \nu_\alpha - 1, \ldots, \nu - 1$ are equal.

If $\nu_\alpha - 1 = 0$, then $\mathcal{S}_\alpha^{II}(X)$ is a left-open interval with the lower bound $-\infty$.

Let $\nu_\alpha - 1 > 0$. For the Set 1, we defined confidence levels as internal points in intervals $\gamma_i \in (\beta_{i-1}, \beta_i)$. Using these definitions of γ_i, we can express $\mathcal{S}_\alpha^{II}(X)$ as the interval:

$$\mathcal{S}_\alpha^{II}(X) = \left[\sum_{i=\nu_\alpha-1}^{\nu-1} q_i VaR_{\gamma_i}(X), \sum_{i=\nu_\alpha-1}^{\nu-1} q_i VaR_{\gamma_{i+1}}(X)\right]$$

To simplify notations, let us denote: $V_i = VaR_{\gamma_i}(X)$; $C_\gamma = CVaR_\gamma(X)$; and L, U are bounds for $\mathcal{S}_\alpha^{II}(X)$ such that $\mathcal{S}_\alpha^{II}(X) = [L, U]$.

Statistic $\overline{\mathcal{S}}_\alpha(X)$ of the CVaR quadrangle is defined in Lemma 1 (see Equation (1)) as $\overline{\mathcal{S}}_\alpha(X) = \sum_{i=\nu_\alpha}^{\nu} p_i V_i$.

Therefore, we wish to prove that:

$$L = \sum_{i=\nu_\alpha-1}^{\nu-1} q_i V_i \leq \overline{\mathcal{S}}_\alpha(X) \leq \sum_{i=\nu_\alpha-1}^{\nu-1} q_i V_{i+1} = U$$

Let us prove the right inequality for the upper bound.

According to Lemmas 1 and 2, it is valid that:

$$\overline{\mathcal{R}}_\alpha(X) = \sum_{i=\nu_\alpha}^{\nu} p_i C_{\gamma_i} = \sum_{i=\nu_\alpha-1}^{\nu-1} q_i C_{\beta_i}$$

Let $d_i = q_{i-1} - p_i$. Then, from the last equality:

$$\sum_{i=\nu_\alpha}^{\nu} p_i C_{\gamma_i} = \sum_{i=\nu_\alpha}^{\nu} q_{i-1} C_{\beta_{i-1}} = \sum_{i=\nu_\alpha}^{\nu} p_i C_{\beta_{i-1}} + \sum_{i=\nu_\alpha}^{\nu} d_i C_{\beta_{i-1}}$$

$$\sum_{i=\nu_\alpha}^{\nu} d_i C_{\beta_{i-1}} = \sum_{i=\nu_\alpha}^{\nu} p_i \left(C_{\gamma_i} - C_{\beta_{i-1}}\right) \geq 0$$

The last inequality is valid because $p_i > 0$, $\gamma_i > \beta_{i-1}$, and $C_{\gamma_i} \geq C_{\beta_{i-1}}$.

Because $C_{\beta_{i-1}}$ may have arbitrary but ordered values ($C_{\beta_{i-1}} \leq C_{\beta_i}$ due to ordered β_i), $i = \nu_\alpha, \ldots, \nu$, the last inequality is valid for any ordered sequence of values $z_{i-1} \leq z_i$. Therefore, it is valid for VaRs V_i that:

$$\sum_{i=\nu_\alpha}^{\nu} d_i V_i = \sum_{i=\nu_\alpha}^{\nu} (q_{i-1} - p_i) V_i \geq 0$$

This leads to the inequality for the upper bound:

$$U = \sum_{i=\nu_\alpha}^{\nu} q_{i-1} V_i \geq \sum_{i=\nu_\alpha}^{\nu} p_i V_i = \overline{S}_\alpha(X).$$

Let us prove the inequality for the lower bound L:

$$L = \sum_{i=\nu_\alpha}^{\nu} q_{i-1} V_{i-1} = \sum_{i=\nu_\alpha}^{\nu} (p_i + d_i)(V_i - \Delta V_i) = \overline{S}_\alpha(X) - \sum_{i=\nu_\alpha}^{\nu} p_i \Delta V_i + \sum_{i=\nu_\alpha}^{\nu} d_i V_{i-1} \quad \text{(A13)}$$

where $\Delta V_i = V_i - V_{i-1}$.

Let us calculate the upper estimate of L. Let us set $V_{\nu_{\alpha-1}} = V_{\nu_\alpha}$, therefore $\Delta V_{\nu_\alpha} = 0$. This increases the right hand side of Equation (A13). By recalling that $p_i = \delta/(1-\alpha) = \frac{1}{\nu(1-\alpha)}$, $i = \nu_\alpha + 1, \ldots, \nu$, for the Set 1, we have:

$$\sum_{i=\nu_\alpha}^{\nu} p_i \Delta V_i = \delta(V_\nu - V_{\nu_\alpha})/(1-\alpha)$$

Because $\sum_{i=\nu_\alpha}^{\nu} p_i = \sum_{i=\nu_\alpha}^{\nu} q_{i-1}$, then $\sum_{i=\nu_\alpha}^{\nu} d_i = 0$ and $\sum_{i=\nu_\alpha}^{\nu} d_i V_{i-1} = \sum_{i=\nu_\alpha}^{\nu} d_i (V_{i-1} - D)$ for any D.

Because $\sum_{i=\nu_\alpha}^{\nu} d_i z_i \geq 0$ for any increasing ordered z_i, and we can set $z_{\nu_\alpha} = 0$, $z_i = 1$, $i = \nu_\alpha + 1, \ldots, \nu$, then:

$$\sum_{i=\nu_\alpha}^{\nu} d_i z_i = \sum_{i=\nu_{\alpha+1}}^{\nu} d_i = -d_{\nu_\alpha} = p_{\nu_\alpha} - q_{\nu_\alpha - 1} \geq 0$$

Taking into account that $p_{\nu_\alpha}, q_{\nu_\alpha - 1} \geq 0$ and $p_{\nu_\alpha} \leq \frac{\delta}{1-\alpha}$, we have $\sum_{i=\nu_{\alpha+1}}^{\nu} d_i \leq \frac{\delta}{1-\alpha}$. Then:

$$\sum_{i=\nu_\alpha}^{\nu} d_i V_{i-1} = \sum_{i=\nu_\alpha}^{\nu} d_i \left(V_{i-1} - V_{\nu_{\alpha-1}}\right) \leq \sum_{i=\nu_\alpha+1}^{\nu} d_i \left(V_{\nu-1} - V_{\nu_{\alpha-1}}\right) \leq \left(V_{\nu-1} - V_{\nu_{\alpha-1}}\right) \frac{\delta}{1-\alpha}$$

Let us return to estimation L by taking into account that $V_{\nu_{\alpha-1}} = V_{\nu_\alpha}$:

$$\begin{aligned} L \leq \overline{S}_\alpha(X) - \textstyle\sum_{i=\nu_\alpha}^{\nu} p_i \Delta V_i + \sum_{i=\nu_\alpha}^{\nu} d_i V_{i-1} \leq \overline{S}_\alpha(X) - \frac{\delta(V_\nu - V_{\nu_\alpha})}{1-\alpha} + \frac{\delta\left(V_{\nu-1} - V_{\nu_{\alpha-1}}\right)}{1-\alpha} \\ = \overline{S}_\alpha(X) - \frac{\delta(V_\nu - V_{\nu-1})}{1-\alpha} \leq \overline{S}_\alpha(X) \end{aligned}$$

Therefore, we have proved that $L \leq \overline{S}_\alpha(X) \leq U$.

Lemma 3 is proved. □

References

Acerbi, Carlo, and Dirk Tasche. 2002. On the Coherence of Expected Shortfall. *Journal of Banking and Finance* 26: 1487–503. [CrossRef]

Adrian, Tobias, and Markus K. Brunnermeier. 2016. CoVaR. *American Economic Review* 106: 1705–41. [CrossRef]

Basset, Gilbert W., and Hsiu-Lang Chen. 2001. Portfolio Style: Return-based Attribution Using Quantile Regression. *Empirical Economics* 26: 293–305. [CrossRef]

Beraldi, Patrizia, Antonio Violi, Massimiliano Ferrara, Claudio Ciancio, and Bruno Antonio Pansera. 2019. Dealing with complex transaction costs in portfolio management. *Annals of Operations Research*, 1–16. [CrossRef]

Carhart, Mark M. 1997. On Persistence in Mutual Fund Performance. *Journal of Finance* 52: 57–82. [CrossRef]

Case Study. 2014. Style Classification with Quantile Regression. Available online: http://www.ise.ufl.edu/uryasev/research/testproblems/financial_engineering/style-classification-with-quantile-regression/ (accessed on 24 June 2019).

Case Study. 2016. Estimation of CVaR through Explanatory Factors with CVaR (Superquantile) Regression. Available online: http://www.ise.ufl.edu/uryasev/research/testproblems/financial_engineering/on-implementation-of-cvar-regression/ (accessed on 24 June 2019).

Fissler, Tobias, and Johanna F. Ziegel. 2015. Higher order elicitability and Osband's principle. *arXiv*. [CrossRef]

Huang, Wei-Qiang, and Stan Uryasev. 2018. The CoCVaR Approach: Systemic Risk Contribution Measurement. *Journal of Risk* 20: 75–93. [CrossRef]

Koenker, R. 2005. *Quantile Regression*. Cambridge: Cambridge University Press.

Koenker, Roger, and Gilbert Bassett. 1978. Regression Quantiles. *Econometrica* 46: 33–50. [CrossRef]

Portfolio Safeguard. 2018. American Optimal Decisions, USA. Available online: http://www.aorda.com (accessed on 24 June 2019).

Rockafellar, R. Tyrrell, and Johannes O. Royset. 2018. Superquantile/CVaR Risk Measures: Second-order Theory. *Annals of Operations Research* 262: 3–29. [CrossRef]

Rockafellar, R. Tyrrell, and Stan Uryasev. 2000. Optimization of Conditional Value-At-Risk. *Journal of Risk* 2: 21–41. [CrossRef]

Rockafellar, R. Tyrrell, and Stan Uryasev. 2002. Conditional Value-at-Risk for General Loss Distributions. *Journal of Banking and Finance* 26: 1443–71. [CrossRef]

Rockafellar, R. Tyrrell, and Stan Uryasev. 2013. The Fundamental Risk Quadrangle in Risk Management, Optimization and Statistical Estimation. *Surveys in Operations Research and Management Science* 18: 33–53. [CrossRef]

Rockafellar, R. Tyrrell, Stan Uryasev, and Michael Zabarankin. 2008. Risk Tuning with Generalized Linear Regression. *Mathematics of Operations Research* 33: 712–29. [CrossRef]

Rockafellar, R. Terry, Johannes O. Royset, and Sofia I. Miranda. 2014. Superquantile Regression with Applications to Buffered Reliability, Uncertainty Quantification and Conditional Value-at-Risk. *European Journal Operations Research* 234: 140–54. [CrossRef]

Sharpe, William F. 1992. Asset Allocation: Management Style and Performance Measurement. *Journal of Portfolio Management (Winter)* 18: 7–19. [CrossRef]

Ziegel, Johanna F. 2014. Coherence and elicitability. *arXiv*. [CrossRef]

5

Capital Structure and Firm Performance in Australian Service Sector Firms: A Panel Data Analysis

Rafiuddin Ahmed [1,*] and Rafiqul Bhuyan [2]

[1] Program of Accounting and Finance, James Cook University, Douglas, QLD 4814, Australia
[2] Department of Accounting and Finance, Alabama A&M University, Normal, AL 35762, USA; rafiqul.bhuyan@aamu.edu
* Correspondence: rafiuddin.ahmed@jcu.edu.au

Abstract: Using cross-sectional panel data over eleven years (2009–2019), or 1001 firm-year observations, this study examines the relationship between capital structure and firm performance of service sector firms from Australian stock market. Unlike other studies, in this study directional causalities of all performance measures were used to identify the cause of firm performance. The study finds that long-term debt dominates debt choices of Australian service sector companies. Although the finding is to some extent similar to trends in debt financed operations observed in companies in developed and developing countries, the finding is unexpected because the sectoral and institutional borrowing rules and regulations in Australia are different from those in other parts of the world.

Keywords: firm performance; causality tests; leverage; long-term debt; capital structure

1. Introduction

Capital structure is one of the most perplexing puzzles in the financial literature that deals with solutions to optimal mix of debt and equity. The seminal work of Modigliani and Miller (1958) initiated this body of work, other researchers later developed theories along the MM, and empirical researchers validated the assumptions underlying the theoretical body of the literature by examining different dimensions such as firm characteristics, time or industry sector category. A mirror image of capital structure choice is essentially a decision to fund capital from the cheapest sources to maximize income after taxes (Yazdanfar 2012). The seminal work of Jensen and Meckling (1976) posits managerial behavior in the best interest of the shareholders which is to borrow at a level that will maximize shareholder value and firm profitability. Since the work of Jensen and Meckling (1976) several researchers have examined the relationship between leverage and profitability. The findings of these studies are contradictory and mixed, some suggesting a positive relationship (Ghosh et al. 2000; Hadlock and James 2002; Roden and Lewellen 1995; Taub 1975) and some suggesting a negative relationship (Fama and French 1998; Gleason et al. 2000) between leverage and profitability (El-Sayed Ebaid 2009). There are many studies on capital structure in the context of service sectors in Europe, USA, the Middle East and other parts of the world (Chakrabarti and Chakrabarti 2019; Choi et al. 2018; Park and Jang 2018; Sardo et al. 2020; Sermpinis et al. 2019; Szemán 2017). Compared to other sectors, service sector capital structure research is at a nascent stage. Further research needs to be done to enrich the understanding of the drivers of financial performance of this sector.

The key aim of this paper is to empirically examine the relationship between debt financing and firm performance of service sector companies listed in the Australian Stock Exchange. The service sector is chosen to reflect the changing configuration of the Australian economy from a resource-based economy to a service-based economy. Over the last one and a half decades, the Australian service sector contributed between 60–70% and is a major employer (Australian Bureau of Statistics 2019, 2020). This trend is expected to continue in the foreseeable future and it is important to get some insights into the effect of capital structure on this sector firms. Four performance measures are used to capture firm performance: (a) return on asset, (b) return on equity, (c) return on capital employed and (d) operating margin. The paper finds: (a) portability (measured by return on equity, ROE) and leverage (measured by a ratio of short-term debt to total assets) is positively associated, (b) profitability (measured by return on assets) and leverage (measured by short-term debt) is positively associated and (c) no significant association between either ROE and ROA and long-term or total debt. The main contribution of this paper is that it has extended the current body of literature on capital structure by adding the Australian service industry context from very recent data. Australia's move from a resource-based economy to a service-based economy means the sector is growing, so the findings of this paper are expected to shed light on this emerging frontier of capital structure practices of service sector firms. The remainder of the paper is organized as follows. In Section 2, the literature is discussed. In the third section, the empirical literature is reviewed, followed by three sections on data, results and discussions. The final section concludes the paper with some possible directions for future research.

2. Literature Review

Numerous theories have been developed following the initial development of capital structure theory by Modigliani and Miller (1958). These theories were later classified by their assumptions about how they affect firm value in the financial market. The first of these theories is the Trade-off theory of capital structure. This theory precedes some initial refinements in 1963 (Modigliani and Miller 1963) of Modigliani and Miller's (1958) initial work, in which taxes are added to theorize the effect of taxes on a firm's tax payable amount, increase in after tax income and its market value. This development was later labelled as trade-off theory, a theory which states that a firm's optimal leverage is achieved by minimizing taxes, costs of financial distress and agency costs. Baxter (1967) argued that increased debt levels increases the chances of bankruptcy and increases interest payable to the debtholders. A firm's optimal leverage is where tax advantage from debt exactly equals the cost of debt. Kraus and Litzenberger (1973) argue that a firm's market value declines if its debt obligations are greater than its earnings. DeAngelo and Masulis (1980) propose the static trade off theory and include other tax minimizing offsets such as depreciation and investment tax credits. They argue that firms weigh tax advantages of debt against business risk (a cost). Their theoretical model proposes that a firm's optimum debt level is where the present value of tax savings from debt equals the present value of costs of distress.

Myers (1984), in his theoretical explanation of the asymmetric information hypothesis, proposes different information held by firms' internal and external stakeholders. Managers hold real information about firms' income distribution plans (Ross 1977). Thus, firm's leverage level signals its confidence levels, suggesting lower leverage as a poor signal about income and its distribution potential and vice versa. Pettit and Singer (1985) discuss the problems of asymmetric information and possible agency costs affecting firms' demand and supply of credit. They argue that small firms possess a higher level of asymmetric information due to financial constraints for sufficient disclosure of financial information to outsiders. This theory has laid the foundation for Pecking Order Theory (POT). Donaldson (1984) proposes the concepts and ideas of Pecking Order Theory (POT) which was later refined by Myers (1984) and Myers and Majluf (1984). The fundamental premise of this theory is that firms' preferences for funding is stacked by a pecking order of risk preferences and corresponding costs. Thus, firms use the cheapest source of internal funds such as retained earnings, debt, convertible debt and preference shares) and external equity (Myers 1984). The cost of sourcing extra funding is dependent on the extent of information asymmetries of risk perceptions emanating from differential information needs held by inside management and potential investors. In addition to a firm's desire to source the cheapest fund to finance its needs, other factors, such as the stage of development of a firm (a startup, a mature firm etc.) influence the supply of funds (Macan Bhaird and Lucey 2010).

Agency theory (Jensen and Meckling 1976) addresses the fundamental problem of managing a firm's capital structure from the cheapest source of funds. While common equity is an expensive source of funds, its use results in suboptimal firm value when equity holders insist on risk reduction from lower leverage usage. If managers' and shareholders' interests are not aligned, it is highly unlikely that optimal firm value is ever going to eventuate from managerial actions. The debtholders' risk perceptions encourage them to ask for debt covenants or other costly debt shielding instruments. The tensions between the two subgroups of owners impose increased risk of monitoring by management, resulting in costly monitoring and hence, agency costs. A number of remedial measures can be implemented such as reduction in consumption of resources when debt and bankruptcy risks increase (Grossman and Hart 1982), increasing the stake of managers in a firm or increasing the leverage (Jensen 1986), commonly packed as 'free cash flow hypothesis'. Free cash flow hypothesis proposes adoption of measures to reduce free cash flow at managers' disposal by increasing leverage (Stulz 1990) so that less cash flow is available for desired investment choices.

The theories above are prevalent in different country specific studies. An empirical study by El-Sayed Ebaid (2009) on Egyptian firms suggest a negative relationship between profitability and shorter-term or total debt when return on asset is used to measure profitability. The results also suggest no significant relationship between short-term or long-term debt and profitability when return on equity or gross margin is used as a measure of profitability. Salim and Yadav's (2012) study on 237 listed Malaysian companies from 1995–2001 found a negative relationship between short-term and long-term debts and all measures of profitability, return on assets, return on equity and earnings per share. Ahmed Sheikh and Wang (2011) examined 240 listed Pakistani non-financial companies during the 2004–2009 period. Three statistical tests, fixed effects, random effects and ordinary least squares found negative relationships between debt and return on assets. Weill (2008) used the maximum likelihood estimation method to analyze the effect of financial leverage on the performance of 11,836 firms from seven European countries over a three-year time period, 1998–2000. The results indicate that the long-term debt ratio is positively related at statistically significant level in Spain and Italy but negatively related at statistically significant level in Germany, France, Belgium and Norway, and insignificantly in Portugal. Goddard et al. (2005) used the generalized methods of moments system to test the determinants of profitability of manufacturing and service firms in Belgium, France, Italy and the U.K. from 1993–2001. They found a negative relationship between the sample firms' gearing ratio and profitability, and higher profitability in more liquid firms. Abor (2007) used a generalized least squares regression to study a sample of 160 Ghanaian and 200 South African Small and Medium Enterprises (SMEs) from 1998–2003 and found a negative relationship between longer-term and total debt ratios and profitability. Yazdanfar and Öhman's (2015) study used 15,897 Swedish SMEs from five different sectors from 2009–2012 to examine the effect of three different forms of debt ratios, trade credit, short-term debt and long-term debt on profitability. The results suggest a negative relationship between all types of debts and profitability, suggesting an increased use of equity capital to finance Swedish SMEs.

There are not many Australian studies on the relationship between capital structure and profitability. Li and Stathis (2017) examined the determinants of the capital structure of Australian manufacturing listed traded firms. The study used eight factors: profitability, log of assets, median industry leverage, industry growth, market-to-book ratio, tangibility, capital expenditure and investment tax credits. They found weak support for the pecking order hypothesis and increasing support for the trade-off theory in Australia. Qiu and La (2010) examined the relationship between firm characteristics and capital structure of 367 Australian firms over a 15 year period. Their study identified the role of debt on profitability, tangibility, growth prospects and risk of these firms. They concluded that profitability has the potential to reduce debt levels of Australian firms, implying debt reduction through increased profits was possible in Australian firms. Barth et al. (2001) examined the relationship between capital structure and profitability of 107 countries including Australia. They tested for regulatory power, supervision, and other factors affecting the relationship between profitability and leverage across the countries studied. Rashid and Islam (2009) examined 60 companies in the Australian Financial services sector during the years 2002–2003. The results suggest that profitability is negatively affected by leverage, and positively affected by board size, liquid markets and information efficiency (all control variables).

Firm performance as a measure of the impact of different proxies for capital structure has added new insights in recent times. Some country-specific studies have examined the direct effect of using different types of debts on firm performance. Most of these studies reported a significant negative relationship between debts and firm performance. Chakrabarti and Chakrabarti (2019) examined firm-specific and macro-economic variables on 18 Indian non-insurance firms for seven years. They found a positive relationship between low insurance, low input costs, low inflation rates, higher return on investment, liquidity and profitability. Dalci (2018) examined the impact of capital structure on 1503 listed manufacturing firms in the Chinese stock exchange between the years 2008–2016. They found an inverted U-shaped relationship between capital structure and profitability and provided the causes of a negative and positive relationship between financial leverage (as a measure of capital structure) and profitability. This is a major study that highlighted the importance of the developments of credit market policies and rules for the advancement of different-sized Chinese manufacturing firms.

Dave et al. (2019) examined the impact of capital structure and profitability of firms in the Indian Steel industry and observed a significant negative relationship between long-term and short-term debts as a ratio of total assets and profitability. Helmy et al. (2020) examined the impact of capital structure, internal governance mechanism, and firm-performance of 183 Bursa-listed Malaysian companies for the years 2007–2010. They found a positive impact of capital structure on firm performance. Gharaibeh and Bani Khaled (2020) examined the factors that played key roles in the profitability of 46 Jordanian service sector companies between the years 2014–2018. They found that debt as a portion of total assets and tangible assets have significantly negative relationships with profitability whereas tangible size and business risk had a positive relationship with profitability.

Hussein et al. (2019) examined listed Jordanian firms between 2005–2017. Using three measures of firm performance, return on assets, Tobin's Q and return on assets, and total and short-term debt as a proxy for capital structure, they observed a positively significant relationship between firm size, asset growth, significant negative relationship between short-term debt and long-term debt and return on assets. However, they did not find any significant negative relationship between short-term and long-term debts and return on equity measure of firm performance. Lastly, Yazdanfar examined 15,897 firms working in five SME sectors of the Swedish economy between 2009–2012. They found debt ratios (trade-credit, short-term and long-term debts) negatively affected firm profitability.

Capital structure studies that examined the relationship between different proxies for capital structure and firm performance used a variety of measures to define profitability. Some studies used a single measure (see, for example, Arifin 2017; Negasa 2016) while others used multiple measures such as return on Equity (ROE), return on assets (ROA), and return on capital invested (ROCE) (see, for example, Gharaibeh and Bani Khaled 2020; Musah and Kong 2019). In these studies, different types of debts are used as proxies for capital structure and different control variables are used to measure the collective impacts on firm performance. The relationship between firm performance and capital structure is assumed to be unidirectional in most of the studies reviewed above. However, some recent studies validated the causal relationship between capital structure and firm performance (Arifin 2017; M'ng et al. 2017). Finally, the studies above showed a negative, positive, and mixed relationship between capital structure measures and firm performance.

The studies are from diverse sectors and cover a wide range of firm year cross sectional observations. There is a limited number of studies that examine the linkage between different measures of firm performance (or profitability) and capital structure. Studies covering the services sector are hardly noteworthy in the Australian and global contexts. Moreover, the directional causal relationship between different types of borrowings and firm performance is hardly examined in detail in the studies reviewed above. This study contributes to the growing body of literature in the study of capital structure in the under-researched domains of service sector firms in Australia and internationally.

3. Data

We consider a comprehensive database from the Australian service sector (as classified by Australian Bureau of Statistics) for the period 2009 to 2019. The data was collected from Datanalysis database-a database that publishes financial data of companies in different Australian sectors. Although our initial sample was much larger than what we have included in the study, due to matching inconsistency in variable definition and the availability of all variables of all companies, we have truncated the data to 91 companies that have same data set for the entire time period. These companies are all listed in the Australian Stock Exchange.

Table 1 shows that a total of nine sectors are considered to conduct research for the period 2009 to 2019. Based on the availability of the entire data set with chosen variables, some sectors had the most samples and others had only a few companies. The percentage column shows the degree of weight from each sector of our sample. Based on the literature surveyed, we consider several variables shown in Table 2 to investigate our research question.

To avoid spurious regression estimates in our empirical analysis, variables under consideration should ideally be stationary. To confirm this, we used the panel unit root test of Levin et al. (2002). Table 3 shows that the unit-roots hypothesis is rejected by all variables at the 1% level of significance. Following (Canarella and Miller 2018; Köksal and Orman 2015; Khan et al. 2018; M'ng et al. 2017), we also checked for stationarity using a unit root test and observed that all variables were stationary with respect to the dependent variables (Return on Equity (ROE), Operating Margin (op_margin), Return on Asset (ROA) and Return on Invested Capital (ROIC)), confirmed by the tests for heteroscedasticity and autocorrelation diagnostics.

The panel regressions were run for four dependent variables (return on equity, return on assets, return on invested capital, and operating margin), two treatment variables (leverage and long-term debt to total assets ratio), and five control variables (size, liquidity, revenue growth for three years, tangibility and depreciation tax shield). A series of regressions were run for these variables and diagnostic tests were conducted to confirm the appropriateness of fixed or random effects panel regressions models.

For each of the dependent variables, outputs for two models are presented, after eliminating the inappropriate models using Hausman tests. The Breausch Pagan test was employed to confirm the outputs of the Hausman test for this purpose. Earlier studies in capital structure used the Hausman test to identify the appropriate panel data model from two available models: fixed effect model and random effect model (Dalci 2018; Mayuri and Kengatharan 2019; Sivalingam and Kengatharan 2018; Suntraruk and Liu 2017). Breausch Pagan Lagrange Multiplier tests were used for confirming the appropriateness of the random effects model (see for example, Dalci 2018; Ghasemi et al. 2018; Khan et al. 2018). The tables below present the outputs of these models.

Table 1. Table shows the various Australian service sector companies considered for this research.

Sector-ID	Frequency	Percent	Cumulative Percentage
Utilities	242	24.18	24.18
Construction	33	3.30	27.47
Retail trade	132	13.19	40.66
Transport	143	14.29	54.95
Communication services	308	30.77	85.71
Consumer discretionary	22	2.20	87.91
Commercial services	121	12.09	100.00
Total	1001	100.00	

(Source: Authors' compilation).

Table 2. List of dependent and explanatory variables.

Variables	Calculated as	Sources
Return on assets (ROA)	Earnings before Interest and tax (EBIT)/Total assets	(Dalci 2018; Gharaibeh and Bani Khaled 2020; Goddard et al. 2005; Nunes et al. 2009)
Return on equity (ROE)	EBIT/Total Equity	(Arifin 2017; Dalci 2018; Gharaibeh and Bani Khaled 2020)
Operating margin (op_margin)	EBIT/Operating revenue	(Gharaibeh and Bani Khaled 2020)
Return on Invested Capital (ROIC)	EBIT/Invested capital	(Musah and Kong 2019)
Leverage	Total liabilities/total assets	(Gharaibeh and Bani Khaled 2020; Nunes et al. 2009)
Long-term debt to total asset (LTd_TA)	Long-term debt/total assets	(Yazdanfar and Öhman 2015)
Liquidity	Current assets/Current liabilities	(Nunes et al. 2009)
Tangibility	Fixed assets/Total assets	(Fitim et al. 2019; Gharaibeh and Bani Khaled 2020; Nunes et al. 2009; Shalini and Biswas 2019)
Tax shield	Depreciation/Total assets	(Fitim et al. 2019; Shalini and Biswas 2019; Yazdanfar and Öhman 2015)
Operating revenue (size)	Log of Operating revenue	(Fitim et al. 2019; Shalini and Biswas 2019)
Revenue growth (3-year)	% of revenue growth (3 yearly average, given)	(Chadha and Sharma 2015; Chakrabarti and Chakrabarti 2019)

Table 3. Unit root tests results.

ROIC	ROE	ROA	OP_MARGIN	LTD_TA	LEVERAGE	TAXSHIELD	TANBIBILITY	LIQUIDITY	LNSALES
−6.4783 (0.0000)	−6.04 (0.0000)	−9.75 (0.0000)	−7.0134 (0.0000)	−9.1691 (0.0000)	−3.2673 (0.0000)	−5.5693 (0.0000)	−12.2987 (0.0000)	−5.9927 (0.0000)	−11.4026 (0.0000)

4. Results

In the following section, we have presented the results of our analysis. Four different measures of financial performance and six explanatory variables were analyzed to identify the important explanatory variables affecting firm performance of Australian service sector firms between the years 2009 and 2019. In each table below, two models are presented: Model 1 and Model 2. In Model 1, leverage is used as a treatment variable and in Model 2, long-term debt is used as the treatment variable. Size, depreciation tax-shield, revenue growth for 3 years, operating revenue (measure of size), liquidity and tangibility are used as control variables in measuring firm performance.

As shown in Table 4, the fixed effect model (FEM) in Model 1 identified leverage, tangibility, liquidity and operating revenue as important predictors of operating margin. The random effect model (REM) in Model 1 identified leverage, tangibility, operating revenue and revenue growth for three years as significant predictors of operating margin. The constant is also important at 1% level of significance. The Granger causality test shows a unidirectional relationship between leverage and operating margin. This relationship is positive as evidenced by a significant positive coefficient of leverage.

Table 4. Panel regression outputs for operating margin (dependent variable).

Dependent Variable	Model 1		Model 2	
	FEM	REM	FEM	REM
leverage	0.5972 * (4.48)	0.2411 ** (2.23)		
LTD_TA			−0.0447 (−1.21)	−0.0295 (−0.88)
Depreciation tax shield	−0.0041 (−0.10)	−0.1265 (−0.36)	−0.0208 (−0.49)	−0.0179 (−0.51)
Tangibility	−0.04682 ** (−1.21)	−0.0584 *** (−1.80)	−0.0640 (−1.63)	−0.0648 ** (−1.97)
Liquidity	0.1589 ** (2.15)	0.0191 (0.32)	−0.0983 *** (−1.69)	−0.0893 *** (−1.78)
Operating revenue	0.0075 * (2.39)	0046 * (2.84)	0.0086 * (2.68)	0.0059 * (3.68)
Revenue growth (3-year)	−0.0081 (−2.66)	−0.0074 ** (−2.45)	−0.0091 * (−2.95)	−0.0080 * (−2.65)
Constant	0.0454 (0.22)	0.5986 * (3.92)	0.9164 * (9.34)	0.9457 * (12.37)
F-test	6.49 (0.0000)		3.32 (0.0031)	
Hausman test		24.85 (0.0000)		3.80 (0.7033)
Breusch-Pagan test		445.16 (0.0000)		480.54 (0.0000)

*, ** and *** are used for 1%, 5% and 10% level of significance.

In Model 2, the fixed effect model identified liquidity, operating revenue and revenue growth as significant predictors of operating margin. In the random effects model, tangibility, liquidity and operating revenue are significant predictors of operating margin. In both models, the constants are significant at 1% level. The Granger causality test revealed a unidirectional relationship between long-term debt to total asset and operating margin. This relationship is negative but not significant at any level.

In Table 5, the random effect model of Model 1 identified leverage, operating revenue, revenue growth as significant predictors of return on assets. In the fixed effect model of Model 1, leverage, operating revenue and growth are significant predictors of return on assets. The Granger causality test indicates a bi-directional causality between leverage and return on asset. Leverage significantly pulls return on assets down, as evidenced by the negative coefficient in the equations in Model 1.

Table 5. Panel regression outputs for Return on Asset (ROA).

Variables	Model 1		Model 2	
	REM	FEM	REM	FEM
leverage	−1.1143 * (−3.75)	−1.6487 * (−4.58)		
LTD_TA			0.2475 ** (2.49)	0.1819 ** (1.98)
Depreciation tax shield	0.01242 (0.13)	−0.1814 (−1.60)	−0.1288 (−1.13)	0.0429 (0.44)
Tangibility	−0.01436 (−0.16)	0.05435 (0.52)	0.0810 (0.77)	0.0045 (0.05)
Liquidity	0.2029 (1.24)	0.1243 (0.62)	0.9096 * (5.82)	0.7347 * (5.35)
Operating revenue	0.02992 * (6.61)	0.01709 ** (2.01)	0.0120 (1.39)	0.0232 * (5.19)
Revenue growth (3-year)	0.0168 ** (2.06)	0.01639 ** (1.98)	0.0193 ** (2.32)	0.01966 ** (2.38)
Constant	0.3210 (0.76)	1.2644 ** (2.03)	−1.2744 * (−4.82)	−1.3402 * (−6.41)
F-test		10.06 (0.0000)		7.48 (0.0000)
Hausman test	26.13 (0.0002)		21.04 (0.0016)	
Breusch-Pagan test	544.07 (0.0000)		532.38 (0.0000)	

*, ** and *** are used for 1%, 5% and 10% level of significance.

The constant is also significant at the 5% level. We also observe that in both fixed effect model and random effect model of Model 2, long-term debt to total assets, liquidity, operating revenue and revenue growth for three years are significant predictors of return on assets. The constant is also significant at the 1% level.

In the Table 6, the fixed effect model in Model 1 (leverage as treatment variable), leverage, tax shield, tangibility and operating revenue are significant explanatory variables of return on capital invested. In the random effects model, leverage, tax shield, tangibility and operating revenue are significant predictors of return on invested capital. In both models, the constants are significant at the 10% level of significance. The Granger causality test indicates a bi-directional relationship between leverage and return on invested capital. This relationship is negative, as evidenced by the negative coefficient of leverage.

In Model 2 (long-term debt as a treatment variable), long-term debt, tax shield, tangibility, liquidity and operating revenue are significant predictors of return on capital employed. In the random effects model in Model 2, long-term debt, tax shield, tangibility, liquidity and operating revenue are significant predictors of return on invested capital. The constant is also significant at the 10% level. When we run the Granger's causality test, we only observe a unidirectional relationship between long-term debt to total assets and return on invested capital. That is, return on invested capital is largely influenced by debt positively. Granger causality test indicates a uni-directional relationship between long-term debt and return on invested capital. This relationship is positive, as evidenced by the coefficient of long-term debt above.

Table 6. Return on invested capital (RoIC).

Dependent Variable	**Model 1**		**Model 2**	
	FEM	**REM**	**FEM**	**REM**
Leverage	−7.6385 ** (−2.17)	−7.509 ** (−2.42)		
LTD_TA			2.1416 ** (2.22)	1.6844 *** (1.83)
Depreciation tax shield	−3.5081 * (−3.16)	−2.4426 ** (−2.45)	−3.2105 * (−2.9)	−2.2261 ** (−2.23)
Tangibility	−2.4543 ** (−2.40)	−3.152 * (−3.41)	−2.4969 ** (−2.43)	−3.0967 * (−3.33)
Liquidity	−0.4048 (−0.21)	0.09087 (0.05)	3.8359 ** (2.53)	4.0090 * (2.86)
Operating revenue	0.2143 ** (2.58)	0.2100 * (4.09)	0.17597 ** (2.09)	0.1623 * (3.15)
Revenue growth (3-year)	0.05104 (0.63)	0.0623 (0.78)	0.06777 (0.84)	0.0803 (1.00)
Constant	8.9454 *** (1.66)	8.1662 *** (1.83)	−3.893121 (−1.52)	−3.6585 *** (−1.70)
F-test	4.52 (0.0000)		4.56 (0.000)	
Hausman test		13.67 (0.00)		16.76 (0.0102)
Breusch-Pagan test		1149.89 (0.00)		1166.45 (0.0000)

*, ** and *** are used for 1%, 5% and 10% level of significance.

In Table 7 below, in Model 1, the random effect regression identified leverage and tangibility as two important explanatory variables of return on equity. The fixed effect model identified tangibility, size and liquidity as significant explanatory variables at the 5% level of significance. The Granger causality test indicates a unidirectional relationship between leverage and return on equity. This relationship is positive in the REM regression of Model 1 above.

In model 2 (long-term debt as a second measure of debt level), long-term debt to total assets, tangibility, liquidity and operating revenues were identified as significant explanatory variables of return on equity at the 1% level of significance. In the random effects model, long-term debt, tangibility and operating revenues were identified at the 5%, 1% and 10% levels of significance, respectively. The Granger causality test suggests a bi-directional relationship between long-term debt and return on equity. The positive coefficient of long-term debt to total assets in the equations in Model 2 in Table 7 indicates long-term debt to finance the purchase of assets for operations is beneficial to Australian service sector firms.

Table 7. Panel regression output for Return on Equity (ROE).

Dependent Variable	Model 1		Model 2	
	REM	FEM	FEM	REM
Leverage	3.8214 * (2.76)	3.7031 (1.41)		
LTD_TA			3.3723 * (4.73)	1.2797 ** (2.45)
Depreciation tax shield	0.1380 (0.29)	−0.9583 (−1.15)	−0.8636 (−1.05)	0.2167 (0.45)
Tangibility	−1.3156 * (−3.01)	−1.5104 ** (1.97)	−2.2271 * (−2.93)	−1.7410 * (3.89)
Liquidity	1.1551 (1.54)	3.1308 ** (2.15)	3.7468 * (3.34)	0.9349 (1.28)
Operating revenue	0.02765 (1.56)	−0.1364 ** (−2.20)	−0.1853 * (−2.97)	0.0335 *** (1.91)
Revenue growth (3-year)	−0.0013 (−0.02)	0.03421 (0.56)	0.03896 (0.65)	−0.0113 (−0.20)
Constant	−4.1381 ** (−2.17)	−1.596 (−0.40)	−0.14843 (−0.08)	−1.2106 (−1.08)
F-test		3.44 (0.0023)	6.90 (0.0000)	
Hausman test	17.72 (0.0000)			38.67 (0.0000)
Breusch-Pagan test	0.62 (0.2150)			2.88 (0.0449)

*, ** and *** are measured for 1%, 5% and 10% level of significance.

5. Discussion

The relationship between capital structure and firm performance can be summarized in two different ways: leverage and firm performance, and long-term debt and firm performance. These themes are discussed in the following section.

In the Table 8 above, leverage is significantly associated with operating margin but has significant negative association with two measures of firm performance: return on assets and return on invested capital. In Table 9, other control variables have influenced profitability in positive and negative ways. Tangibility has affected operating margin, return on vested capital and return on equity negatively, suggesting that Australian firms are overinvesting on fixed assets. Revenue growth has also affected operating margin and return on assets significantly. Depreciation tax shield is observed to have affected operating margin negatively and liquidity has affected return on equity positively. The constant is significant in both operating margin and return on invested capital, suggesting a guaranteed minimum return from the presence of service sector firms in the economy. However, return on assets and return on equity, not assumed as constants, are not significant at any level of confidence.

In the table below, all relevant regression models are summarized to demonstrate the effect of long-term debt being used to finance total assets in order to improve firms' performance in the Australian services sectors.

Table 8. Leverage and its effect of firm performance.

Variables	Performance Measures Used			
	Operating Margin	**Return on Assets**	**Return on Invested Capital**	**Return on Equity**
	REM	**REM**	**REM**	**FEM**
Leverage	0.2411 ** (−2.23)	−1.1143 * (−3.75)	−7.509 ** (−2.42)	3.7031 (−1.41)
Depreciation tax shield	1265 (−0.36)	0.01242 (−0.13)	−2.4426 ** (−2.45)	−0.9583 (−1.15)
Tangibility	−0.0584 *** (−1.80)	−0.01436 (−0.16)	−3.152 * (−3.41)	−1.5104 ** (−1.97)
Liquidity	0.0191 (−0.32)	0.2029 (−1.24)	0.09087 (−0.05)	3.1308 ** (−2.15)
Operating revenue	0.0046 * (−2.84)	0.02992 * (−6.61)	0.2100 * (−4.09)	−0.1364 ** (−2.20)
Revenue growth (3-year)	−0.0074 ** (−2.45)	0.0168 ** (−2.06)	0.0623 (−0.78)	0.03421 −0.56
contant	0.5986 * (3.92)	0.321 (−0.76)	8.1662 *** (−1.83)	−1.596 (−0.40)
F-test				3.44 (−0.0023)
Hausman test	24.85 (0.0000)	26.13 (−0.0002)	35.01 (0.000)	
Breusch-Pagan test	445.16 (0.0000)	544.07 (0.0000)	13.67 (0.000)	

*, ** and *** are measured for 1%, 5% and 10% level of significance.

In the Table 9 above, long-term debt to finance assets is significantly associated (positively) with all measures of firm performance, except operating margin. Depreciation tax shield significantly influenced return on invested capital negatively while tangibility has negatively affected, at different levels of statistical significance, all measures of firm performance except return on assets. Liquidity has affected firm performance in all instances, natively in operating profit, and positively (at different levels of statistical significance) in improving return on assets and return on invested capital. Operating revenue positively influenced all measures of firm performance except return on assets. Finally, revenue growth for three years was a significantly influential negative factor in affecting operating margin but a positive factor in improving return on assets.

In light of the discussions above, we can say, 'capital structure matters.' It enhances the performance of the service sectors in Australia not only through improved operating margin and higher return on invested capital; it also increases shareholder value by improving return on equity and return on assets. As observed in our data analysis, we have used four different measures of firm performance. Three of these measures relate to balance sheets and the other one relates to profit and loss in the short-term. Leverage in all measures of performance was significant at the 1% level except when return on invested capital was used to measure firm performance. Even in the presence of other variables (used in the literature as explanatory variables) that showed significant influence in firm performance, leverage remains significant in shaping the performance of service sector firms in Australia. The positive and significant relationship between leverage and operating margin implies that the service sectors in Australia can greatly benefit by increasing the debt level in its capital structure. The negative and significant relationship with tangibility also makes economic sense and implies that tying funds in fixed assets can be detrimental to operating profits as the company will have lees funds available for generating revenues.

Table 9. Long-term debt and its effect on firm performance.

Variables	Performance Measures Used			
	Operating Margin	Return on Assets	Return on Invested Capital	Return on Equity
	REM	REM	REM	REM
LTD_TA	−0.0295 (−0.88)	0.2475 ** (−2.49)	1.6844 *** (−1.83)	1.2797 ** (−2.45)
Depreciation tax shield	−0.0179 (−0.51)	−0.1288 (−1.13)	−2.2261 ** (−2.23)	0.2167 (−0.45)
Tangibility	−0.0648 ** (−1.97)	0.081 (−0.77)	−3.0967 * (−3.33)	−1.7410 * (−3.89)
Liquidity	−0.0893 *** (−1.78)	0.9096 * (−5.82)	4.0090 * (−2.86)	0.9349 (−1.28)
Operating revenue	0.0059 * (−3.68)	0.012 (−1.39)	0.1623 * (−3.15)	0.0335 *** (−1.91)
Revenue growth (3-year)	−0.0080 * (−2.65)	0.0193 ** (−2.32)	0.0803 (−1.000)	−0.0113 (−0.20)
Constant	0.9457 * (−12.37)	−1.2744 * (−4.82)	−3.6585 *** (−1.70)	−1.2106 (−1.08)
Hausman test	3.800 (−0.7033)	21.04 (−0.0016)	16.76 (−0.0102)	38.67 (0.000)
Breusch-Pagan test	480.54 (0.0000)	532.38 (0.0000)	1166.45 (0.0000)	2.88 (−0.0449)

*, ** and *** are measured for 1%, 5% and 10% level of significance.

Long-term debt to total assets does not play any role in operating margin as the service sectors can generate frequent cash flows and turnover rate is very high. As such, need for long-term debt is irrelevant. The economic impact is different for return on assets where our analysis shows that both leverage and long-term debt to total assets are positively significant in impacting return on assets under both REM and FEM. It further strengthens our earlier arguments that service sectors greatly benefit from increased debt level in its capital structure. Revenue growth for three years also shows promising impact on return on asset which also remain positively significant. When we tested our model for ROIC, we find that leverage is negatively significant whereas long-term debt to total asset remains positively significant. The negative relationship with leverage implies that firms face challenges in return on invested capital if too much of the funds are tied with short-term borrowing as they are payable quickly. In addition, just like in operating margin, tangibility remains negatively significant implying that firms are adversely affected by increased tangible assets holding. Finally, when analyzing the relationship with return on equity, just like ROA, we observe that both leverage and long-term debt to total assets remain positively significant. Unlike ROA, tangibility remains negatively significant with ROE. In conclusion, we can assert that in the case of service sector firms in Australia, a high level of leverage and a high level of long-term debt in capital structure is beneficial to increasing shareholders' wealth.

6. Conclusions

This paper has examined firm level characteristics and firm performance (or profitability) of service sector firms listed in the Australian Stock Exchange (ASX). Using a panel regression approach on data collected over an eleven-year period (2009–2019), the effect of capital structure and leverage was examined. Four measures of firm performance were used: return on assets, return on equity, operating margin ratio and return on capital employed. The analysis of data reveals a significant association between return on equity and leverage levels. Leverage affects firm performance at a statistically significant level in these service sector firms. For every dollar of increase in leverage, operating margin improves by 0.24 times, return on assets reduces by 1.11 times and return on invested capital reduces by 7.59 times (all statistically significant at the 1% and 5% level), suggesting that Australian services sector firms are not benefitting much from the use of debts to finance their operations. This finding is in sharp contrast to asymmetric information theory that suggests that lower debt levels hide firm performance (Myers 1984). In fact, they are overburdened with debts. When long-term debt is used to finance total assets, the picture changes dramatically. Return on assets, return on invested capital and return on equity changes by 0.24 times (significant at 5% level), return on capital employed increases by 1.68 times (significant at 10% level) and return on equity improves by 1.27 times (significant at the 5% level), suggesting the positive value adding contributions of the use of long-term debt. The directional causality tests, as captured in the Granger causality test, indicated a positive unidirectional association between leverage and operating margin, bi-directional causality with return on assets (negative) and return on invested capital, with return on assets (negative) and a unidirectional (positive) causality between leverage and return on equity. The test also identified a bidirectional causality between long-term debt to total assets and operating margin, and a bidirectional relationship with return on assets, return on invested capital and return on equity. The presence of unidirectional and bidirectional causality between different types of debts to finance operations mean significant interdependencies and negative effects of debt on service sector firms in Australia.

The study has the inherent limitations of any research project. The sample size may be questioned for two reasons: the number of firms from the Australian service sector and the years included in the data. Due to unavailability of data, only three years of data are used. Inclusion of more years can be a possibility for the extension of the current research project. Interested researchers may consider a robust dataset, encompassing industries from all sectors of the Australian economy. The current study has examined a limited number of constructs to reflect on the profitability of Australian service sector listed firms. The influence of extra-organizational factors may contribute to the profitability of Australian service sector companies. Researchers willing to pursue the line of inquiry in this paper may include economic factors such as inflation, interest rate and GDP in future research. Finally, service sector heterogeneity may be partially responsible for poor reflection of profitability. So the inclusion of industry effects may be worthwhile before a conclusion can be reached about the industry sector effect on Australian service sector performance.

Author Contributions: R.A. developed the research idea and conducted review of literature. R.A. collected the data under the guidance and recommendation of R.B. and run the statistical analysis which is also suggested by R.B. R.B. provided the insights of the statistical results and analysis. Both authors have read and agreed to the published version of the manuscript.

References

Abor, Joshua. 2007. Corporate governance and financing decisions of Ghanaian listed firms. *Corporate Governance: The International Journal of Business in Society* 7: 83–92. [CrossRef]

Ahmed Sheikh, Nadeem, and Zongjun Wang. 2011. Determinants of capital structure: An empirical study of firms in manufacturing industry of Pakistan. *Managerial Finance* 37: 117–33. [CrossRef]

Arifin, Agus Zainul. 2017. Interactions between capital structure and profitability: Evidence from Indonesia stock exchange. *International Journal of Economic Perspectives* 11: 117–21.

Australian Bureau of Statistics. 2019. *Australian System of National Accounts, 2018–19 5204.0*. Canberra: Australian Bureau of Statistics.

Australian Bureau of Statistics. 2020. *Labour Force, Australia, Detailed, Quarterly, May 2020. (6291.0.55.003)*. Canberra: Australian Bureau of Statistics.

Barth, James Richard, Gerard Caprio Jr., and Ross Levine. 2001. Banking Systems around the Globe: Do Regulation and Ownership Affect Performance and Stability? In *Prudential Supervision: What Works and What Doesn't*. Chicago: University of Chicago Press, pp. 31–96.

Baxter, Nevins D. 1967. Leverage, risk of ruin and the cost of capital. *The Journal of Finance* 22: 395–403.

Canarella, Giorgio, and Stephen Michael Miller. 2018. The determinants of growth in the U.S. information and communication technology (ICT) industry: A firm-level analysis. *Economic Modelling* 70: 259–71. [CrossRef]

Chadha, Saurabh, and Anil Kumar Sharma. 2015. Capital Structure and Firm Performance: Empirical Evidence from India. *Vision* 19: 295–302. [CrossRef]

Chakrabarti, Anindita, and Ahindra Chakrabarti. 2019. The capital structure puzzle—Evidence from Indian energy sector. *International Journal of Energy Sector Management* 13: 2–23. [CrossRef]

Choi, Serin, Seoki Lee, Kyuwan Choi, and Kyung-A Sun. 2018. Investment–cash flow sensitivities of restaurant firms: A moderating role of franchising. *Tourism Economics* 24: 560–75. [CrossRef]

Dalci, Ilhan. 2018. Impact of financial leverage on profitability of listed manufacturing firms in China. *Pacific Accounting Review* 30: 410–32. [CrossRef]

Dave, Ashvin, Ashwin Parwani, Ashish Joshi, and Tejas Dave. 2019. A study of capital structure and profitability of Indian steel sector companies. *International Journal of Advanced Science and Technology* 28: 866–73.

DeAngelo, Harry, and Ronald W. Masulis. 1980. Optimal capital structure under corporate and personal taxation. *Journal of Financial Economics* 8: 3–29. [CrossRef]

Donaldson, Gordon. 1984. *Managing Corporate Wealth*. New York: Praeger.

El-Sayed Ebaid, Ibrahim. 2009. The impact of capital-structure choice on firm performance: Empirical evidence from Egypt. *The Journal of Risk Finance* 10: 477–87. [CrossRef]

Fama, Eugene Frank, and Kenneth Richard French. 1998. Value versus growth: The international evidence. *Journal of Finance* 53: 1975–99. [CrossRef]

Fitim, Deari, Matsuk Zoriana, and Lakshina Valeriya. 2019. Leverage and Macroeconomic Determinants: Evidence from Ukraine. *Studies in Business and Economics* 14: 5–19. [CrossRef]

Gharaibeh, Omar Krishna, and Marie Hal Bani Khaled. 2020. Determinants of profitability in Jordanian services companies. *Investment Management and Financial Innovations* 17: 277–90. [CrossRef]

Ghasemi, Maziar, Nazrul Hisyam Ab Razak, and Komeli Dehghani. 2018. Determinants of debt structure in ACE Market Bursa Malaysia: A panel data analysis. *Journal of Social Sciences Research* 2018: 390–95. [CrossRef]

Ghosh, Chinmoy, Raja Nag, and C. F. Sirmans. 2000. The pricing of seasoned equity offerings: Evidence from REITs. *Real Estate Economics* 28: 363–84. [CrossRef]

Gleason, Kimberley Clark, Linette Knowles Mathur, and Ike Mathur. 2000. The interrelationship between culture, capital structure, and performance: Evidence from European retailers. *Journal of Business Research* 50: 185–91. [CrossRef]

Goddard, John, Manouche Tavakoli, and John Oliver Wilson. 2005. Determinants of profitability in European manufacturing and services: Evidence from a dynamic panel model. *Applied Financial Economics* 15: 1269–82. [CrossRef]

Grossman, Sanford John, and Oliver David Hart. 1982. Corporate financial structure and managerial incentives. In *The Economics of Information and Uncertainty*. Chicago: University of Chicago Press, pp. 107–40.

Hadlock, Charles Jones, and Christopher M. James. 2002. Do banks provide financial slack? *The Journal of Finance* 57: 1383–419. [CrossRef]

Helmy, Muhammad Harith Zulqarnain Bin Noor, Goh Chin Fei, Tan Owee Kowang, Ong Choon Hee, Tan Seng Teck, Lim Kim Yew, and Wong Chee Hoo. 2020. Capital structure, internal governance mechanisms and firm performance. *International Journal of Psychosocial Rehabilitation* 24: 7313–21. [CrossRef]

Hussein, Mohammed Jalal, Alrabba Hussein, Muhannad Akram Ahmad, and Mashhoor Hamadneh. 2019. Capital structure and firm performance: Evidence from Jordanian listed companies. *International Journal of Scientific and Technology Research* 8: 364–75.

Jensen, Michael Clark. 1986. Agency cost of free cash flow, corporate finance, and takeovers. *Corporate Finance, and Takeovers. American Economic Review* 76: 323–29.

Jensen, Michael Clark, and William H. Meckling. 1976. Theory of the firm: Managerial behavior, agency costs and ownership structure. *Journal of Financial Economics* 3: 305–60. [CrossRef]

Khan, Tasneem, Mohd Shamim, and Jatin Goyal. 2018. Panel data analysis of profitability determinants: Evidence from Indian telecom companies. *Theoretical Economics Letters* 8: 3581–93. [CrossRef]

Köksal, Bülent, and Cüneyt Orman. 2015. Determinants of capital structure: Evidence from a major developing economy. *Small Business Economics* 44: 255–82. [CrossRef]

Kraus, Alan, and Robert H. Litzenberger. 1973. A state-preference model of optimal financial leverage. *The Journal of Finance* 28: 911–22. [CrossRef]

Levin, Andrew, Chien-Fu Lin, and Chia-Shang James Chu. 2002. Unit root tests in panel data: Asymptotic and finite-sample properties. *Journal of Econometrics* 108: 1–24. [CrossRef]

Li, Hui, and Petros Stathis. 2017. Determinants of Capital Structure in Australia: An Analysis of Important Factors. *Managerial Finance* 43: 881–97. [CrossRef]

M'ng, Jacinta Chan Phooi, Mahfuzur Rahman, and Selvam Sannacy. 2017. The determinants of capital structure: Evidence from public listed companies in Malaysia, Singapore and Thailand. *Cogent Economics and Finance* 5: 1418609. [CrossRef]

Macan Bhaird, Ciarán, and Brian Lucey. 2010. Determinants of capital structure in Irish SMEs. *Small Business Economics* 35: 357–75. [CrossRef]

Mayuri, Tulsi, and Lingesiya Kengatharan. 2019. Determinants of Capital Structure: Evidence from Listed Manufacturing Companies in Sri Lanka. *SCMS Journal of Indian Management* 16: 43–56.

Modigliani, Franco, and Merton H. Miller. 1958. The cost of capital, corporation finance and the theory of investment. *The American Economic Review* 48: 261–97.

Modigliani, Franco, and Merton H. Miller. 1963. Corporate income taxes and the cost of capital: A correction. *The American Economic Review* 53: 433–43.

Musah, Mohammed, and Yusheng Kong. 2019. The relationship between capital structure and the financial performance of non-financial firms listed on the Ghana Stock Exchange (GSE). *International Journal of Research in Social Sciences* 9: 92–123.

Myers, Stewart Charles. 1984. The capital structure puzzle. *The Journal of Finance* 39: 574–92. [CrossRef]

Myers, Stewart C., and Nicholas Samual Majluf. 1984. Corporate financing and investment decisions when firms have information that investors do not have. *Journal of Financial Economics* 13: 187–221. [CrossRef]

Negasa, Tran. 2016. The Effect of Capital Structure on Firms' Profitability (Evidenced from Ethiopian). *Preprints*, 2016070013. [CrossRef]

Nunes, Paulo J. Maçãs, Zélia M. Serrasqueiro, and Tiago N. Sequeira. 2009. Profitability in Portuguese service industries: A panel data approach. *The Service Industries Journal* 29: 693–707. [CrossRef]

Park, Kwangmin, and SooCheong Jang. 2018. Is franchising an additional financing source for franchisors? A Blinder–Oaxaca decomposition analysis. *Tourism Economics* 24: 541–59. [CrossRef]

Pettit, Richardson, and Ronald F. Singer. 1985. Small business finance: A research agenda. *Financial Management* 14: 47–60. [CrossRef]

Qiu, Mei, and Bo La. 2010. Firm characteristics as determinants of capital structures in Australia. *International journal of the Economics of Business* 17: 277–87. [CrossRef]

Rashid, Kashif, and Sardar Islam. 2009. Capital structure and firm performance in the developed financial market. *Corporate Ownership and Control* 7: 189–201. [CrossRef]

Roden, Dianne M., and Wilbur Grand Lewellen. 1995. Corporate Capital Structure Decisions: Evidence from Leveraged Buyouts. *Financial Management* 24: 76–87. [CrossRef]

Ross, Stephen Allen. 1977. The determination of financial structure: The incentive-signalling approach. *The Bell Journal of Economics* 8: 23–40. [CrossRef]

Salim, Mahfuzah, and Raj Yadav. 2012. Capital Structure and Firm Performance: Evidence from Malaysian Listed Companies. *Procedia-Social and Behavioral Sciences* 65: 156–66. [CrossRef]

Sardo, Filipe, Zélia Serrasqueiro, and Elisabete GS Félix. 2020. Does Venture Capital affect capital structure rebalancing? The case of small knowledge-intensive service firms. *Structural Change and Economic Dynamics* 53: 170–79. [CrossRef]

Sermpinis, Georgios, Serafeim Tsoukas, and Ping Zhang. 2019. What influences a bank's decision to go public? *International Journal of Finance and Economics* 24: 1464–85. [CrossRef]

Shalini, Ramaswamy, and Mahua Biswas. 2019. Capital structure determinants of SandP BSE 500: A panel data research. *International Journal of Recent Technology and Engineering* 8: 377–80. [CrossRef]

Sivalingam, Logavathani, and Lingesiya Kengatharan. 2018. Capital Structure and Financial Performance: A Study on Commercial Banks in Sri Lanka. *Asian Economic and Financial Review* 8: 586–98. [CrossRef]

Stulz, René. 1990. Managerial discretion and optimal financing policies. *Journal of Financial Economics* 26: 3–27. [CrossRef]

Suntraruk, Phassawan, and Xiaoxing Liu. 2017. The impacts of institutional characteristics on capital structure: Evidence from listed commercial banks in China. *Afro-Asian Journal of Finance and Accounting* 7: 337–50. [CrossRef]

Szemán, Judith. 2017. Relevance of Capital Structure Theories in the Service Sector. *Theory, Methodology, Practice* 13: 53–64. [CrossRef]

Taub, Allan Jay. 1975. Determinants of the firm's capital structure. *The Review of Economics and Statistics* 57: 410–16. [CrossRef]

Weill, Laurent. 2008. Leverage and corporate performance: Does institutional environment matter? *Small Business Economics* 30: 251–65. [CrossRef]

Yazdanfar, Darush. 2012. The Impact of Financing Pattern on Firm Growth: Evidence from Swedish Micro Firms. *International Business Research* 5: 16. [CrossRef]

Yazdanfar, Darush, and Peter Öhman. 2015. Debt financing and firm performance: An empirical study based on Swedish data. *The Journal of Risk Finance* 16: 102–18. [CrossRef]

6

An Empirical Analysis of the Volatility Spillover Effect between World-Leading and the Asian Stock Markets: Implications for Portfolio Management

Imran Yousaf [1,*], Shoaib Ali [1] and Wing-Keung Wong [2,3,4]

[1] Air University School of Management, Air University, Islamabad 44000, Pakistan; ShoaibAli@mail.au.edu.pk
[2] Department of Finance, Fintech Center, and Big Data Research Center, Asia University, Taichung 41354, Taiwan; wong@asia.edu.tw
[3] Department of Medical Research, China Medical University Hospital, Taichung 40402, Taiwan
[4] Department of Economics and Finance, The Hang Seng University of Hong Kong, Hong Kong 999077, China
* Correspondence: imranyousaf.fin@gmail.com

Abstract: This study employs the Vector Autoregressive-Generalized Autoregressive Conditional Heteroskedasticity (VAR-AGARCH) model to examine both return and volatility spillovers from the USA (developed) and China (Emerging) towards eight emerging Asian stock markets during the full sample period, the US financial crisis, and the Chinese Stock market crash. We also calculate the optimal weights and hedge ratios for the stock portfolios. Our results reveal that both return and volatility transmissions vary across the pairs of stock markets and the financial crises. More specifically, return spillover was observed from the US and China to the Asian stock markets during the US financial crisis and the Chinese stock market crash, and the volatility was transmitted from the USA to the majority of the Asian stock markets during the Chinese stock market crash. Additionally, volatility was transmitted from China to the majority of the Asian stock markets during the US financial crisis. The weights of American stocks in the Asia-US portfolios were found to be higher during the Chinese stock market crash than in the US financial crisis. For the majority of the Asia-China portfolios, the optimal weights of the Chinese stocks were almost equal during the Chinese stock market crash and the US financial crisis. Regarding hedge ratios, fewer US stocks were required to minimize the risk for Asian stock investors during the US financial crisis. In contrast, fewer Chinese stocks were needed to minimize the risk for Asian stock investors during the Chinese stock market crash. This study provides useful information to institutional investors, portfolio managers, and policymakers regarding optimal asset allocation and risk management.

Keywords: return spillover; volatility spillover; shock spillover; US financial crisis; Chinese stock market crash

JEL Classification: G10; G11; G12; G15

---

1. Introduction

Information transmissions from both return and volatility across national equity markets are of greater interest to both investors and policymakers, with increasing financial integration in the stock markets all over the world (Yousaf et al. 2020). If, for example, asset volatility is transmitted from one market to another during turmoil or crisis period (Forbes and Rigobon 2002; Diebold and Yilmaz 2009), then portfolio managers need to adjust their asset allocations (Baele 2005; Engle et al. 2012) and financial policymakers need to adapt their policies in order to mitigate the contagion risk. Changes in linkages between national equity markets, especially during a crisis, can also have important implications for asset allocations, business valuation, risk management, and access to finance.

Several studies have examined linkages between the equity markets during the 1997 Asian financial crisis (In Francis et al. 2001; Wan and Wong 2001; Yang et al. 2003), and the last 2008 global financial crisis (Yilmaz 2010; Cheung et al. 2007; Kim et al. 2015; Li and Giles 2015; Lean et al. 2015; Vieito et al. 2015; Zhu et al. 2019) and some studies, see, for example, Fung et al. (2011) and Guo et al. (2017), develop theories to explain that crisis. However, the linkages between equity markets during the Chinese stock market crash of 2015have been rarely examined. The Chinese stock market experienced a major crash in 2015 (Zhu et al. 2017; Yousaf and Hassan 2019; Yousaf et al. 2020; Yousaf and Ali 2020). The CSI 300 index increased before reaching 5178 points in mid-June of 2015. Then, it took a roller-coaster ride and dropped by up to 34% in just 20 days; Chinese stock market also lost 1000 points within just one week. Around 50% of Chinese stocks lost more than half of their pre-crash market value. The Chinese stock market crash affected many other commodities and financial markets, including Asian (Allen 2015) and the US stock markets (The causes and consequences of China's market crash 2015).

Despite the importance of the Chinese crash for international portfolio managers, few studies have examined how it was transmitted to other national financial markets. Xiong et al. (2018) investigate the time-varying correlation between economic policy uncertainty and Chinese stock market returns during the Chinese crash of 2015, while Yousaf and Hassan (2019) examine the linkages between crude oil and emerging Asian stock markets during this crisis. However, research on the linkages between stock markets has not been investigated yet for the 2015 Chinese crash. Therefore, this study focuses on providing useful insights about this issue for the Asian region, which has attracted considerable attention from finance practitioners and academics due to its position as the center of global economic activity in the 21st century[1]. While using the US and Chinese equity markets as the indicators of global markets, we explore whether global investors can get the maximum benefit of diversification by adding emerging Asian market stocks in their portfolios. In literature, several studies have examined the linkages between the global (US and China) and emerging Asian equity markets during the Asian financial crisis, and the US financial crisis (Yang et al. 2003; Beirne et al. 2013; Jin 2015; Li and Giles 2015), but not in the Chinese stock market crash.

We address the above-mentioned literature gap by examining the return and volatility spillover from the US and China to the emerging Asian equity markets during the Chinese stock market crash by using the VAR-AGARCH model that was developed by Ling and McAleer (2003). Moreover, we examine the ability of spillovers during the full sample period and the 2008 US financial crisis to provide comparative insights to investors about whether the impact of the Chinese crash on equity market spillovers was different from those in the other sample periods. Our findings show that return spillover was observed from the US and China to the Asian stock market during the US financial crisis and the Chinese stock market crash. Volatility was also transmitted from the US to the majority of the Asian stock markets during the Chinese stock market crash. However, volatility was transmitted from China to the majority of the Asian stock markets during the US financial crisis. Overall, as the return and volatility transmission vary across pairs of stock markets and financial crises, investors have to adjust their asset allocations from time to time to improve their profits. Therefore, we also estimate the optimal weights and hedge ratios during the full sample period, the US financial crisis, and the Chinese stock market crash. Our findings imply that fewer US stocks were required to minimize the risk for Asian stock investors during the US financial crisis compared to during the Chinese crash. In contrast, fewer Chinese stocks were needed to minimize the risk for Asian stock investors during the Chinese stock market crash as compared to during the US crisis. Overall, our findings draw several important implications for risk management and portfolio diversification that could be useful for investors and policymakers related to the US and Asian stock markets.

---

[1] Source: https://www.ft.com/content/520cb6f6-2958-11e9-a5ab-ff8ef2b976c7.

The rest of the paper is organized as follows: Section 2 provides the literature review. Section 3 describes the data and methodology. Section 4 reports the findings, and Section 5 concludes the whole discussion.

2. Literature Review

The analysis of both return and volatility spillover between stock markets is crucial for investors in designing optimal portfolios. According to modern portfolio theory, the gains of international portfolio diversification decrease when the correlation of security returns increases and vice versa. Michaud et al. (1996) discuss the advantages of a low correlation between the developed and emerging markets for international portfolio diversification. Due to this trend, investors can benefit by investing in emerging markets that are weakly interconnected with developed markets. However, this correlation becomes higher during an economic crisis, suggesting low diversification benefits when diversification is most required.

2.1. Linkages between US, China, and Asian Stock Markets

Many studies have been conducted to investigate the link between different stock markets during the last three decades. Liu and Pan (1997) examine the mean and the volatility spillover from the US and Japan to Singapore, Hong Kong, Thailand, and Taiwan. The results show that the US market is more dominant than the Japanese stock market in transmitting return and volatility effects to four Asian stock markets. Huang et al. (2000) investigate the link between the US, Japan, and South China growth triangle. The US stock market significantly and dominantly affects the south Chinese growth triangle compared to the impact of Japan on China's stock market. The return spillover has been also found to be significant from the US to Hong Kong and Taiwan, and from Hong Kong to the Taiwanese stock market. Miyakoshi (2003) estimates the return and volatility spillover between the US, Japan, and seven Asian stock markets (South Korea, Taiwan, Singapore, Thailand, Indonesia, and Hong Kong). It finds a significant return spillover from the US to Asian markets, whereas no return spillover is found from Japan to Asian stock markets. Moreover, the volatility spillover from Japan to other Asian stock markets is observed to be dominant as compared to the volatility spillover from the US to Asian stock markets.

Johansson and Ljungwall (2009) examine the association between stock markets of China, Hong Kong, and Thailand. It reports a significant return spillover from Taiwan to China and the Hong Kong stock market. In contrast, volatility spillover runs from Hong Kong to Taiwan and from Taiwan to the Chinese stock market. Zhou et al. (2012) estimate the spillover between Chinese and international (the US, the UK, France, Germany, Japan, India, Hong Kong, Taiwan, South Korea, and Singapore) stock markets from 1996 to 2009. Before 2005, the Chinese stock market was affected by spillover from other international markets. After 2005, volatility spillover was significantly transmitted from China to most of the other international stock markets. Chien et al. (2015) report on the significant financial integration between China and the ASEAN-5 (Indonesia, Malaysia, the Philippines, Singapore, and Thailand) stock markets. Huo and Ahmed (2017) provide significant evidence of both return and volatility effects from China to the Hong Kong equity market.

2.2. Linkages between US, China, and Asian Stock Markets during Crisis

Many studies have examined the linkages between markets during crisis periods. Yang et al. (2003) investigate the short and long-run relationship between the US, Japan, and ten Asian stock markets, mainly focusing on the Asian financial crisis of 1997–1998. This study reports a strengthened long-run co-integration among these stock markets during the Asian financial crises. The degree of integration is found to change during crises and non-crisis periods. Beirne et al. (2013) look at the volatility spillover from developed to emerging stock markets during periods of turbulence in mature stock markets. It finds that volatility in mature markets affects the conditional variances in emerging stock markets.

Moreover, the spillover effect from developed to emerging markets is also changed during times of turbulence in mature markets.

Jin (2015) examines the mean and volatility spillover between China, Taiwan, and Hong Kong. It reveals that financial crises have a substantial and positive effect on expected conditional variances, but also that the size and dynamics of impacts vary from market to market. Li and Giles (2015) investigate the volatility spillover across the US, Japan, and four Asian developing economies during the Asian financial crisis of 1997 and the US financial crises of 2008. The results revealed that there is a presence of a volatility spillover effect from the USA to Asian developing economies and Japan. This study also finds a bidirectional volatility spillover between US and Asian markets that occurred during the Asian financial crisis. Gkillas et al. (2019) explore integration and co-movement between 68 international stock markets (including in the Asian region) during the US financial crisis.

Overall, several studies have examined the return and volatility spillover from the US to Asian markets during the Asian financial crisis of 1997 and the US financial crisis of 2008. However less has been done on both return and volatility transmission from China to the emerging Asian stock markets during the US financial crisis and the Chinese stock market crash. Moreover, no study has examined return and volatility spillovers from the US to the emerging Asian stock markets during the Chinese crash. Therefore, this study addresses these above-mentioned literature gaps.

3. Data and Methodology

3.1. Data

We based our empirical investigation on daily data of accepted benchmark stock indices of nine Asian countries and the US. The Emerging Asian stock markets include China, India, South Korea, Indonesia, Pakistan, Malaysia, the Philippines, Thailand, and Taiwan. The emerging Asian economies were selected from the list of countries, including the MSCI (Morgan Stanley Capital International) emerging market index. The data of stock indices were taken from the Data Stream database. The index is assumed to be the same on non-trading days (holidays except weekends) as on the previous trading day, as suggested by Malik and Hammoudeh (2007) and many others.[2]

This study used the full sample period from 1 January 2000 to 30 June 2018 and studies the following two sub-samples: the first sub-period from 1 August 2007 to 31 July 2010 presenting the period with the US financial crisis; and the second sub-period from 1 June 2015 to 30 May 2018 presenting the period with the Chinese Stock market crash. We note that Yousaf and Hassan (2019) also use similar time frames for the US financial crisis and the Chinese stock market crash. This study followed He (2001) and many others to use three-year data for each crisis for short-run analysis. Changes in market correlations take place continuously not only as a result of crises but also due to the consequences of many financial, economic, and political events. Moreover, Arouri et al. (2015) have also used the daily data covering periods shorter than three years to estimate the return and volatility spillover between gold and Chinese stock markets in US financial crisis by applying the VAR-GARCH model. The difference in the opening time of US and Asian stock markets was adjusted by using lags where necessary.

---

2 In time-series data, if there are missing values, there are two ways to deal with the incomplete data: (a) omit the entire record that contains information, (b) Impute the missing information. We used 10 series in this paper and if we wanted to omit the missing data for one series then the data of all other nine series needed to be removed as well for that specific day. So, if we omitted the data for days where values are missing at specific days, then we lost the data for many days, which is not good for getting realistic results. Therefore, we followed many studies, for example, Malik and Hammoudeh (2007), and imputed the missing data by using previous day data. Indeed, there are many methods used to impute the missing data and every method has pros and cons, but we used this imputation method following past literature. Moreover, our missing observations were less than one percent of overall data, therefore the imputation method should not create a larger effect than that on results.

3.2. Methodology

This study estimated the return and volatility transmissions using the Vector Autoregressive-Generalized Autoregressive Conditional Heteroskedasticity (VAR-AGARCH) model proposed by McAleer et al. (2009). Several studies have previously used the VAR-GARCH and VAR-AGARCH model to estimate spillover between different asset classes (Arouri et al. 2011; Arouri et al. 2012; Jouini 2013; Yousaf and Hassan 2019). This model includes the Constant Conditional Correlation (CCC-GARCH) model of Bollerslev (1990) as a special case. The selection of the model was based on three reasons. First, the most commonly used multivariate models are the BEKK (Baba, Engle, Kraft, and Kroner) model and the DCC (dynamic conditional correlation) model. These models often suffer from unreasonable parameter estimates and data convergence problems (Bouri 2015). The VAR-AGARCH model overcomes these problems regarding parameters and data convergence. Second, it incorporates asymmetry into the model. Third, this model can be used to calculate the optimal weights and hedge ratios.

Ling and McAleer (2003) propose the multivariate VAR-GARCH Model to estimate the return and volatility transmission between the different series. For two series, the VAR-GARCH model has the following specifications for the conditional mean equation[3]:

$$R_t = \mu + FR_{t-1} + e_t \text{ with } e_t = D_t^{1/2}\eta_t, \tag{1}$$

in which R_t represents a 2×1 vector of daily returns[4] on the stocks x and y at time t, μ denotes a 2×1 vector of constants, F is a 2×2 matrix of parameters measuring the impacts of own lagged and cross mean transmissions between two series, e_t is the residual of the mean equation for the two stocks returns series at time t, η_t indicates a 2×1 vector of independently and identically distributed random vectors, and $D_t^{1/2} = diag\ (\sqrt{h_t^x}, \sqrt{h_t^y})$, where h_t^x and h_t^y representing the conditional variances of the returns for stocks x and y, respectively, are given as

$$h_t^x = C_x + a_{11}\left(e_{t-1}^x\right)^2 + a_{21}\left(e_{t-1}^y\right)^2 + b_{11}h_{t-1}^x + b_{21}h_{t-1}^y, \tag{2}$$

$$h_t^y = C_y + a_{12}\left(e_{t-1}^x\right)^2 + a_{22}\left(e_{t-1}^y\right)^2 + b_{12}h_{t-1}^x + b_{22}h_{t-1}^y. \tag{3}$$

Equations (2) and (3) reveal how shock and volatility are transmitted over time and across the markets under investigation. Furthermore, the conditional covariance between returns from two different stock markets can be estimated as follows:

$$h_t^{x,y} = p \times \sqrt{h_t^x} \times \sqrt{h_t^y}. \tag{4}$$

In the above equation, $h_t^{x,y}$ refers to the conditional covariance between the returns of two stock markets $(x,\ y)$ at time t. Moreover, p indicates the constant conditional correlation between the returns of two stock markets $(x,\ y)$.

The VAR–GARCH model assumes that positive or negative shocks have the same impact on the conditional variance. To estimate the spillover between different markets, we estimated spillover between two stock markets by using the VAR–AGARCH Model proposed by the McAleer et al. (2009).

---

[3] Several studies, for example, Hammoudeh et al. (2009), Arouri et al. (2011), and Dutta et al. (2018) have applied the VAR for the conditional mean equation.

[4] $Stock\ Returns_t = ln\left(\frac{Stock\ Index_t}{Stock\ Index_{t-1}}\right)$.

The VAR AGARCH model incorporates asymmetry in the model as well. Specifically, instead of using Equations (2) and (3), the conditional variance of the VAR AGARCH model was defined as follows:

$$h_t^x = C_x + a_{11}A\left(e_{t-1}^x\right)^2 + a_{21}A\left(e_{t-1}^y\right)^2 + b_{11}h_{t-1}^x + b_{21}h_{t-1}^y + a_{11}B\left[\left(e_{t-1}^x\right)\left(\left(e_{t-1}^x\right) < 0\right)\right], \quad (5)$$

$$h_t^y = C_y + a_{12}A\left(e_{t-1}^x\right)^2 + a_{22}A\left(e_{t-1}^y\right)^2 + b_{12}h_{t-1}^x + b_{22}h_{t-1}^y + a_{22}B\left[\left(e_{t-1}^y\right)\left(\left(e_{t-1}^y\right) < 0\right)\right]. \quad (6)$$

In the above equations, $A\left(e_{t-1}^x\right)^2$ and $B\left[\left(e_{t-1}^x\right)\left(\left(e_{t-1}^x\right) < 0\right)\right]$ as well as $A\left(e_{t-1}^y\right)^2$ and $B\left[\left(e_{t-1}^y\right)\left(\left(e_{t-1}^y\right) < 0\right)\right]$ reveal the relationships between a market's volatility and both positive and negative own lagged returns, respectively (Lin et al. 2014). Equations (5) and (6) show that the conditional variance of each market depends upon its past shock and past volatility, as well as the past shock and past volatility of other markets. In Equation (5), $\left(e_{t-1}^x\right)^2$ and $\left(e_{t-1}^y\right)^2$ explain how the past shocks of both x and y affect the current conditional volatility of x. Moreover, h_{t-1}^x and h_{t-1}^y measure how the past volatilities of both x and y affect the current conditional volatility of x. The parameters of the VAR-AGARCH model can be estimated by using the Quasi-Maximum Likelihood estimation (QMLE) and using the BFGS algorithm.[5]

The estimates of the VAR-AGARCH model can be used to calculate optimal portfolio weights. This study followed Kroner and Ng (1998) to calculate the optimal portfolio weights for the pairs of the stock market (x, y) as:

$$w_{xy,t} = \frac{h_{y,t} - h_{xy,t}}{h_{x,y} - 2h_{xy,t} + h_{y,t}} \quad (7)$$

$$w_{xy,t} = \begin{cases} 0 & if\ W_{xy,t} < 0 \\ w_{xy,t} & if\ 0 \leq w_{xy,t} \leq 1, \\ 1 & if\ w_{xy,t} > 1 \end{cases}$$

where $w_{xy,t}$ is the weight of stock(x) in a \$1 stock(x)-stock(y) portfolio at time t, $h_{xy,t}$ is the conditional covariance between the two stock markets, $h_{x,t}$ and $h_{y,t}$ are the conditional variance of stock(x) and stock(y), respectively, and 1-$w_{xy,t}$ is the weight of stock(y) in a \$1 stock(x)-stock(y) portfolio.

It is also essential to estimate the risk-minimizing optimal hedge ratios for the portfolio of different stocks. The estimates of the VAR-AGARCH model can also be used to calculate optimal hedge ratios. This study followed Kroner and Sultan (1993) to calculate the optimal hedge ratios as:

$$\beta_{xy,t} = \frac{h_{xy,t}}{h_{y,t}}, \quad (8)$$

where $\beta_{xy,t}$ represents the hedge ratio. This shows that a short position in the stock (y) market can hedge a long position in the stock (x). Lastly, RATS 10.0 software is used for estimations.

4. Empirical Results

4.1. Descriptive Analysis

Table 1 reports the summary statistics of the daily returns for the US, China, and eight emerging Asian stock markets, namely India, Korea, Indonesia, Pakistan, Malaysia, the Philippines, Thailand, and Taiwan. The average returns of the Pakistani stock market are the highest out of these markets, whereas the lowest returns are found in the US stock market during the full sample period. The unconditional volatility is lower in Malaysia and the US market and is highest in the

---

[5] Arouri et al. (2011), Sadorsky (2012), and Allen et al. (2013) use the Quasi-Maximum Likelihood estimation (QMLE) and use the BFGS algorithm to estimate the parameters in the VAR-GARCH model.

Chinese stock market. The skewness is negative in all cases, kurtosis is higher than 3 for all stocks, and Jarque–Bera statistics do not accept the hypothesis of the normality for all stocks. Moreover, we applied the Ljung–Box Q test for autocorrelation to the standardized residuals and squared standardized residuals. The coefficients both Q(12) and $Q^2(12)$ were found to be signifcant for all series. ARCH effects were also statistically significant for all series.[6]

Table 1. Summary Statistics.

	Mean	Median	Max	Min	Std. Dev.	Skewness	Kurtosis	Jarque-Bera	Q-Stat	ARCH Test
USA	0.00016	0.00055	0.10958	−0.09470	0.01200	−0.20353	11.57202	14802.7 [a]	37.24 [a]	206.42 [a]
CHN	0.00045	0.00096	0.09401	−0.09256	0.01570	−0.31725	8.21506	5547.4 [a]	54.64 [a]	180.10 [a]
IND	0.00050	0.00094	0.15990	−0.11809	0.01472	−0.22234	10.54239	11474.1 [a]	84.62 [a]	283.89 [a]
INDO	0.00046	0.00113	0.07623	−0.10954	0.01357	−0.85402	10.92376	13206.3 [a]	154.0 [a]	457.66 [a]
KOR	0.00028	0.00080	0.11284	−0.12805	0.01509	−0.57337	9.64860	9149.3 [a]	24.06 [a]	210.02 [a]
MYS	0.00020	0.00041	0.04503	−0.09979	0.00816	−0.85496	13.33067	22038.9 [a]	226.8 [a]	267.36 [a]
PAK	0.00081	0.00109	0.08507	−0.07741	0.01359	−0.34875	6.83764	3058.01 [a]	165.5 [a]	594.62 [a]
PHL	0.00038	0.00055	0.16178	−0.13089	0.01309	0.23024	19.78304	56658.3 [a]	96.40 [a]	161.15 [a]
TAIW	0.00018	0.00070	0.06525	−0.09936	0.01356	−0.27454	6.59593	2659.6 [a]	77.68 [a]	201.54 [a]
THA	0.00044	0.00064	0.10577	−0.16063	0.01316	−0.70520	12.86191	19948.5 [a]	70.19 [a]	656.27 [a]

Notes: [a] indicates the statistical significance at 1% level.

4.2. Return, Shock and Volatility Spillover Analysis

4.2.1. Stock Market Linkages between the USA and Asia from the Full Sample Period

Table 2 represents the return and volatility spillover between US and Asian stock markets during the full sample period. The lagged stock returns were found to significantly affect the current stock returns in all studied Asian stock markets except for Korea. This highlights the possibility of short-term predictions of current returns through past returns in the Asian stock markets. Moreover, the autoregressive term of the USA stock market was found to be significant as well. This depicts that past returns help to predict current returns in the American stock market.

The estimate of return spillover from one market to another market can be estimated by using the coefficient of lagged return of one market (i.e., the US) onto another market (i.e., India) and vice versa. The return spillover from the USA to all Asian stock markets is significant. This implies that US stock market prices play an important role in predicting the prices of all Asian stock markets during the full sample period. These results are in line with the findings of Huyghebaert and Wang (2010), which find a significant return spillover from the USA to Asian markets. This shows that the effect of the returns of the American stock market are significantly transmitted to the Asian stock markets. However, the return spillover from all Asian stock markets to the USA was found to be insignificant. This implies that Asian stock market prices are not helpful in predicting the prices of the US stock market during the full sample period.

---

6 We applied both Augmented Dickey–Fuller (ADF), and Phillip–Perron (PP) tests to examine the stationarity of all returns series and found that all returns series are stationary, but we do not report these results in Table form for the sake of brevity.

Table 2. Estimates of bivariate Vector Autoregressive-Generalized Autoregressive Conditional Heteroskedasticity (VAR-AGARCH) for the USA and Asian stock markets during a full sample period.

	IND	USA	INDO	USA	KOR	USA	MYS	USA	PAK	USA	PHL	USA	TAIW	USA	THA	USA
Panel A: Mean Equation																
Constant	$4.90 \times 10^{-4\,a}$ (0.001)	$2.48 \times 10^{-4\,c}$ (0.023)	$5.13 \times 10^{-4\,a}$ (0.001)	$2.40 \times 10^{-4\,b}$ (0.022)	1.55×10^{-4} (0.252)	$2.08 \times 10^{-4\,c}$ (0.059)	1.08×10^{-4} (0.195)	$2.46 \times 10^{-4\,b}$ (0.026)	$8.79 \times 10^{-4\,a}$ (0.000)	$2.07 \times 10^{-4\,b}$ (0.048)	$3.53 \times 10^{-4\,b}$ (0.030)	$2.20 \times 10^{-4\,a}$ (0.001)	1.28×10^{-4} (0.296)	$2.64 \times 10^{-4\,b}$ (0.016)	$6.11 \times 10^{-4\,a}$ (0.000)	$2.71 \times 10^{-4\,c}$ (0.010)
r^{o}_{t-1}	0.100^{a} (0.000)	-7.65×10^{-4} (0.933)	0.136^{a} (0.000)	−0.010 (0.260)	0.017 (0.241)	0.016 (0.126)	0.168^{a} (0.000)	−0.022 (0.232)	0.168^{a} (0.000)	-4.52×10^{-3} (0.648)	0.118^{a} (0.000)	-1.87×10^{-3} (0.132)	0.054^{a} (0.000)	0.014 (0.232)	0.093^{a} (0.000)	-8.99×10^{-3} (0.423)
r^{u}_{t-1}	0.231^{a} (0.000)	-0.041^{a} (0.009)	0.301^{a} (0.000)	-0.032^{b} (0.031)	0.420^{a} (0.000)	-0.037^{b} (0.017)	0.201^{a} (0.000)	-0.034^{b} (0.035)	$6.40 \times 10^{-3\,b}$ (0.017)	-0.030^{b} (0.040)	0.421^{a} (0.000)	-0.030^{b} (0.014)	0.386^{a} (0.000)	-0.037^{b} (0.021)	0.224^{a} (0.000)	-0.037^{b} (0.021)
Panel B: Variance Equation																
Constant	$2.68 \times 10^{-6\,a}$ (0.000)	$1.72 \times 10^{-6\,a}$ (0.000)	$1.46 \times 10^{-5\,a}$ (0.000)	$1.47 \times 10^{-6\,a}$ (0.000)	$1.40 \times 10^{-6\,a}$ (0.000)	$1.71 \times 10^{-6\,a}$ (0.000)	$6.67 \times 10^{-7\,a}$ (0.000)	$1.70 \times 10^{-6\,a}$ (0.000)	$7.61 \times 10^{-6\,a}$ (0.000)	$2.30 \times 10^{-6\,a}$ (0.000)	$3.82 \times 10^{-5\,a}$ (0.000)	$-1.46 \times 10^{-6\,a}$ (0.000)	$1.11 \times 10^{-6\,a}$ (0.000)	$1.64 \times 10^{-6\,a}$ (0.000)	$4.90 \times 10^{-6\,a}$ (0.000)	$1.72 \times 10^{-6\,a}$ (0.000)
$\left(e^{o}_{t-1}\right)^{2}$	0.056^{a} (0.000)	0.038^{a} (0.000)	0.118^{a} (0.000)	0.057^{a} (0.000)	0.053^{a} (0.000)	0.020^{a} (0.001)	0.090^{a} (0.000)	$7.04 \times 10^{-3\,a}$ (0.002)	0.100^{a} (0.000)	$-1.84 \times 10^{-3\,a}$ (0.000)	0.125^{a} (0.000)	0.041^{a} (0.000)	0.037^{a} (0.000)	0.035^{a} (0.000)	0.080^{a} (0.000)	8.93×10^{-3} (0.315)
$\left(e^{u}_{t-1}\right)^{2}$	$8.17 \times 10^{-3\,a}$ (0.001)	-5.68×10^{-3} (0.401)	2.30×10^{-3} (0.538)	-0.011^{c} (0.052)	$6.81 \times 10^{-3\,b}$ (0.032)	−0.011 (0.108)	4.97×10^{-3} (0.524)	-0.011^{c} (0.073)	$5.98 \times 10^{-3\,a}$ (0.000)	$-6.98 \times 10^{-3\,c}$ (0.083)	$-4.96 \times 10^{-3\,a}$ (0.000)	-0.014^{a} (0.000)	6.71×10^{-3} (0.110)	-8.59×10^{-3} (0.228)	$7.89 \times 10^{-3\,b}$ (0.012)	-9.45×10^{-3} (0.119)
h^{o}_{t-1}	0.877^{a} (0.000)	-0.030^{a} (0.000)	0.579^{a} (0.000)	0.085 (0.130)	0.906^{a} (0.000)	−0.018 (0.290)	0.853^{a} (0.000)	2.77×10^{-3} (0.378)	0.791^{a} (0.000)	1.71×10^{-4} (0.369)	0.506^{a} (0.000)	−0.029 (0.142)	0.925^{a} (0.000)	-0.030^{a} (0.000)	0.813^{a} (0.000)	5.47×10^{-3} (0.599)
h^{u}_{t-1}	$-4.98 \times 10^{-3\,b}$ (0.031)	0.896^{a} (0.000)	6.82×10^{-4} (0.915)	0.908^{a} (0.000)	-2.59×10^{-3} (0.407)	0.900^{a} (0.000)	0.010 (0.355)	0.894^{a} (0.000)	$-7.80 \times 10^{-3\,a}$ (0.000)	0.892^{a} (0.000)	0.036^{a} (0.000)	0.903^{a} (0.000)	2.08×10^{-4} (0.968)	0.897^{a} (0.000)	$-4.67 \times 10^{-3\,c}$ (0.076)	0.900^{a} (0.000)
Asymmetry	0.098^{a} (0.000)	0.172^{a} (0.000)	0.199^{a} (0.000)	0.164^{a} (0.000)	0.068^{a} (0.000)	0.173^{a} (0.000)	0.060^{a} (0.000)	0.183^{a} (0.000)	0.143^{a} (0.000)	0.193^{a} (0.000)	0.143^{a} (0.000)	0.167^{a} (0.000)	0.051^{a} (0.000)	0.167^{a} (0.000)	0.157^{a} (0.000)	0.175^{a} (0.000)
Panel C: Constant Conditional Correlation																
$p^{0,u}$	0.198^{a} (0.000)		0.104^{a} (0.000)		0.185^{a} (0.000)		0.082^{a} (0.000)		0.031^{b} (0.030)		0.054^{a} (0.000)		0.142^{a} (0.000)		0.134^{a} (0.000)	
Panel D: Diagnostic Tests																
LogL	30500.4		30678.873		30692.048		33342.351		30597.4		30633.0		30931.2		30683.3	
AIC	−9.520		−9.912		−10.060		−11.074		−10.044		−10.537		−10.030		−10.174	
SIC	−9.126		−9.518		−9.671		−10.680		−9.040		−10.143		−9.780		−9.779	

Table 2. *Cont.*

	IND	USA	INDO	USA	KOR	USA	MYS	USA	PAK	USA	PHL	USA	TAIW	USA	THA	USA
Panel D: Diagnostic Tests																
JB	533.870 [a] (0.000)	381.111 [a] (0.000)	688.360 [a] (0.000)	528.060 [a] (0.000)	870.230 [a] (0.000)	391.040 [a] (0.000)	1921.200 [a] (0.000)	500.410 [a] (0.000)	3442.14 [a] (0.000)	508.010 [a] (0.000)	10842.7 [a] (0.000)	438.390 [a] (0.000)	295.380 [a] (0.000)	462.680 [a] (0.000)	20331.6 [a] (0.000)	560.990 [a] (0.000)
Q (12)	16.143 (0.185)	22.788 (0.303)	9.012 (0.702)	15.021 (0.240)	5.829 (0.924)	18.049 (0.114)	13.179 (0.356)	14.149 (0.291)	55.870 [c] (0.097)	15.244 (0.228)	21.845 [b] (0.039)	15.008 (0.241)	18.533 [c] (0.100)	19.566 [c] (0.076)	33.450 [c] (0.087)	16.279 (0.179)
$Q^2(12)$	13.026 (0.367)	9.413 (0.667)	19.566 [c] (0.076)	8.416 (0.752)	5.242 (0.949)	9.919 (0.623)	14.799 (0.253)	10.027 (0.614)	5.970 (0.918)	0.000 [a] (0.000)	1.823 (0.923)	0.000 [a] (0.000)	16.456 (0.171)	10.976 (0.531)	0.928 (0.965)	11.617 (0.477)

Notes: The number of lags for VAR was decided using SIC (Schwartz information criterion) and AIC (Akaike information criterion) criteria. JB, Q(12), and $Q^2(12)$ indicate the empirical statistics of the Jarque–Bera test for normality, while Ljung–Box Q statistics with order 12 for autocorrelation were applied to the standardized residuals and squared standardized residuals, respectively. USA, United States of America; IND, India; INDO, Indonesia; KOR, South Korea; MYS, Malaysia; PAK, Pakistan; PHL, the Philippines; TAIW, Taiwan; THA, Thailand. Values in parentheses are the *p*-values. [a], [b], [c] indicate the statistical significance at 1%, 5%, and 10%, respectively.

ARCH coefficient captures the shock dependence, while the GARCH coefficient captures the persistence of volatility in conditional variance equations. The findings reveal that the sensitivity of past own shocks (ARCH term) is significantly positive for all Asian Stock Markets in the short run. In addition, the sensitivity of past own volatility (the GARCH term) was found to be significant for all stock markets (including the Asian and American Markets), thus the ARCH (1) volatility model was determined to be more appropriate in this case. The coefficient of past own volatility was than the coefficients of past own shocks in all Asian stock markets, implying that past own volatilities are more critical for prediction of future volatility as compared to past own shocks.

The conditional volatility of India's, South Korea's, the Philippines', Pakistan's, and Thailand's stock markets was found to be significantly affected by shocks in the American stock market. These results are similar to the findings of Syriopoulos et al. (2015), which show that past shocks in the American market significantly affect the market volatility of India, Brazil, and Russia. Therefore, this implies that shock in the American stock market leads to an increase in the volatility of the majority of Asian markets. The past volatility of the American stock market significantly influenced the conditional volatility of India's, The Philippines', Pakistan's, and Thailand's stock markets. These results confirm the previous findings of Li and Giles (2015), which finds a significant volatility spillover from the USA to emerging Asian stock markets. Further, Syriopoulos et al. 2015 found a significant volatility spillover from the USA to India. In addition, the past volatility of the majority of Asian Markets (Except for India and Taiwan) has not been significantly transmitted to the American stock market. The asymmetric coefficients of all Asian stock markets are significant and positive, showing that negative news (or unexpected shocks) for the American stock market has more ability to increase the volatility of all Asian Stock markets as compared to positive news.

Besides, the asymmetric coefficient of the American stock market is positively significant, demonstrating that negative unexpected shocks in Asian Stock markets will increase the volatility more in the American Stock market as compared to a positive shock. Constant conditional correlation (CCC) is positively significant for all pairs of stock markets. However, cross-market correlation is weak in almost all pairs, indicating that investors can get substantial gains by having these pairs in the same portfolio.

4.2.2. Stock Market Linkages between China and Asia from the Full Sample Period

Table 3 reports the return and volatility spillover between the Chinese and other Asian stock markets during the full sample period. The current stock returns of Asian stock markets are significantly affected by their own lagged stock returns. This highlights the possibility of short-term predictions of current returns through past returns in the Asian stock markets. Moreover, Chinese stock returns are also significantly influenced by their own single period lagged returns. These findings depict that stock prices can be predicted in the short term in the Chinese stock market.

The return spillover is not significant from China to the majority of other Asian markets except for the Indian, Philippines, and Thai stock markets. Besides, the return transmission from Asian markets to the Chinese market is insignificant except for in the case of the Indian Stock market. Moreover, there is a presence of bi-directional return transmission between the Indian and Chinese stock markets. This implies that Chinese (Indian) stock market prices play an important role in predicting the prices of Indian (Chinese) stock markets during the full sample period. The coefficient of past own shock of all Asian markets (including China) was found to be significant; thus, past shocks affect current conditional volatility in Asian stock markets. Besides, the sensitivity of past own volatility for all Asian markets was found to be significant as well.

Table 3. Estimates of bivariate VAR–AGARCH for China's and other Asian stock markets during the full sample period.

	IND	CHN	INDO	CHN	KOR	CHN	MYS	CHN	PAK	CHN	PHL	CHN	TAIW	CHN	THA	CHN
Panel A: Mean Equation																
Constant	5.06×10^{-4} [a] (0.000)	3.25×10^{-4} [c] (0.053)	5.37×10^{-4} [a] (0.000)	3.50×10^{-4} [b] (0.022)	2.71×10^{-4} [c] (0.065)	3.26×10^{-4} [b] (0.043)	1.69×10^{-4} [b] (0.036)	3.74×10^{-4} [b] (0.023)	8.64×10^{-4} [a] (0.000)	3.39×10^{-4} [b] (0.029)	3.32×10^{-4} [b] (0.031)	3.50×10^{-4} [b] (0.030)	2.18×10^{-4} [c] (0.089)	3.51×10^{-4} [b] (0.027)	5.05×10^{-4} [a] (0.000)	2.12×10^{-4} (0.139)
r^{o}_{t-1}	0.137 [a] (0.000)	0.036 [a] (0.004)	0.152 [a] (0.000)	0.017 (0.237)	0.083 [a] (0.000)	0.011 (0.345)	0.191 [a] (0.000)	0.018 (0.456)	0.167 [a] (0.000)	6.27×10^{-3} (0.576)	0.144 [a] (0.000)	−0.028 [b] (0.048)	0.100 [a] (0.000)	0.017 (0.241)	0.125 [a] (0.000)	0.018 (0.183)
r^{c}_{t-1}	−0.018 [c] (0.096)	0.049 [a] (0.001)	-6.05×10^{-3} (0.603)	0.050 [a] (0.000)	−0.013 (0.269)	0.055 [a] (0.000)	7.11×10^{-3} (0.295)	0.053 [a] (0.000)	0.010 (0.272)	0.055 [a] (0.000)	0.025 [b] (0.043)	0.055 [a] (0.000)	2.26×10^{-3} (0.841)	0.051 [a] (0.000)	−0.020 (0.103)	0.049 [a] (0.001)
Panel B: Variance Equation																
Constant	2.14×10^{-6} [a] (0.000)	1.14×10^{-6} [a] (0.001)	4.30×10^{-6} [a] (0.000)	1.19×10^{-6} [a] (0.002)	1.00×10^{-6} [a] (0.000)	1.37×10^{-6} [a] (0.000)	5.84×10^{-7} [a] (0.000)	1.26×10^{-6} [a] (0.000)	7.29×10^{-6} [a] (0.000)	1.15×10^{-6} [a] (0.002)	1.36×10^{-5} [a] (0.000)	1.19×10^{-6} [b] (0.048)	9.17×10^{-7} [a] (0.000)	1.21×10^{-6} [a] (0.000)	1.01×10^{-6} [a] (0.001)	1.19×10^{-6} [a] (0.001)
$\left(\varepsilon^{o}_{t-1}\right)^2$	0.051 [a] (0.000)	-1.95×10^{-3} (0.413)	0.074 [a] (0.000)	-1.77×10^{-4} (0.950)	0.031[a] (0.000)	7.03×10^{-3} (0.241)	0.074 [a] (0.000)	3.80×10^{-4} (0.668)	0.100 [a] (0.000)	-9.47×10^{-4} (0.509)	0.053 [a] (0.000)	8.20×10^{-3} [b] (0.016)	0.033 [a] (0.000)	3.74×10^{-3} [b] (0.034)	0.070 [a] (0.000)	0.023 [a] (0.000)
$\left(\varepsilon^{c}_{t-1}\right)^2$	-5.03×10^{-3} [c] (0.081)	0.069 [a] (0.000)	9.29×10^{-3} [c] (0.099)	0.066 [a] (0.000)	-1.32×10^{-3} (0.551)	0.070 [a] (0.000)	-6.92×10^{-3} (0.422)	0.069 [a] (0.000)	3.56×10^{-3} (0.453)	0.064 [a] (0.000)	8.77×10^{-4} (0.754)	0.068 [a] (0.000)	9.94×10^{-3} [a] (0.000)	0.069 [a] (0.000)	7.51×10^{-3} [b] (0.025)	0.062 [a] (0.000)
h^{o}_{t-1}	0.876 [a] (0.000)	9.48×10^{-3} [a] (0.001)	0.840 [a] (0.000)	0.011 [b] (0.041)	0.929 [a] (0.000)	-6.06×10^{-3} [a] (0.000)	0.871 [a] (0.000)	2.91×10^{-3} [b] (0.022)	0.792 [a] (0.000)	1.39×10^{-3} (0.458)	0.797 [a] (0.000)	2.22×10^{-3} (0.656)	0.922 [a] (0.000)	-1.34×10^{-3} (0.508)	0.878 [a] (0.000)	-8.97×10^{-3} [a] (0.001)
h^{c}_{t-1}	7.61×10^{-3} [b] (0.033)	0.920 [a] (0.000)	-4.86×10^{-3} (0.482)	0.923 [a] (0.000)	2.24×10^{-3} (0.314)	0.918 [a] (0.000)	0.010 (0.278)	0.921 [a] (0.000)	-1.29×10^{-3} (0.816)	0.923 [a] (0.000)	1.53×10^{-3} (0.816)	0.921 [a] (0.000)	-8.00×10^{-3} [a] (0.002)	0.921 [a] (0.000)	-3.06×10^{-3} (0.497)	0.919 [a] (0.000)
Asymmetry	0.105 [a] (0.000)	0.016 [b] (0.048)	0.096 [a] (0.000)	0.012 (0.138)	0.067 [a] (0.000)	0.019 [b] (0.023)	0.073 [a] (0.000)	0.016 [c] (0.055)	0.139 [a] (0.000)	0.019 [b] (0.012)	0.096 [a] (0.000)	0.018 [b] (0.035)	0.072 [a] (0.000)	0.016 [b] (0.050)	0.073 [a] (0.000)	0.030 [a] (0.001)
Panel C: Constant Conditional Correlation																
$p^{0,c}$	0.158 [a] (0.000)		0.164 [a] (0.000)		0.211 [a] (0.000)		[a] (0.000)		0.047 [a] (0.001)		0.131 [a] (0.000)		0.218 [a] (0.000)		0.146 [a] (0.000)	
Panel D: Diagnostic Tests																
LogL	28559.5		28690.5		28584.3		31339.5		28776.2		28564.8		28882.9		28821.945	
AIC	9.191		−9.543		−9.621		−10.762		−9.824		−9.920		−9.766		−9.770	
SIC	−8.797		−9.149		−9.227		−10.368		−9.430		−9.520		−9.372		−9.375	

Table 3. *Cont.*

	IND	CHN	INDO	CHN	KOR	CHN	MYS	CHN	PAK	CHN	PHL	CHN	TAIW	CHN	THA	CHN
Panel D: Diagnostic Tests																
JB	574.750 [a] (0.000)	1348.92 [a] (0.000)	1310.07 [a] (0.000)	1391.71 [a] (0.000)	1385.22 [a] (0.000)	705.970 [a] (0.000)	1651.18 [a] (0.000)	1428.96 [a] (0.000)	3079.52 [a] (0.000)	1424.88 [a] (0.000)	1248.20 [a] (0.000)	1408.79 [a] (0.000)	464.070 [a] (0.000)	1238.55 [a] (0.000)	56026.9 [a] (0.000)	1380.10 [a] (0.000)
Q (12)	17.399 (0.135)	48.803 [a] (0.000)	10.709 (0.554)	45.754 [a] (0.000)	9.252 (0.681)	46.376 [a] (0.000)	15.229 (0.229)	44.283 [a] (0.000)	54.789 [a] (0.000)	51.513 [a] (0.000)	14.166 (0.290)	50.473 [a] (0.000)	24.449 [b] (0.018)	39.022 [a] (0.000)	31.740 [a] (0.002)	48.687 [a] (0.000)
$Q^2(12)$	11.249 (0.508)	11.647 (0.474)	17.033 (0.148)	9.322 (0.675)	5.203 (0.951)	11.672 (0.472)	20.075 [c] (0.066)	11.915 (0.453)	6.182 (0.907)	12.065 (0.440)	4.715 (0.967)	10.218 (0.597)	11.938 (0.451)	13.649 (0.324)	2.396 (0.999)	10.561 (0.567)

Notes: The number of lags for VAR was decided using SIC (Schwartz information criterion) and AIC (Akaike information criterion) criteria. JB, Q(12) and $Q^2(12)$ indicated the empirical statistics of the Jarque–Bera test for normality, while Ljung–Box Q statistics of order 12 for autocorrelation applied to the standardized residuals and squared standardized residuals, respectively. CHN, China; IND, India; INDO, Indonesia; KOR, South Korea; MYS, Malaysia; PAK, Pakistan; PHL, the Philippines; TAIW, Taiwan; THA, Thailand. Values in parentheses are the *p*-values. [a], [b], [c] indicate the statistical significance at 1%, 5%, and 10%, respectively.

The conditional volatility of India, Indonesia, Taiwan, and Thailand is significantly affected by shocks in the Chinese market. Also, the conditional volatility of the Chinese market is significantly impacted by the shocks in the Philippines, Taiwanese, and Thai stock markets. The past volatility of the Chinese stock market has not influenced the conditional volatility of the most of the Asian stock markets except for the Indian and Taiwanese stock markets. These findings corroborate with the results of Zhou et al. (2012), which report a significant spillover from China to the Taiwanese stock market. However, the past volatility of the majority Asian markets (except for Pakistan, the Philippines, and Taiwan) significantly affected the conditional volatility of the Chinese stock market.

The asymmetric coefficients of all Asian stock markets were found to be significant and positive, showing that negative news of the Chinese stock market has more of an ability to increase the volatility of all Asian stock markets as compared to positive news. Moreover, the asymmetric coefficient of the Chinese stock market is significant and positive, showing that negative news in Asian markets (except in Indonesia) has a greater ability to increase the volatility of the Chinese market as compared to positive news. Constant conditional correlation is positively significant for all pairs of stock markets, but CCC is weak in majority pairs.

4.2.3. Stock Market Linkages between the USA and Asia from the US Financial Crisis

Table 4 shows the mean and volatility spillover between the USA and Asian stock markets during the US financial crisis. In Asian Stock markets (except for South Korea), past lagged returns significantly influenced the current returns. This highlights the possibility of short-term prediction of current returns through past returns in the Asian stock markets. Moreover, the American stock returns were also significantly influenced by their own single period lagged returns in the majority of cases.

The return spillover effect from the USA to all Asian markets was seen to be significant during the US financial crisis. This implies that US stock market prices played an important role in predicting the prices of all Asian stock markets during the US financial crisis. These results confirm the previous findings of Glick and Hutchison (2013), who reported a significant impact of American equity returns on Asian equity returns during the US financial crisis. Moreover, no single Asian stock market transmitted the return effect to the American market during the US financial crisis. The sensitivity of past own shock was significant for the majority of Asian markets other than Indonesia, Korea, and Taiwan. The coefficient of past own shocks of the American stock market was insignificant in the majority estimations. Besides, the coefficient of own past volatility in all Asian markets was significant except in the Philippines.

The past shocks in the American stock market significantly influenced the conditional volatility of Korea, the Philippines, and Taiwan during the US financial crisis. However, past shocks in most of the Asian stock markets (Except India) have not affected the conditional volatility of the American stock market. The effect of past volatility in the USA on conditional volatility of the Asian stock markets (except Korea) was found to be insignificant. These results match with the findings of Li and Giles (2015), which observe an absence of volatility spillover from the USA to emerging Asian stock markets during the US financial crisis. Moreover, the past volatility of majority Asian stock has not significantly affected American stock market volatility. The asymmetric coefficient of all Asian markets is significant and positive. Moreover, the asymmetric coefficient of the US market is significant and positive in all cases. Constant conditional correlation is positively significant for all pairs of stock markets, but CCC is weak in majority pairs.

Table 4. Estimates of bivariate VAR–AGARCH for the American and Asian stock markets during the US Financial Crisis.

	IND	USA	INDO	USA	KOR	USA	MYS	USA	PAK	USA	PHL	USA	TAIW	USA	THA	USA
Panel A: Mean Equation																
Constant	7.60 × 10^{-4} (0138)	3.49 × 10^{-4} (0.936)	−2.0 × 10^{-5} (0957)	8.03 × 10^{-5} (0.834)	8.5 × 10^{-4} [c] (0.080)	−7.01 × 10^{-5} (0.865)	3.72 × 10^{-4} (0.123)	−1.43 × 10^{-4} (0.742)	9.53 × 10^{-4} [a] (0.005)	3.57 × 10^{-4} (0.262)	9.83 × 10^{-4} [a] (0.001)	4.79 × 10^{-4} [a] (0.066)	1.90 × 10^{-4} (0.684)	−2.64 × 10^{-5} (0.955)	6.00 × 10^{-4} (0.194)	−6.12 × 10^{-5} (0.886)
r^{o}_{t-1}	0.081 [b] (0.031)	0.016 (0.510)	0.031 [b] (0.036)	−0.024 (0.445)	0.044 (0.141)	−0.010 (0.713)	0.145 [a] (0.000)	−0.046 (0.409)	0.125 [a] (0.001)	0.015 (0.648)	0.101 [a] (0.000)	0.016 (0.523)	0.063 [c] (0.076)	0.020 (0.565)	0.065 [b] (0.027)	−0.012 (0.663)
r^{u}_{t-1}	0.293 [a] (0.000)	−0.093 [b] (0.016)	0.423 [a] (0.000)	−0.062 [a] (0.091)	0.325 [a] (0.000)	−0.079 [b] (0.041)	0.187 [a] (0.000)	−0.080 [b] (0.042)	0.175 [a] (0.000)	0.016 (0.706)	0.489 [a] (0.000)	1.60 × 10^{-3} (0.963)	0.383 [a] (0.000)	−0.103 [b] (0.011)	0.249 [a] (0.000)	−0.077 [b] (0.034)
Panel B: Variance Equation																
Constant	2.91 × 10^{-6} (0.161)	3.39 × 10^{-6} [a] (0.001)	1.33 × 10^{-5} [a] (0.000)	5.10 × 10^{-6} [a] (0.000)	3.98 × 10^{-5} [a] (0.000)	4.34 × 10^{-6} [a] (0.000)	1.80 × 10^{-6} [b] (0.038)	3.40 × 10^{-6} [a] (0.001)	4.22 × 10^{-6} [a] (0.007)	3.93 × 10^{-6} [a] (0.000)	3.87 × 10^{-5} [a] (0.000)	2.98 × 10^{-6} [a] (0.000)	1.15 × 10^{-6} (0.352)	3.73 × 10^{-6} [a] (0.009)	3.72 × 10^{-5} [a] (0.000)	4.59 × 10^{-8} (0.984)
$(e^{o}_{t-1})^2$	0.086 [a] (0.000)	0.075 [b] (0.016)	0.198 [a] (0.000)	0.001 (0.877)	0.012 [a] (0.000)	0.077 [c] (0.078)	0.119 [a] (0.000)	−7.24 × 10^{-4} (0.879)	0.036 [b] (0.034)	7.53 × 10^{-3} (0.439)	0.283 (0.140)	0.051 [b] (0.044)	4.98 × 10^{-3} (0.761)	0.029 (0.230)	0.151 [a] (0.000)	0.015 (0.678)
$(e^{u}_{t-1})^2$	6.65 × 10^{-3} [a] (0.053)	−0.01 × 10^{-3} (0.337)	0.011 (0.138)	−0.029 [b] (0.014)	7.71 × 10^{-3} [b] (0.052)	0.001 (0.917)	0.051 (0.213)	−0.013 (0.361)	0.018 (0.146)	−0.060 [a] (0.001)	−4.01 × 10^{-3} [b] (0.033)	−0.061 [a] (0.000)	0.026 [c] (0.056)	−0.022 (0.113)	−5.15 × 10^{-3} (0.690)	−6.17 × 10^{-3} (0.661)
h^{o}_{t-1}	−0.868 [a] (0.000)	−0.048 [b] (0.012)	0.750 [a] (0.000)	0.060 (0.208)	0.426 [a] (0.000)	−0.012 (0.330)	0.755 [a] (0.000)	0.013 [c] (0.090)	0.854 [a] (0.000)	8.26 × 10^{-3} (0.470)	0.099 (0.196)	0.043 (0.171)	0.951 [a] (0.000)	−0.029 [c] (0.062)	0.449 [a] (0.000)	0.110 [b] (0.024)
h^{u}_{t-1}	2.17 × 10^{-3} [c] (0.087)	0.894 [a] (0.000)	5.62 × 10^{-4} (0.765)	0.945 [a] (0.000)	−0.176 (0.394)	0.906 [a] (0.000)	−0.035 (0.312)	0.915 [a] (0.000)	−0.021 (0.322)	0.907 [a] (0.000)	8.49 × 10^{-3} (0.115)	0.906 [a] (0.000)	−0.017 (0.382)	0.910 [a] (0.000)	0.031 (0.299)	0.905 [a] (0.000)
Asymmetry	0.064 [c] (0.090)	0.179 [a] (0.000)	0.265 [a] (0.000)	0.159 [a] (0.000)	0.446 [a] (0.000)	0.161 [a] (0.000)	0.163 [a] (0.004)	0.154 [a] (0.000)	0.101 [a] (0.004)	0.237 [a] (0.000)	0.523 [a] (0.000)	0.233 [a] (0.000)	0.075 [a] (0.000)	0.169 [a] (0.000)	0.187 [a] (0.000)	0.147 [a] (0.000)
Panel C: Constant Conditional Correlation																
$p^{0,u}$	0.172 [a] (0.000)		0.280 [a] (0.000)		0.220 [a] (0.000)		0.195 [a] (0.000)		0.016 [c] (0.086)		0.065 [b] (0.044)		0.187 [a] (0.000)		0.218 [a] (0.000)	
Panel D: Diagnostic Tests																
LogL	4230.27		4320.18		4436.36		4843.96		4850.88		6628.56		4391.36		4399.67	
AIC	−10.066		−10.502		−10.394		−11.480		−11.499		−11.518		−10.601		−10.710	
SIC	-9.773		−10.209		−10.101		−11.187		−11.205		−11.224		−10.308		−10.417	

Table 4. *Cont.*

	IND	USA	INDO	USA	KOR	USA	MYS	USA	PAK	USA	PHL	USA	TAIW	USA	THA	USA
Panel D: Diagnostic Tests																
JB	513.31 [a] (0.000)	372.23 [a] (0.000)	679.01 [a] (0.000)	512.65 [a] (0.000)	843.871 [a] (0.000)	399.862 [a] (0.000)	321.220 [a] (0.000)	627.410 (0.000)	676.080 [a] (0.000)	654.370 [a] (0.000)	1598.28 [a] (0.000)	629.470 [a] (0.000)	352.580 [a] (0.000)	635.280 [a] (0.000)	330.040 [a] (0.000)	643.320 [a] (0.000)
Q (12)	14.588 (0.264)	8.939 (0.708)	11.706 (0.469)	6.002 (0.915)	12.006 (0.445)	7.401 (0.829)	11.897 (0.454)	6.280 (0.901)	17.579 (0.129)	9.780 (0.635)	9.776 (0.636)	7.878 (0.795)	14.047 (0.298)	5.797 (0.926)	13.234 (0.352)	5.468 (0.941)
$Q^2(12)$	7.779 (0.802)	13.979 (0.301)	5.303 (0.947)	14.983 (0.242)	6.336 (0.898)	13.673 (0.322)	5.422 (0.942)	14.540 (0.268)	6.417 (0.894)	32.448 [a] (0.001)	2.113 (0.999)	20.023 [c] (0.067)	6.158 (0.908)	13.940 (0.305)	8.596 (0.737)	14.294 (0.282)

Notes: The number of lags for VAR was decided using SIC (Schwartz information criterion) and AIC (Akaike information criterion) criteria. JB, Q(12) and $Q^2(12)$ indicate the empirical statistics of the Jarque–Bera test for normality, while Ljung–Box Q statistics of order 12 for autocorrelation were applied to the standardized residuals and squared standardized residuals, respectively. USA, United States of America; IND, India; INDO, Indonesia; KOR, South Korea; MYS, Malaysia; PAK, Pakistan; PHL, the Philippines; TAIW, Taiwan; THA, Thailand. Values in parentheses are the *p*-values. [a], [b], [c] indicate the statistical significance at 1%, 5%, and 10%, respectively.

4.2.4. Stock Market Linkages between China and Asia during the US Financial Crisis

Table 5 reports the return and volatility spillover between China and Asian stock markets during the US financial crisis. The current stock returns of the majority of Asian stock markets (Except in South Korea) are significantly affected by their own lagged stock returns. This highlights the possibility of short-term prediction of current returns through past returns in the Asian stock markets. However, Chinese stock returns were not significantly affected by their lagged returns during the US financial crisis. Therefore, there is no evidence of Chinese stock price prediction being possible through lagged values during the US financial crisis.

The return transmission effect from China to all Asian markets was insignificant during the US financial US crisis. However, most of the Asian markets did not transmit the return effect to the Chinese stock market other than India, Indonesia, and Malaysia. The coefficient of past own shock was found to be significant in the majority of Asian markets except for Indonesia, Korea, and Pakistan. The sensitivity to past own shocks from Chinese stock markets was found to be insignificant in the majority of markets during the US financial crisis. Moreover, the sensitivity of past own volatility in all Asian markets was significant. The past shocks of China did not influence the conditional volatility of the majority of Asian stock markets (except India) during the US financial crisis. The conditional volatility of the Chinese stock market was not affected by shocks in most of the Asian stock markets (except for Indonesia and Thailand).

There is no significant evidence of volatility spillover from Chinese to Asian stock markets except in India and Taiwan. Besides, the volatility spillover was insignificant in the majority of Asian markets (except Indonesia, Pakistan, and The Philippines) to the Chinese stock market. The asymmetric coefficient of all Asian markets is significant and positive. Moreover, the asymmetric coefficient of China is asymmetric, showing that that negative news of all Asian stock markets (except Pakistan) has more ability to increase the volatility of the Chinese stock market as compared to positive news. Constant conditional correlation is positively significant for all pairs of stock markets. However, CCC has a medium level in the majority of pairs.

4.2.5. Stock Market Linkages between the USA and Asia from the Chinese Stock Market Crash

Table 6 reports the mean and volatility spillover between the USA and Asian stock markets during the Chinese stock market crash. The autoregressive term of Asian market returns (Except Korea, the Philippines, and Taiwan) can be seen to be significant in the majority of stock markets. This shows the short-term predictability in stock price changes in the Asian stock markets. In addition, stock returns of the American stock market were significantly influenced by their lagged returns during the Chinese Crisis.

Table 5. Estimates of bivariate VAR–AGARCH for China and Asian stock markets during the US Financial Crisis.

	IND	CHN	INDO	CHN	KOR	CHN	MYS	CHN	PAK	CHN	PHL	CHN	TAIW	CHN	THA	CHN
Panel A: Mean Equation																
Constant	6.04 × 10^{-4} (0.2680	−2.82 × 10^{-4} (0.671)	6.72 × 10^{-4} (0.220)	−2.28 × 10^{-4} (0.712)	−7.49 × 10^{-5} (0.867)	−2.18 × 10^{-5} (0.974)	4.34 × 10^{-4} (0.146)	−5.07 × 10^{-5} (0.943)	7.61 × 10^{-4} [b] (0.017)	1.87 × 10^{-4} (0.684)	2.57 × 10^{-4} (0.615)	−8.84 × 10^{-5} (0.895)	2.65 × 10^{-4} (0.580)	3.46 × 10^{-5} (0.954)	7.98 × 10^{-4} [c] (0.091)	9.09 × 10^{-5} (0.876)
r^{ρ}_{t-1}	0.147 [a] (0.000)	0.109 [a] (0.002)	0.146 [a] (0.000)	0.136 [a] (0.001)	0.052 (0.211)	8.15 × 10^{-3} (0.861)	0.216 [a] (0.000)	0.196 [a] (0.008)	0.117 [a] (0.000)	−0.032 (0.451)	0.138 [a] (0.001)	5.75 × 10^{-4} (0.990)	0.102 [a] (0.007)	0.017 (0.689)	0.128 [a] (0.000)	0.045 (0.268)
r^{c}_{t-1}	−0.044 (0.135)	0.022 (0.576)	−9.00 × 10^{-3} (0.728)	0.019 (0.614)	0.011 (0.675)	0.067 [c] (0.086)	9.47 × 10^{-3} (0.534)	0.030 (0.432)	0.029 (0.259)	0.031 (0.432)	−7.18 × 10^{-3} (0.773)	0.043 (0.282)	0.015 (0.570)	0.052 (0.183)	−0.025 (0.233)	0.054 (0.147)
Panel B: Variance Equation																
Constant	5.43 × 10^{-6} [c] (0.064)	1.43 × 10^{-5} [a] (0.002)	3.74 × 10^{-5} [a] (0.001)	3.62 × 10^{-6} (0.389)	3.52 × 10^{-6} (0.101)	1.61 × 10^{-5} [a] (0.004)	2.63 × 10^{-6} [c] (0.087)	1.48 × 10^{-5} [a] (0.004)	−2.77 × 10^{-6} (0.365)	3.62 × 10^{-6} [c] (0.064)	5.69 × 10^{-5} [a] (0.000)	1.70 × 10^{-5} [b] (0.043)	1.84 × 10^{-6} (0.223)	1.10 × 10^{-5} [a] (0.050)	1.50 × 10^{-5} [a] (0.007)	8.77 × 10^{-6} (0.144)
$(e^{\rho}_{t-1})^2$	0.067 [a] (0.001)	7.41 × 10^{-3} (0.446)	5.31 × 10^{-3} (0.821)	−0.025 [b] (0.013)	−0.016 (0.271)	2.75 × 10^{-3} (0.636)	0.112 [a] (0.000)	−6.75 × 10^{-4} (0.819)	−0.018 (0.269)	5.44 × 10^{-3} (0.489)	0.070 [a] (0.001)	1.27 × 10^{-3} (0.870)	0.022 (0.108)	7.35 × 10^{-3} (0.229)	0.055 [b] (0.019)	−0.011 [b] (0.044)
$(e^{c}_{t-1})^2$	−7.19 × 10^{-3} [b] (0.017)	−2.74 × 10^{-3} (0.870)	−6.98 × 10^{-3} (0.748)	0.027 [c] (0.070)	0.015 (0.398)	9.48 × 10^{-3} (0.587)	0.100 (0.113)	0.020 (0.265)	−7.46 × 10^{-3} (0.613)	0.025 [b] (0.047)	1.48 × 10^{-3} (0.864)	0.017 (0.391)	−0.018 (0.184)	0.013 (0.388)	−6.48 × 10^{-3} (0.704)	0.012 (0.468)
h^{ρ}_{t-1}	0.893 [a] (0.000)	−0.016 (0.210)	0.623 [a] (0.000)	0.073 [c] (0.049)	0.919 [a] (0.000)	−7.47 × 10^{-4} (0.924)	0.789 [a] (0.000)	3.24 × 10^{-3} (0.615)	0.844 [a] (0.000)	0.064 [a] (0.015)	0.726 [a] (0.000)	−0.042 [b] (0.023)	0.936 [a] (0.000)	−5.07 × 10^{-3} (0.540)	0.834 [a] (0.000)	4.85 × 10^{-3} (0.637)
h^{c}_{t-1}	0.030 [a] (0.002)	0.865 [a] (0.000)	0.060 (0.157)	0.896 [a] (0.000)	1.15 × 10^{-3} (0.952)	0.879 [a] (0.000)	7.69 × 10^{-3} (0.916)	0.865 [a] (0.000)	2.14 × 10^{-3} (0.909)	0.952 [a] (0.000)	0.016 (0.504)	0.867 [a] (0.000)	0.041 [c] (0.099)	0.885[a] (0.000)	0.044 (0.128)	0.876 [a] (0.000)
Asymmetry	0.091 [a] (0.003)	0.151 [a] (0.000)	0.361 [a] (0.000)	0.054 [b] (0.044)	0.152 [a] (0.000)	0.120 [a] (0.001)	0.149 [a] (0.006)	0.109 [a] (0.001)	0.167 [a] (0.000)	0.010 (0.451)	0.153 [a] (0.005)	0.126 [a] (0.000)	0.059 [a] (0.005)	0.116 [a] (0.000)	0.124 [a] (0.001)	0.130 [a] (0.000)
Panel C: Constant Conditional Correlation																
$p^{0,c}$	0.307 [a] (0.000)		0.290 [a] (0.000)		0.401 [a] (0.000)		0.287 [a] (0.000)		0.103 [a] (0.001)		0.270 [a] (0.000)		0.373 [a] (0.000)		0.269 [a] (0.000)	
Panel D: Diagnostic Tests																
LogL	3980.79		4085.73		4161.31		4570.93		4646.76		4104.40		4125.88		4144.48	
AIC	−9.451		−9.746		−9.840		−10.871		−11.412		−9.940		−9.990		−10.001	
SIC	−9.158		−9.453		−9.547		−10.578		−11.118		−9.650		−9.770		−9.707	

Table 5. *Cont.*

	IND	CHN	INDO	CHN	KOR	CHN	MYS	CHN	PAK	CHN	PHL	CHN	TAIW	CHN	THA	CHN
Panel D: Diagnostic Tests																
JB	501.810 ^{a} (0.000)	693.810 ^{a} (0.000)	541.320 ^{a} (0.000)	686.040 ^{a} (0.000)	374.450 ^{a} (0.000)	673.900 ^{a} (0.000)	541.990 ^{a} (0.000)	694.160 (0.000)	590.770 ^{a} (0.000)	756.610 ^{a} (0.000)	1903.28 ^{a} (0.000)	678.760 ^{a} (0.000)	356.750 ^{a} (0.000)	671.310 ^{a} (0.000)	358.350 ^{a} (0.000)	686.450 ^{a} (0.000)
Q (12)	14.715 (0.257)	7.454 (0.826)	11.152 (0.516)	7.664 (0.811)	12.457 (0.410)	8.501 (0.745)	11.203 (0.512)	6.933 (0.862)	18.800 ^{c} (0.093)	9.775 (0.636)	11.248 (0.508)	7.940 (0.790)	13.853 (0.310)	8.794 (0.720)	12.923 (0.375)	7.812 (0.800)
Q^2(12)	8.980 (0.705)	13.615 (0.326)	5.438 (0.942)	14.090 (0.295)	6.297 (0.900)	14.103 (0.294)	6.502 (0.889)	14.401 (0.276)	6.331 (0.899)	20.886 ^{b} (0.052)	27.233 ^{a} (0.007)	13.701 (0.320)	5.748 (0.928)	15.255 (0.228)	6.937 (0.862)	14.507 (0.270)

Notes: The number of lags for VAR was decided using SIC (Schwartz information criterion) and AIC (Akaike information criterion) criteria. JB, Q(12) and Q^2(12) indicate the empirical statistics of the Jarque–Bera test for normality, while Ljung–Box Q statistics of order 12 for autocorrelation were applied to the standardized residuals and squared standardized residuals, respectively. USA, United States of America; IND, India; INDO, Indonesia; KOR, South Korea; MYS, Malaysia; PAK, Pakistan; PHL, the Philippines; TAIW, Taiwan; THA, Thailand. Values in parentheses are the *p*-values. ^{a}, ^{b}, ^{c} indicate the statistical significance at 1%, 5%, and 10%, respectively.

Table 6. Estimates of bivariate VAR–AGARCH for American and Asian stock markets during the Chinese Stock Market Crash.

	IND	USA	INDO	USA	KOR	USA	MYS	USA	PAK	USA	PHL	USA	TAIW	USA	THA	USA
Panel A: Mean Equation																
Constant	3.17 × 10^{-4} (0.227)	5.16 × 10^{-4} ^{b} (0.020)	−8.21 × 10^{-5} ^{a} (0.743)	4.52 × 10^{-4} ^{b} (0.047)	2.08 × 10^{-4} (0.395)	5.26 × 10^{-4} ^{b} (0.013)	−5.44 × 10^{-5} (0.721)	4.01 × 10^{-4} ^{b} (0.055)	3.33 × 10^{-4} (0.225)	4.55 × 10^{-4} ^{b} (0.030)	−3.09 × 10^{-4} (0.341)	4.77 × 10^{-4} ^{b} (0.029)	−3.28 × 10^{-6} (0.988)	5.05 × 10^{-4} ^{b} (0.011)	1.19 × 10^{-4} (0.557)	5.36 × 10^{-4} ^{a} (0.005)
r^{o}_{t-1}	0.077 ^{b} (0.032)	−0.044 (0.166)	0.107 ^{a} (0.001)	−0.028 (0.359)	−0.020 (0.560)	−0.018 (0.588)	0.089 ^{b} (0.013)	0.022 (0.685)	0.284 ^{a} (0.000)	−0.020 (0.367)	0.032 (0.342)	−0.032 (0.188)	0.031 (0.359)	0.076 (0.120)	0.147 ^{a} (0.000)	−0.024 (0.519)
r^{u}_{t-1}	0.233 ^{a} (0.000)	−0.082 ^{b} (0.024)	0.241 ^{a} (0.000)	−0.069 ^{c} (0.065)	0.351 ^{a} (0.000)	−0.079 ^{b} (0.016)	0.234 ^{a} (0.000)	−0.070 ^{c} (0.080)	0.088 ^{b} (0.016)	−0.077 ^{b} (0.044)	0.416 ^{a} (0.000)	−0.077 ^{b} (0.046)	0.405 ^{a} (0.000)	−0.081 ^{b} (0.030)	0.148 ^{a} (0.000)	−0.054 (0.173)
Panel B: Variance Equation																
Constant	3.75 × 10^{-6} ^{b} (0.020)	4.42 × 10^{-6} ^{a} (0.000)	2.05 × 10^{-6} ^{a} (0.005)	4.16 × 10^{-6} ^{a} (0.000)	1.11 × 10^{-5} ^{a} (0.001)	−1.41 × 10^{-6} (0.670)	3.61 × 10^{-7} ^{c} (0.092)	3.28 × 10^{-6} ^{a} (0.004)	4.09 × 10^{-6} ^{a} (0.000)	4.13 × 10^{-6} ^{a} (0.000)	4.62 × 10^{-6} ^{b} (0.025)	1.00 × 10^{-6} (0.714)	2.60 × 10^{-6} ^{a} (0.008)	3.85 × 10^{-6} ^{a} (0.000)	1.12 × 10^{-6} ^{a} (0.001)	7.03 × 10^{-7} (0.518)
$\left(e^{o}_{t-1}\right)^2$	−0.034 ^{b} (0.037)	0.079 ^{a} (0.000)	2.40 × 10^{-3} (0.911)	0.086 ^{a} (0.003)	0.149 ^{a} (0.001)	0.073 (0.270)	0.088 ^{a} (0.000)	0.018 (0.110)	−0.018 (0.402)	0.011 (0.425)	0.032 (0.277)	0.063 (0.130)	−0.020 (0.271)	0.071 (0.160)	−0.023 (0.146)	0.020 (0.311)
$\left(e^{u}_{t-1}\right)^2$	1.73 × 10^{-3} (0.915)	0.051 (0.102)	−6.47 × 10^{-4} (0.950)	0.065 ^{c} (0.064)	9.22 × 10^{-3} (0.738)	0.020 (0.534)	−0.077 ^{a} (0.000)	0.046 (0.216)	4.99 × 10^{-3} (0.296)	0.015 (0.640)	−0.014 (0.145)	0.037 (0.343)	0.030 (0.162)	0.068 ^{c} (0.059)	0.061 ^{b} (0.046)	0.014 (0.650)
h^{o}_{t-1}	0.896 ^{a} (0.000)	−0.066 ^{a} (0.000)	0.891 ^{a} (0.000)	−0.058 (0.202)	0.511 ^{a} (0.000)	0.049 (0.243)	0.865 ^{a} (0.000)	−7.32 × 10^{-3} (0.556)	0.856 ^{a} (0.000)	−0.020 (0.155)	0.870 ^{a} (0.000)	−0.051 ^{a} (0.004)	0.890 ^{a} (0.000)	−0.054 ^{a} (0.001)	0.913 ^{a} (0.000)	−0.017 (0.219)

Table 6. *Cont.*

	IND	USA	INDO	USA	KOR	USA	MYS	USA	PAK	USA	PHL	USA	TAIW	USA	THA	USA
Panel B: Variance Equation																
h^u_{t-1}	−0.012 (0.644)	0.754 [a] (0.000)	0.021 (0.306)	0.696 [a] (0.000)	0.204 (0.149)	0.672 [a] (0.000)	0.303 [a] (0.002)	0.642 [a] (0.000)	−0.012 [a] (0.002)	0.797 [a] (0.000)	0.070 (0.166)	0.711 (0.290)	−0.018 (0.515)	0.726 [a] (0.000)	0.157 [c] (0.072)	0.645 [a] (0.000)
Asymmetry	0.126 [a] (0.002)	0.270 [a] (0.000)	0.123 [a] (0.001)	0.314 [a] (0.000)	−0.079 (0.183)	0.346 [a] (0.000)	0.040 (0.231)	0.317 [a] (0.000)	0.253 [a] (0.000)	0.258 [a] (0.000)	0.081 [b] (0.035)	0.338 [a] (0.000)	0.131 [a] (0.001)	0.296 [a] (0.000)	0.164 [a] (0.000)	0.343 [a] (0.000)
Panel C: Constant Conditional Correlation																
$p^{0,u}$	0.324 [a] (0.000)		0.123 [a] (0.000)		0.281 [a] (0.000)		0.146 [a] (0.000)		0.059 [a] (0.078)		0.103 [a] (0.001)		0.214 [a] (0.000)		0.224 [a] (0.000)	
Panel D: Diagnostic Tests																
LogL	5580.558		5500.464		5640.723		5895.388		5471.88		5390.25		5605.08		5699.52	
AIC	−13.712		−13.614		−14.028		−14.525		−13.330		−13.434		−13.877		−13.881	
SIC	−13.418		−13.321		−13.734		−14.232		−13.037		−13.140		−13.583		−13.587	
JB	72.640 [a] (0.000)	204.890 [a] (0.000)	58.520 [a] (0.000)	412.790 [a] (0.000)	134.540 [a] (0.000)	168.440 [a] (0.000)	22.680 [a] (0.000)	442.380 [a] (0.000)	30.880 [a] (0.000)	428.120 [a] (0.000)	27.070 [a] (0.000)	422.700 [a] (0.000)	84.480 [a] (0.000)	230.350 [a] (0.000)	68.060 [a] (0.000)	154.390 [a] (0.000)
Q (12)	16.833 (0.156)	8.760 (0.723)	12.383 (0.415)	9.384 (0.670)	4.583 (0.970)	10.350 (0.585)	7.511 (0.822)	9.705 (0.642)	11.341 (0.500)	10.270 (0.592)	9.383 (0.670)	12.352 (0.418)	20.039 [c] (0.066)	13.680 (0.322)	6.178 (0.907)	7.190 (0.845)
Q^2(12)	7.383 (0.831)	3.713 (0.988)	18.894 [c] (0.091)	8.117 (0.776)	8.144 (0.774)	5.426 (0.942)	8.099 (0.777)	5.282 (0.948)	7.532 (0.821)	5.161 (0.952)	12.838 (0.381)	7.290 (0.838)	7.271 (0.839)	8.802 (0.720)	24.298 [b] (0.019)	5.675 (0.932)

Notes: The number of lags for VAR was decided using SIC (Schwartz information criterion) and AIC (Akaike information criterion) criteria. JB, Q(12) and Q^2(12) indicate the empirical statistics of the Jarque–Bera test for normality, while Ljung–Box Q statistics of order 12 for autocorrelation were applied to the standardized residuals and squared standardized residuals, respectively. USA, United States of America; IND, India; INDO, Indonesia; KOR, South Korea; MYS, Malaysia; PAK, Pakistan; PHL, the Philippines; TAIW, Taiwan; THA, Thailand. Values in parentheses are the *p*-values. [a], [b], [c] indicate the statistical significance at 1%, 5%, and 10%, respectively.

The return spillover from the USA to all Asian Stock markets was significant during the Chinese crisis. This implies that US stock market prices played an important role in predicting the prices of all Asian stock markets during the Chinese stock market crash. Moreover, the return spillover from Asia to the US stock market was insignificant. The coefficient of past shock was insignificant in the majority of Asian markets except for in India, Korea, and Malaysia. The sensitivity of past own shocks of the USA was insignificant in most of the cases. In addition, the coefficient of past own volatility significantly affected the conditional volatility of all Asian markets.

The conditional volatility of the majority of Asian stock markets (except Malaysia and Thailand) was not significantly affected by the shocks in the US stock market. In addition, past shocks in most of the Asian stock markets (Except in India and Indonesia) did not influence the conditional volatility of the US stock market. The volatility transmission from the USA to most of the Asian stock markets (except Malaysia, Pakistan, and Thailand) was found to be insignificant during the Chinese Crisis. On the other hand, volatility spillover from most of the Asian stock markets to the USA stock market was evidently insignificant.

The asymmetric coefficients of all Asian stock markets (except Korea and Malaysia) were significant and positive, showing that negative news from the US stock market has a greater ability to increase the volatility of all Asian Stock markets as compared to positive news. However, the asymmetric coefficient of the US stock market is significant and positive. Constant conditional correlation was positively significant for all pairs of stock markets. However, CCC was weak in the majority of pairs.

4.2.6. Stock Market Linkages between China and Asia from the Chinese Stock Market Crash

Table 7 reports the return and volatility spillover between Chinese and Asian stock markets during the Chinese stock market crash. There is significant evidence that lagged returns influence the current stock returns of Asian Stock markets (Except in Korea, the Philippines, and Taiwan). This shows the short-term predictability in stock price changes in the Asian stock markets. Moreover, Chinese stock market returns were not affected by their lags during the Chinese stock market crash.

The return spillover was found to be insignificant from China to all Asian markets. However, the return spillover was found to be insignificant from the majority of Asian markets to the Chinese market, except for India and Taiwan, during the Chinese stock market crash. The coefficient of past own shock did not significantly influence the conditional variance of the most of the Asian stock markets except in India, Malaysia, and Thailand. Moreover, the sensitivity to past own shock of the Chinese stock market was insignificant during the Chinese crash. However, the sensitivity of past own volatility was found to be significant for all Asian stock markets.

The conditional volatility of India, Indonesia, Taiwan, and Thailand was significantly affected by the shocks in the Chinese stock market. However, the shocks in the majority of Asian stock markets (except India and the Philippines) did not influence the Chinese stock market. The past volatility of China significantly impacted the conditional volatility of the stock markets of India, Indonesia, Taiwan, and Thailand. However, volatility spillover was not found from most of the Asian stock markets (except India, Taiwan, and Thailand) to the Chinese stock market during the Chinese stock market crash.

The asymmetric coefficients of all Asian stock markets (except Malaysia and the Philippines) were significant and positive, showing that negative news of the US stock market has a greater ability to increase the volatility of Asian stock markets as compared to positive news. Asymmetric coefficients of China were significant and positive in all pairs, demonstrating that negative news for any Asian markets except for India had a greater ability to increase the volatility of Chinese stock markets as compared to positive news during the Chinese crash. Constant conditional correlation was positively significant for all pairs of stock markets, but CCC was weak in the majority of pairs.

Table 7. Estimates of bivariate VAR–AGARCH for Chinese and Asian stock markets during the Chinese Stock Market Crash.

	IND	CHN	INDO	CHN	KOR	CHN	MYS	CHN	PAK	CHN	PHL	CHN	TAIW	CHN	THA	CHN
Panel A: Mean Equation																
Constant	3.32×10^{-4} (0.168)	5.74×10^{-5} (0.835)	1.10×10^{-5} (0.964)	1.39×10^{-4} (0.615)	3.28×10^{-4} (0.147)	8.49×10^{-5} (0.715)	-1.35×10^{-5} (0.933)	9.62×10^{-5} (0.715)	4.13×10^{-4} (0.145)	1.16×10^{-4} (0.663)	5.57×10^{-5} (0.875)	1.07×10^{-4} (0.699)	2.52×10^{-4} (0.310)	9.50×10^{-5} (0.714)	1.73×10^{-4} (0.359)	1.56×10^{-4} (0.561)
r^{o}_{t-1}	0.167 [a] (0.000)	0.106 [b] (0.022)	0.144 [a] (0.000)	−0.034 (0.500)	0.084 (0.350)	−0.014 (0.715)	0.125 [a] (0.000)	0.062 (0.322)	0.288 [a] (0.000)	−0.012 (0.660)	0.056 (0.141)	-1.15×10^{-4} (0.997)	0.086 (0.280)	0.106 [b] (0.014)	0.155 [a] (0.000)	0.021 [c] (0.620)
r^{c}_{t-1}	−0.029 (0.185)	0.037 (0.356)	−0.012 (0.550)	0.069 (0.330)	−0.019 (0.361)	0.062 [c] (0.077)	0.021 (0.222)	0.053 (0.181)	0.014 (0.452)	0.062 (0.120)	0.033 (0.274)	0.062 (0.113)	5.37×10^{-3} (0.836)	0.021 (0.580)	-8.57×10^{-4} (0.965)	0.070 [b] (0.052)
Panel B: Variance Equation																
Constant	3.77×10^{-6} [a] (0.000)	1.79×10^{-6} [b] (0.012)	4.23×10^{-6} [c] (0.010)	-4.44×10^{-8} (0.930)	3.43×10^{-6} (0.139)	9.89×10^{-7} (0.425)	1.42×10^{-6} (0.126)	-2.42×10^{-7} (0.768)	4.07×10^{-6} [a] (0.000)	8.60×10^{-7} [a] (0.005)	6.97×10^{-6} [b] (0.032)	1.31×10^{-6} (0.259)	9.22×10^{-6} [a] (0.000)	3.22×10^{-6} [b] (0.017)	2.64×10^{-6} [a] (0.001)	-8.98×10^{-7} [b] (0.031)
$(e^{o}_{t-1})^{2}$	−0.063 [a] (0.000)	-5.50×10^{-3} [a] (0.002)	0.052 (0.119)	5.85×10^{-4} (0.901)	0.024 (0.413)	5.44×10^{-3} (0.107)	0.102 [a] (0.003)	-8.63×10^{-4} (0.677)	−0.013 (0.560)	-2.66×10^{-5} (0.986)	0.058 (0.144)	0.018 [b] (0.022)	−0.028 (0.209)	5.81×10^{-3} (0.347)	−0.039 [b] (0.021)	-2.55×10^{-3} (0.364)
$(e^{c}_{t-1})^{2}$	0.070 [a] (0.000)	0.035 [b] (0.031)	−0.018 [a] (0.007)	0.012 (0.567)	8.96×10^{-3} (0.618)	0.023 (0.290)	−0.013 (0.780)	9.71×10^{-3} (0.611)	5.15×10^{-3} (0.180)	8.77×10^{-3} (0.656)	0.021 (0.137)	0.020 (0.342)	0.062 [a] (0.008)	0.027 (0.176)	−0.033 [a] (0.000)	0.017 (0.355)
h^{o}_{t-1}	0.880 [a] (0.000)	0.016 [a] (0.000)	0.789 [a] (0.000)	8.85×10^{-3} (0.176)	0.879 [a] (0.000)	-2.40×10^{-3} (0.665)	0.768 [a] (0.000)	9.26×10^{-3} (0.283)	0.841 [a] (0.000)	-1.18×10^{-3} (0.517)	0.824 [a] (0.000)	-9.03×10^{-3} (0.137)	0.659 [a] (0.000)	0.021 [b] (0.038)	0.818 [a] (0.000)	0.015 [b] (0.028)
h^{c}_{t-1}	−0.096 [a] (0.000)	0.945 [a] (0.000)	0.039 [b] (0.025)	0.940 [a] (0.000)	−0.014 (0.742)	0.944 [a] (0.000)	0.100 (0.343)	0.934 [a] (0.000)	-6.00×10^{-3} (0.109)	0.949 [a] (0.000)	−0.026 (0.341)	0.943 [a] (0.000)	−0.118 [b] (0.017)	0.944 [a] (0.000)	0.112 [a] (0.000)	0.933 [a] (0.000)
Asymmetry	0.171 [a] (0.000)	0.029 (0.145)	0.163 [a] (0.000)	0.059 [b] (0.021)	0.044 [c] (0.077)	0.045 [c] (0.058)	0.087 (0.112)	0.065 [b] (0.015)	0.263 [a] (0.000)	0.058 [c] (0.010)	0.061 (0.210)	0.048 [b] (0.045)	0.276 [a] (0.000)	0.047 [b] (0.040)	0.229 [a] (0.000)	0.049 [b] (0.029)
Panel C: Constant Conditional Correlation																
$p^{0,c}$	0.204 [a] (0.000)		0.151 [a] (0.000)		0.282 [a] (0.000)		0.166 [a] (0.000)		0.109 [a] (0.003)		0.179 [a] (0.000)		0.319 [a] (0.000)		0.202 [a] (0.000)	
Panel D: Diagnostic Tests																
LogL	5216.702		5157.83		5258.48		5520.00		5143.34		5026.77		5246.59		5354.32	
AIC	−12.108		−12.093		−12.437		−12.929		−11.856		−11.866		−12.268		−12.364	
SIC	−11.814		−11.799		−12.143		−12.635		−11.562		−11.573		−11.975		−12.070	

Table 7. *Cont.*

	IND	CHN	INDO	CHN	KOR	CHN	MYS	CHN	PAK	CHN	PHL	CHN	TAIW	CHN	THA	CHN
JB	245.340 [a]	365.250 [a]	212.590 [a]	369.930 [a]	325.490 [a]	305.710 [a]	214.270 [a]	342.420 [a]	156.230 [a]	422.300 [a]	216.090 [a]	338.660 [a]	301.330 [a]	269.370 [a]	227.030 [a]	329.200 [a]
	(0.000)	(0.000)	(0.000)	(0.000)	(0.000)	(0.000)	(0.000)	(0.000)	(0.000)	(0.000)	(0.000)	(0.000)	(0.000)	(0.000)	(0.000)	(0.000)
Q (12)	31.405 [a]	16.802	10.948	15.522	27.305 [a]	16.590	15.069	16.940	46.439 [a]	20.284 [c]	26.154 [c]	15.115	13.912	11.123	24.373 [b]	15.957
	(0.002)	(0.157)	(0.533)	(0.214)	(0.007)	(0.166)	(0.238)	(0.152)	(0.000)	(0.062)	(0.010)	(0.235)	(0.306)	(0.518)	(0.018)	(0.193)
$Q^2(12)$	55.871 [a]	17.976	47.212 [a]	21.484 [b]	48.551 [a]	17.416	70.480 [a]	21.463 [b]	70.632 [a]	30.028 [a]	50.180 [a]	23.783 [b]	44.260 [a]	16.310	88.957 [a]	31.358 [a]
	(0.000)	(0.116)	(0.000)	(0.044)	(0.000)	(0.135)	(0.000)	(0.044)	(0.000)	(0.003)	(0.000)	(0.022)	(0.000)	(0.177)	(0.000)	(0.002)

Notes: The number of lags for VAR was decided using SIC (Schwartz information criterion) and AIC (Akaike information criterion) criteria. JB, Q(12) and $Q^2(12)$ indicate the empirical statistics of the Jarque–Bera test for normality, while Ljung–Box Q statistics of order 12 for autocorrelation were applied to the standardized residuals and squared standardized residuals, respectively. USA, United States of America; IND, India; INDO, Indonesia; KOR, South Korea; MYS, Malaysia; PAK, Pakistan; PHL, the Philippines; TAIW, Taiwan; THA, Thailand. Values in parentheses are the *p*-values. [a], [b], [c] indicate the statistical significance at 1%, 5%, and 10%, respectively.

4.3. Optimal Weights and Hedge Ratio Portfolio Implications

Table 8 indicates the optimal weights and hedge ratios for the pairs of Asia-US stock portfolios during the full sample period, US financial crisis, and the Chinese stock market crash.[7] The range of optimal weights is 0.37 for IND/USA to 0.68 for MYS/USA during the period of the full sample, indicating that for a $1 India-USA portfolio, 37 cents should be invested in Indian stocks and the remaining 63 cents in the US stock market. The average optimal portfolio weights vary from 0.38 for IND/USA to 0.80 for MYS/USA during the US financial crisis and range from 0.37 for PHL/USA to 0.69 for MYS/USA during the Chinese stock market crash. Overall, the optimal weights of US stock in Asia-USA portfolios are higher during the Chinese stock market crash compared to the US financial crisis. This implies that investors should have maintained more US stocks in their portfolio of Asia-USA during the Chinese stock market crash compared to the Asian stocks during the US financial crisis.

Table 9 presents the optimal weights and hedge ratios for the pairs of Asia-China stock portfolio during the full sample period, US financial crisis, and the Chinese stock market crash.[8] The range of optimal weights is from 0.56 for IND/CHN, KOR/CHN, PHL/CHN to 0.81 for MYS/CHN during the full sample period. The average optimal portfolio weights vary from 0.53 for IND/CHN to 0.90 for MYS/CHN during the US financial crisis and range from 0.52 for PHL/CHN to 0.82 for MYS/CHN during the Chinese stock market crash. Overall, for the majority of Asia-China portfolios, the optimal weights of Chinese stocks were almost equal or higher during the Chinese stock market crash and the US financial crisis. This suggests that portfolio managers and investors should have maintained almost the same investment in Chinese stock in their majority of the portfolio of Asia-China during both the Chinese crash and the US financial crisis.

Table 8 presents the optimal hedge ratios for the pairs of Asia-USA stock portfolio during the full sample period, US financial crisis, and the Chinese stock market crash. Regarding the hedge ratio, the range of average hedge ratio is 0.04 for PAK/USA to 0.27 for IND/USA during the period of the full sample, showing that a long position of $1 in Pakistani stocks can be hedged for a short position of 4 cents in US stocks. During the US financial crisis, the average optimal hedge ratios varied from 0.08 for PAK/USA to 0.36 for IND/USA. The average optimal hedge ratio ranged from 0.09 for PAK/USA to 0.36 for IND/USA during the Chinese stock market crash. For the majority of pairs of Asia-USA, the hedge ratios were lower in the US financial crisis compared with the Chinese stock market crash. This suggests that few US stocks were required to minimize the risk of Asian stock investors during the US financial crisis as compared to during the Chinese crash.

Table 9 provides the optimal hedge ratios for the pairs of a Asia-China stock portfolio during the full sample period, US financial crisis, and the Chinese stock market crash. The range of average hedge ratio is 0.04 for PAK/CHN to 0.21 for KOR/CHN during the period of the full sample. During the US financial crisis, the average optimal hedge ratios varied from 0.03 for PAK/CHN to 0.32 for KOR/CHN. The average optimal hedge ratio ranged from 0.09 for MYS/CHN to 0.26 for TAIW/CHN during the Chinese stock market crash. Overall, for the Asia-China pairs, the hedge ratio was lower during the Chinese stock market crash compared to the hedge ratios in the US financial crisis. This implies that fewer Chinese stocks were needed to minimize the risk for Asian stock investors during the Chinese stock market crash as compared to during the US crisis.

---

7 We calculated the optimal weights by using both VAR-GARCH and VAR-AGARCH models, but we reported the optimal weights only from the VAR-AGARCH model for the purpose of brevity.

8 We calculated the optimal weights by using both VAR-GARCH and VAR-AGARCH models, but we reported the optimal weights only from the VAR-AGARCH model for the purpose of brevity.

Table 8. Optimal Weights and Hedge Ratios for Asia/USA pairs.

	IND/USA	INDO/USA	KOR/USA	MYS/USA	PAK/USA	PHL/USA	TAIW/USA	THA/USA
				Full Sample Period				
w_t^{SU}	0.37	0.41	0.40	0.68	0.41	0.42	0.43	0.41
β_t^{SU}	0.27	0.13	0.24	0.06	0.04	0.06	0.17	0.17
				US Financial Crisis				
w_t^{SU}	0.38	0.51	0.54	0.80	0.52	0.57	0.52	0.52
β_t^{SU}	0.36	0.18	0.21	0.11	0.08	0.09	0.15	0.22
				Chinese Stock Market Crash				
w_t^{SU}	0.46	0.44	0.51	0.69	0.44	0.37	0.49	0.53
β_t^{SO}	0.36	0.15	0.29	0.11	0.09	0.14	0.22	0.22

Note: w_t^{SU} and β_t^{SU} refer to the optimal weights and hedge ratios, respectively.

Table 9. Optimal Weights and Hedge Ratios for Asia/China pairs.

	IND/CHN	INDO/CHN	KOR/CHN	MYS/CHN	PAK/CHN	PHL/CHN	TAIW/CHN	THA/CHN
				Full Sample Period				
w_t^{SC}	0.56	0.57	0.56	0.81	0.57	0.56	0.59	0.59
β_t^{SC}	0.15	0.15	0.21	0.09	0.04	0.12	0.20	0.13
				US Financial Crisis				
w_t^{SC}	0.53	0.63	0.68	0.90	0.64	0.64	0.66	0.67
β_t^{SC}	0.31	0.24	0.32	0.13	0.03	0.22	0.30	0.19
				Chinese Stock Market Crash				
w_t^{SC}	0.65	0.61	0.66	0.82	0.58	0.52	0.66	0.73
β_t^{SC}	0.17	0.13	0.23	0.09	0.11	0.18	0.26	0.14

Note: w_t^{SC} and β_t^{SC} refer to the optimal weights and hedge ratios, respectively.

5. Conclusions

In this paper, we extend the previous work by examining the return and volatility transmissions from the US and China to the eight emerging Asian stock markets including India, Indonesia, Korea, Malaysia, Pakistan, the Philippines, Taiwan, and Thailand during the Chinese stock market crash by using the VAR-AGARCH model. Moreover, we also examine the spillovers during the full sample period and the 2008 US financial crisis to provide comparative insights to investors about whether the impact of the Chinese crash on equity market spillovers is different from the crashes in other sample periods. Lastly, we also estimate the optimal weights and hedge ratios during the full period and all sub-periods.

Our comprehensive analysis reveals that both return and volatility spillover vary across different pairs of stock markets and during financial crises. The findings of return spillover indicate a significant spillover from the USA to Asian stock markets during the full sample period, the US financial crisis, and the Chinese stock market crash. This implies that US stock market prices play an important role in predicting the prices of the majority of Asian stock markets during the full period and all the sub-periods. However, the return spillover is not significant from China to emerging Asian stock markets during the US financial crisis and the Chinese stock market crash, implying that Chinese stock prices cannot be used for predicting the prices of the majority of Asian stock markets during any of the crisis periods in our study.

Our volatility spillover analysis reveals that the volatility was transmitted from the US to the majority of Asian markets during the full sample period and the Chinese stock market crash, but such a conclusion cannot be drawn for during the US financial crisis. This implies that portfolio investors of Asian stock markets could have gotten the maximum benefits of diversification by holding US stocks in their portfolio during the US financial crisis. However, the volatility spillover was transmitted from China to a majority of Asian markets during the full sample period and US financial crisis, but such a conclusion cannot be reached for during the Chinese crash, implying that portfolio investors of Asian stock markets could have gotten the maximum benefits of diversification by holding Chinese stocks in their portfolio during the Chinese stock market crash.

Based on optimal weights results, the weights of the US stocks in the Asia-USA portfolios are higher during the Chinese crash compared to the US financial crisis, implying that investors should keep more US stocks in their portfolio of the Asia-USA stocks during the Chinese stock market crash, compared to the US financial crisis. For the majority of Asia-China portfolios, the optimal weights of Chinese stocks were almost equal or higher during the Chinese stock market crash and the US financial crisis. This suggests that portfolio managers and investors should have maintained almost the same investment in the Chinese stocks in their portfolio of the Asia-China majority pairs during both the Chinese crash and the US financial crisis.

Regarding the hedge ratios, for most of the Asia-USA pairs, the hedge ratios were smaller in the US financial crisis than in the Chinese stock market crash. This suggests that few US stocks were required to minimize the risk for Asian stock investors during the US financial crisis as compared to during the Chinese crash. In contrast, for the Asia-China pairs, the hedge ratio was smaller during the Chinese stock market crash compared to that in the US financial crisis. This implies that fewer Chinese stocks were needed to minimize the risk for Asian stock investors during the Chinese stock market crash as compared to the US crisis. Overall, our findings provide several important implications for risk management and portfolio diversification that could be useful for investors and for policymakers related to the US and Asian stock markets.

Author Contributions: Conceptualization, estimations, formal analysis, original draft preparation I.Y.; Data collection, methodology writing, and review of draft S.A.; review, editing, and funding W.-K.W. All authors have read and agreed to the published version of the manuscript.

Acknowledgments: The first author gratefully acknowledge Arshad Hassan (department of Management and Social Sciences, Capital University of Science and Technology, Islamabad) for their valuable suggestions. The third author would like to thank Robert B. Miller and Howard E. Thompson for their continuous guidance and encouragement.

References

Allen, Katie. 2015. Why is China's stock market in crisis? *The Guardian*, July 8.

Allen, David E., Ron Amram, and Michael McAleer. 2013. Volatility spillovers from the Chinese stock market to economic neighbours. *Mathematics and Computers in Simulation* 94: 238–57. [CrossRef]

Arouri, Mohamed El Hedi, Amine Lahiani, and Duc Khuong Nguyen. 2011. Return and volatility transmission between world oil prices and stock markets of the GCC countries. *Economic Modelling* 28: 1815–25. [CrossRef]

Arouri, Mohamed El Hedi, Jamel Jouini, and Duc Khuong Nguyen. 2012. On the impacts of oil price fluctuations on European equity markets: Volatility spillover and hedging effectiveness. *Energy Economics* 34: 611–17. [CrossRef]

Arouri, Mohamed El Hedi, Amine Lahiani, and Duc Khuong Nguyen. 2015. World gold prices and stock returns in China: Insights for hedging and diversification strategies. *Economic Modelling* 44: 273–282. [CrossRef]

Baele, Lieven. 2005. Volatility spillover effects in European equity markets. *Journal of Financial and Quantitative Analysis* 40: 373–401. [CrossRef]

Beirne, John, Guglielmo Maria Caporale, Marianne Schulze-Ghattas, and Nicola Spagnolo. 2013. Volatility spillovers and contagion from mature to emerging stock markets. *Review of International Economics* 21: 1060–75. [CrossRef]

Bollerslev, T. 1990. Modelling the coherence in short-run nominal exchange rates: A multivariate generalized ARCH model. *The review of Economics and Statistics* 72: 498–505. [CrossRef]

Bouri, Elie. 2015. Return and volatility linkages between oil prices and the Lebanese stock market in crisis periods. *Energy* 89: 365–71. [CrossRef]

Cheung, Yan-Leung, Yin-Wong Cheung, and Chris C. Ng. 2007. East Asian equity markets, financial crises, and the Japanese currency. *Journal of the Japanese and International Economies* 21: 138–52. [CrossRef]

Chien, Mei-Se, Chien-Chiang Lee, Te-Chung Hu, and Hui-Ting Hu. 2015. Dynamic Asian stock market convergence: Evidence from dynamic cointegration analysis among China and ASEAN-5. *Economic Modelling* 51: 84–98. [CrossRef]

Diebold, Francis X., and Kamil Yilmaz. 2009. Measuring financial asset return and volatility spillovers, with application to global equity markets. *The Economic Journal* 119: 158–71. [CrossRef]

Dutta, Anupam, Elie Bouri, and Md Hasib Noor. 2018. Return and volatility linkages between CO_2 emission and clean energy stock prices. *Energy* 164: 803–10. [CrossRef]

Engle, Robert F., Giampiero M. Gallo, and Margherita Velucchi. 2012. Volatility spillovers in East Asian financial markets: A MEM-based approach. *Review of Economics and Statistics* 94: 222–23. [CrossRef]

Forbes, K., and R. Rigobon. 2002. No contagion, only interdependence: Measuring stock market comovements. *Journal of Finance* 57: 2223–61. [CrossRef]

Fung, Eric S., Kin Lam, Tak-Kuen Siu, and Wing-Keung Wong. 2011. A Pseudo-Bayesian Model for Stock Returns in Financial Crises. *Journal of Risk and Financial Management* 4: 42–72. [CrossRef]

Gkillas, Konstantinos, Athanasios Tsagkanos, and Dimitrios I. Vortelinos. 2019. Integration and risk contagion in financial crises: Evidence from international stock markets. *Journal of Business Research* 104: 350–65. [CrossRef]

Glick, Reuven, and Michael Hutchison. 2013. China's financial linkages with Asia and the global financial crisis. *Journal of International Money and Finance* 39: 186–206. [CrossRef]

Guo, Xu, Michael McAleer, Wing-Keung Wong, and Lixing Zhu. 2017. A Bayesian approach to excess volatility, short-term underreaction and long-term overreaction during financial crises. *North American Journal of Economics and Finance* 42: 346–58. [CrossRef]

Hammoudeh, Shawkat M., Yuan Yuan, and Michael McAleer. 2009. Shock and volatility spillovers among equity sectors of the Gulf Arab stock markets. *The Quarterly Review of Economics and Finance* 49: 829–42. [CrossRef]

He, Ling T. 2001. Time variation paths of international transmission of stock volatility—US vs. Hong Kong and South Korea. *Global Finance Journal* 12: 79–93. [CrossRef]

Huang, Bwo-Nung, Chin-Wei Yang, and John Wei-Shan Hu. 2000. Causality and cointegration of stock markets among the United States, Japan and the South China Growth Triangle. *International Review of Financial Analysis* 9: 281–97. [CrossRef]

Huo, Rui, and Abdullahi D. Ahmed. 2017. Return and volatility spillovers effects: Evaluating the impact of Shanghai-Hong Kong Stock Connect. *Economic Modelling* 61: 260–72. [CrossRef]

Huyghebaert, Nancy, and Lihong Wang. 2010. The co-movement of stock markets in East Asia: Did the 1997–1998 Asian financial crisis really strengthen stock market integration? *China Economic Review* 21: 98–112. [CrossRef]

In Francis, Sangbae Kim, Jai Hyung Yoon, and Christopher Viney. 2001. Dynamic interdependence and volatility transmission of Asian stock markets: Evidence from the Asian crisis. *International Review of Financial Analysis* 10: 87–96.

Jin, Xiaoye. 2015. Volatility transmission and volatility impulse response functions among the Greater China stock markets. *Journal of Asian Economics* 39: 43–58. [CrossRef]

Johansson, Anders C., and Christer Ljungwall. 2009. Spillover effects among the Greater China stock markets. *World Development* 37: 839–51. [CrossRef]

Jouini, Jamel. 2013. Return and volatility interaction between oil prices and stock markets in Saudi Arabia. *Journal of Policy Modeling* 35: 1124–44. [CrossRef]

Kim, Bong-Han, Hyeongwoo Kim, and Bong-Soo Lee. 2015. Spillover effects of the US financial crisis on financial markets in emerging Asian countries. *International Review of Economics & Finance* 39: 192–210.

Kroner, Kenneth F., and Victor K. Ng. 1998. Modeling asymmetric comovements of asset returns. *The Review of Financial Studies* 11: 817–44. [CrossRef]

Kroner, Kenneth F., and Jahangir Sultan. 1993. Time-varying distributions and dynamic hedging with foreign currency futures. *Journal of financial and Quantitative Analysis* 28: 535–51. [CrossRef]

Lean, Hooi Hooi, Michael McAleer, and Wing-Keung Wong. 2015. Preferences of risk-averse and risk-seeking investors for oil spot and futures before, during and after the Global Financial Crisis. *International Review of Economics and Finance* 40: 204–16. [CrossRef]

Li, Yanan, and David E. Giles. 2015. Modelling volatility spillover effects between developed stock markets and Asian emerging stock markets. *International Journal of Finance & Economics* 20: 155–77.

Lin, Boqiang, Presley K. Wesseh Jr., and Michael Owusu Appiah. 2014. Oil price fluctuation, volatility spillover and the Ghanaian equity market: Implication for portfolio management and hedging effectiveness. *Energy Economics* 42: 172–82. [CrossRef]

Ling, Shiqing, and Michael McAleer. 2003. Asymptotic theory for a vector ARMA-GARCH model. *Econometric Theory* 19: 280–310. [CrossRef]

Liu, Y. Angela, and Ming-Shiun Pan. 1997. Mean and volatility spillover effects in the US and Pacific-Basin stock markets. *Multinational Finance Journal* 1: 47–62. [CrossRef]

Malik, Farooq, and Shawkat Hammoudeh. 2007. Shock and volatility transmission in the oil, US and Gulf equity markets. *International Review of Economics & Finance* 16: 357–68.

McAleer, Michael, Suhejla Hoti, and Felix Chan. 2009. Structure and asymptotic theory for multivariate asymmetric conditional volatility. *Econometric Reviews* 28: 422–40. [CrossRef]

Michaud, Richard O., Gary L. Bergstrom, Ronald Frashure, and Brian Wolahany. 1996. Twenty years of international equity investing. *The Journal of Portfolio Management* 23: 9–22. [CrossRef]

Miyakoshi, Tatsuyoshi. 2003. Spillovers of stock return volatility to Asian equity markets from Japan and the US. *Journal of International Financial Markets, Institutions and Money* 13: 383–99. [CrossRef]

Sadorsky, Perry. 2012. Correlations and volatility spillovers between oil prices and the stock prices of clean energy and technology companies. *Energy Economics* 34: 248–55. [CrossRef]

Syriopoulos, Theodore, Beljid Makram, and Adel Boubaker. 2015. Stock market volatility spillovers and portfolio hedging: BRICS and the financial crisis. *International Review of Financial Analysis* 39: 7–18. [CrossRef]

The causes and consequences of China's market crash. 2015. *The Economist*. Available online: https://www.economist.com/news/2015/08/24/the-causes-and-consequences-of-chinas-market-crash (accessed on 24 September 2020).

Vieito, João Paulo, Wing-Keung Wong, and Zhen-Zhen Zhu. 2015. Could The Global Financial Crisis Improve the Performance of The G7 Stocks Markets? *Applied Economics* 48: 1066–80. [CrossRef]

Wan, Henry, Jr., and Wing-Keung Wong. 2001. Contagion or inductance? Crisis 1997 reconsidered. *Japanese Economic Review* 52: 372–80. [CrossRef]

Xiong, Xiong, Yuxiang Bian, and Dehua Shen. 2018. The time-varying correlation between policy uncertainty and stock returns: Evidence from China. *Physica A: Statistical Mechanics and its Applications* 499: 413–19. [CrossRef]

Yang, Jian, James W. Kolari, and Insik Min. 2003. Stock market integration and financial crises: The case of Asia. *Applied Financial Economics* 13: 477–86. [CrossRef]

Yilmaz, Kamil. 2010. Return and volatility spillovers among the East Asian equity markets. *Journal of Asian Economics* 21: 304–13. [CrossRef]

Yousaf, Imran, and Arshad Hassan. 2019. Linkages between crude oil and emerging Asian stock markets: New evidence from the Chinese stock market crash. *Finance Research Letters* 31: 207–217. [CrossRef]

Yousaf, Imran, Shoaib Ali, and Wing-Keung Wong. 2020. Return and Volatility Transmission between World-Leading and Latin American Stock Markets: Portfolio Implications. *Journal of Risk and Financial Management* 13: 148. [CrossRef]

Yousaf, Imran, and Shoaib Ali. 2020. Linkages between gold and emerging Asian stock markets: New evidence from the Chinese stock market crash. *Studies of Applied Economics* 39. [CrossRef]

Zhou, Xiangyi, Weijin Zhang, and Jie Zhang. 2012. Volatility spillovers between the Chinese and world equity markets. *Pacific-Basin Finance Journal* 20: 247–70. [CrossRef]

Zhu, Yanjian, Zhaoying Wu, Hua Zhang, and Jing Yu. 2017. Media sentiment, institutional investors and probability of stock price crash: Evidence from Chinese stock markets. *Accounting & Finance* 57: 1635–70.

Zhu, Zhenzhen, Zhidong Bai, João Paulo Vieito, and Wing-Keung Wong. 2019. The Impact of the Global Financial Crisis on the Efficiency of Latin American Stock Markets. *Estudios de Economía* 46: 5–30. [CrossRef]

7

Determining Distribution for the Quotients of Dependent and Independent Random Variables by using Copulas

Sel Ly [1], Kim-Hung Pho [1,*], Sal Ly [1] and Wing-Keung Wong [2,3,4]

1 Faculty of Mathematics and Statistics, Ton Duc Thang University, Ho Chi Minh City 756636, Vietnam; lysel@tdtu.edu.vn (S.L.); shanlee5611@gmail.com (S.L.)
2 Department of Finance, Fintech Center, and Big Data Research Center, Asia University, Taichung 41354, Taiwan; wong@asia.edu.tw
3 Department of Medical Research, China Medical University Hospital, Taichung 40402, Taiwan
4 Department of Economics and Finance, Hang Seng University of Hong Kong, Shatin, Hong Kong 999077, China
* Correspondence: phokimhung@tdtu.edu.vn

Abstract: Determining distributions of the functions of random variables is a very important problem with a wide range of applications in Risk Management, Finance, Economics, Science, and many other areas. This paper develops the theory on both density and distribution functions for the quotient $Y = \frac{X_1}{X_2}$ and the ratio of one variable over the sum of two variables $Z = \frac{X_1}{X_1+X_2}$ of two dependent or independent random variables X_1 and X_2 by using copulas to capture the structures between X_1 and X_2. Thereafter, we extend the theory by establishing the density and distribution functions for the quotients $Y = \frac{X_1}{X_2}$ and $Z = \frac{X_1}{X_1+X_2}$ of two dependent normal random variables X_1 and X_2 in the case of Gaussian copulas. We then develop the theory on the median for the ratios of both Y and Z on two normal random variables X_1 and X_2. Furthermore, we extend the result of median for Z to a larger family of symmetric distributions and symmetric copulas of X_1 and X_2. Our results are the foundation of any further study that relies on the density and cumulative probability functions of ratios for two dependent or independent random variables. Since the densities and distributions of the ratios of both Y and Z are in terms of integrals and are very complicated, their exact forms cannot be obtained. To circumvent the difficulty, this paper introduces the Monte Carlo algorithm, numerical analysis, and graphical approach to efficiently compute the complicated integrals and study the behaviors of density and distribution. We illustrate our proposed approaches by using a simulation study with ratios of normal random variables on several different copulas, including Gaussian, Student-t, Clayton, Gumbel, Frank, and Joe Copulas. We find that copulas make big impacts from different Copulas on behavior of distributions, especially on median, spread, scale and skewness effects. In addition, we also discuss the behaviors via all copulas above with the same Kendall's coefficient. The approaches developed in this paper are flexible and have a wide range of applications for both symmetric and non-symmetric distributions and also for both skewed and non-skewed copulas with absolutely continuous random variables that could contain a negative range, for instance, generalized skewed-t distribution and skewed-t Copulas. Thus, our findings are useful for academics, practitioners, and policy makers.

Keywords: copulas; dependence structures; quotient of random variables; density functions; distribution functions

1. Introduction

Determining distributions of the functions of random variables is a very crucial task and this problem has been attracted a number of researchers because there are numerous applications in Risk Management, Finance, Economics, Science, and, many other areas, see, for example, (Donahue 1964; Ly et al. 2016; Nadarajah and Espejo 2006; Springer 1979). Basically, the distributions of an algebraic combination of random variables including the sum, product, and quotient are focused on some common distributions along with the assumptions of independence or correlated through Pearson's coefficient or dependence via multivariate normal joint distributions (Arnold and Brockett 1992; Bithas et al. 2007; Cedilnik et al. 2004; Hinkley 1969; Macalos and Arcede 2015; Marsaglia 1965; Matović et al. 2013; Mekićet al. 2012; Nadarajah and Espejo 2006; Nadarajah and Kotz 2006a, 2006b; Pham-Gia et al. 2006; Pham-Gia 2000; Rathie et al. 2016; Sakamoto 1943). Regarding ratio, it often appears in the problems of constructing statistics used in hypothesis testing and estimating issues. Some well-known distributions are results of such quotients. For example, the quotient of a Gaussian random variable divided by a square root of an independent chi-distributed random variable follows the t-distribution while the F-distribution is derived via the ratio of two independent chi-squared distributed random variables. To relax independence assumption, it is necessary to develop a framework for modeling dependence structures of random vectors in more general sense. To do so, Dolati et al. (2017) develop the distribution for X/Y in which both X and Y are positive.

In our paper, we first extend the theory developed by Dolati et al. (2017) to relax the positive assumption for the variables by developing the theory on both density and distribution function (CDF) for the quotient $Y = \frac{X_1}{X_2}$ of two dependent or independent continuous random variables X_1 and X_2 in which X_1 and X_2 could be any real number. Thereafter, we develop a theory on both density and distribution function for the ratio of one variable over the sum of two variables $Z = \frac{X_1}{X_1+X_2}$ of two dependent or independent continuous random variables X_1 and X_2 by using copulas to capture the structures between X_1 and X_2.

Since the density and the CDF formula of the ratios of both Y and Z are in terms of integrals and are very complicated, we cannot obtain the exact forms of the densities and the CDFs. To circumvent the difficulty, in this paper, we propose to use a Monte Carlo algorithm, numerical analysis, and graphical approach to study behavior of density and distribution. We illustrate our proposed approaches by using a simulation study with ratios of standard normal random variables on several different copulas, including Gaussian, Student-t, Clayton, Gumbel, Frank, and Joe Copulas and we find that copulas make big impacts from different Copulas on behavior of distributions, especially on median, spread, skewness and scale effects. For instance, when X_1 and X_2 tend to be more co-monotonic indicated by increasing the parameters of copulas, then the median of Y is shifted to be higher and its shape tends to be more symmetric. In the meantime, the median of Z is equally unchanged one-half and the shape always has symmetry. We note that the approaches developed in this paper are flexible and have a wide range of applications for both symmetric and non-symmetric distributions and also for both skewed and non-skewed copulas with absolutely continuous random variables that could contain a negative range, for instance, generalized skewed-t distribution and skewed-t Copulas.[1] Thus, our findings are useful for academics, practitioners, and policy makers.

The rest of the paper is organized as follows. In Sections 2 and 3, we will briefly discuss the background theory and copula theory related to the theory developed in our paper. In Section 4, we provide main results on the quotients of dependent and independent random variables. Section 5 proposes using the Monte Carlo to deal with complex integrals and estimate some percentiles by using some special copulas, and investigate their effects on the behavior of ratios of two standard normal random variables. The last section provides the conclusions.

---

[1] We would like to thank the anonymous reviewer for giving us helpful comments so that we could draw this conclusion.

2. Background Theory

We first review some previous work on the weighted sum, for example, in constructing portfolio Y_1 that is composed of two dependent assets defined by

$$Y_1 = w_1 X_1 + w_2 X_2, \tag{1}$$

in which the random variables $X_{i,t}$ $(i = 1, 2)$ denotes the rate of return at time t for the asset defined in terms of the following random quotient:

$$X_{i,t} = \frac{P_{i,t} - P_{i,t-1}}{P_{i,t-1}},$$

where $P_{i,t}$ denotes the price of the ith asset at time t. Note that X_i is assumed to be absolutely continuous with the cumulative distribution functions (CDF) F_i. Suppose that (X_1, X_2) follows copula C, then the CDF, $F_{Y_1}(y)$, of Y_1 defined in (1) satisfies:

$$F_{Y_1}(y) = \mathbf{1}_{\{w_2<0\}} + \operatorname{sgn}(w_2) \int_0^1 \frac{\partial}{\partial u} C\left(u, F_2\left(\frac{y - w_1 F_1^{-1}(u)}{w_2}\right)\right) du, \tag{2}$$

where $\operatorname{sgn}(\cdot)$ denotes the sign function such that

$$\operatorname{sgn}(x) = \begin{cases} 1, & \text{if} \quad x > 0, \\ -1, & \text{if} \quad x < 0. \end{cases}$$

Then, the CDF, F_{Y_1}, can be used to estimate the distortion risk measure of the portfolio defined by

$$R_g[Y_1] = \int_0^\infty g(\overline{F}_{Y_1}(y))dy + \int_{-\infty}^0 [g(\overline{F}_{Y_1}(y)) - 1]dy,$$

where g is a *distortion function* and $\overline{F}_{Y_1}(y) = 1 - F_{Y_1}(y)$ is a survival function of Y_1. Readers may refer to Ly et al. (2016) for more detailed information.

In the credit model, the total loss is defined as the aggregation of the product of risk factors. Thus, it is necessary to find the distribution for the product case, for instance, Y_2 given by

$$Y_2 = X_1 X_2. \tag{3}$$

Ly et al. (2019) show that the CDF of Y_2 can be determined by

$$F_{Y_2}(y) = F_1(0) + \int_0^1 \operatorname{sgn}\left(F_1^{-1}(u)\right) \frac{\partial}{\partial u} C\left(u, F_2\left(\frac{y}{F_1^{-1}(u)}\right)\right) du. \tag{4}$$

3. Copulas

In this section, we will briefly discuss the copula theory related to the theory developed in our paper. Readers may refer to (Cherubini et al. 2004; Joe 1997; Nelsen 2007; Tran et al. 2015, 2017) for more information. Let $\mathbb{I} = [0,1]$ be the closed unit interval and $\mathbb{I}^2 = [0,1] \times [0,1]$ be the closed unit square interval. We first state the most basic definition of copula in two dimensions in the following:

Definition 1. ***(Copula)*** *A 2-copula (two-dimensional copula) is a function C: $\mathbb{I}^2 \to \mathbb{I}$ satisfying the following conditions:*

(i) $C(u,0) = C(0,v) = 0$ *for any* $u, v \in \mathbb{I}$;
(ii) $C(u,1) = u$ *and* $C(1,v) = v$ *for any* $u, v \in \mathbb{I}$; *and*

(iii) for any $u_1, u_2, v_1, v_2 \in \mathbb{I}$ *with* $u_1 \leq u_2$ *and* $v_1 \leq v_2$,

$$C(u_2, v_2) + C(u_1, v_1) - C(u_2, v_1) - C(u_1, v_2) \geq 0.$$

In copula theory, Sklar proposed a very important theorem in 1959 called Sklar's Theorem (Cherubini et al. (2004); Joe (1997); Nelsen (2007)), which plays the most important role in this theory. It tells us that given a random vector (X_1, X_2) with absolutely continuous marginal distribution functions F_{X_1} and F_{X_2}, respectively, and its joint distribution function denoted by H, and then there exists a unique copula C such that

$$\begin{aligned} H(x_1, x_2) &= C\left(F_{X_1}(x_1), F_{X_2}(x_2)\right), \\ h(x_1, x_2) &= \frac{\partial^2}{\partial x_1 \partial x_2} H(x_1, x_2) = c\left(F_{X_1}(x_1), F_{X_2}(x_2) f_{X_1}(x_1) f_{X_2}(x_2)\right), \end{aligned} \tag{5}$$

where $c(u,v) := \frac{\partial^2}{\partial u \partial v} C(u,v)$ denotes density of copula C, f_{X_i} is probability density function (PDF) of $X_i, i = 1, 2$, and $h(x_1, x_2)$ is the joint density function of X_1 and X_2. Copula is used to combine several univariate distributions together into bivariate [multivariate] settings so as the copula C can capture the dependence structure of (X_1, X_2) $[(X_1, \cdots, X_n)]$. For any copula C, we have the bounds

$$W(u,v) \leq C(u,v) \leq M(u,v),$$

where the copula $W(u,v) := \max(u + v - 1, 0)$ captures counter-monotonicity structure; that is, $X_2 = f(X_1)$ a.s., where f is strictly decreasing, while the copula $M(u,v) := \min(u,v)$ is used to capture comonotonicity; that is, $X_2 = f(X_1)$ a.s., where f is strictly increasing. In case X_1 and X_2 are independent, they follow copula denoted by $\Pi(u,v) := uv$. Copulas can be used not only to model the dependence structure of the variables, but also capture the correlation between the variables. The Kendall's coefficient τ can be expressed in terms of copulas as shown in the following:

$$\tau(X_1, X_2) = \tau(C) = 4 \int \int_{\mathbb{I}^2} C(u,v) dC(u,v) - 1. \tag{6}$$

In the next section, we will derive the two main propositions regarding formulas that can be used to determine the probability density and probability distribution of the quotient of dependent random variables by using copulas. In addition, we will apply the results to derive some corollaries on PDFs, CDFs, and median of the ratios in case X_1 and X_2 are normal distributed and they follow the Gaussian copulas.

4. Theory

We now develop two propositions on both density and distribution functions for the quotient $Y := \frac{X_1}{X_2}$ and the ratio of one variable over the sum of two variables $Z := \frac{X_1}{X_1 + X_2}$ of two dependent random variables X_1 and X_2 by using copulas. We first develop the proposition on the density and distribution functions for the quotient $Y = \frac{X_1}{X_2}$ as stated in the following:

Proposition 1. *Supposing that* (X_1, X_2) *is a vector of two absolutely continuous random variables* X_1 *and* X_2 *with the marginal distributions* F_1 *and* F_2*, respectively, let* C *be an absolutely continuous copula modeling dependence structure of the random vector* (X_1, X_2)*, and define* Y *as*

$$Y := \frac{X_1}{X_2}. \tag{7}$$

Then, the density $f_Y(y)$ and distribution functions $F_Y(y)$ of Y are

$$f_Y(y) = \int_0^1 |F_2^{-1}(v)| c\Big(F_1\left(yF_2^{-1}(v)\right), v\Big) f_1\left(yF_2^{-1}(v)\right) dv, \tag{8}$$

$$F_Y(y) = F_2(0) + \int_0^1 \operatorname{sgn}\left(F_2^{-1}(v)\right) \frac{\partial}{\partial v} C\left(F_1\left(yF_2^{-1}(v)\right), v\right) dv, \tag{9}$$

respectively, where F_2^{-1} denotes the inverse function of F_2, c is the density of copula C, and $\operatorname{sgn}(\cdot)$ stands for a sign function such that

$$\operatorname{sgn}(x) = \begin{cases} 1, & \text{if} \quad x > 0, \\ -1, & \text{if} \quad x < 0. \end{cases}$$

Proof. Letting

$$\begin{cases} Y_1 := \dfrac{X_1}{X_2}, \\ Y_2 := X_2. \end{cases}$$

We note that since X_2 is absolutely continuous, $P(X_2 = 0) = 0$; that is, $X_2 \neq 0$ almost surely. Hence, the transformation $Y_1 = \frac{X_1}{X_2}$ always exists with probability 1 and we can obtain its inverse transformation by using

$$\begin{cases} X_1 = Y_1 Y_2, \\ X_2 = Y_2, \end{cases}$$

and their corresponding Jacobian

$$J = \begin{vmatrix} Y_2 & Y_1 \\ 0 & 1 \end{vmatrix} = Y_2.$$

Then, we obtain the joint density of Y_1 and Y_2 such that

$$\begin{aligned} h(y_1, y_2) &= f(y_1 y_2, y_2)\, |y_2| \\ &= |y_2|\, c\left(F_1(y_1 y_2), F_2(y_2)\right) f_1(y_1 y_2) f_2(y_2). \end{aligned}$$

This yields the density of Y_1:

$$f_{Y_1}(y_1) = \int_{-\infty}^{\infty} |y_2|\, c\left(F_1(y_1 y_2), F_2(y_2)\right) f_1(y_1 y_2) f_2(y_2)\, dy_2 \tag{10}$$

$$= \int_0^1 \left|F_2^{-1}(v)\right| c\Big(F_1\left(y_1 F_2^{-1}(v)\right), v\Big) f_1\left(y_1 F_2^{-1}(v)\right) dv. \tag{11}$$

As a result, the CDF of Y_1 is determined by

$$F_{Y_1}(t) = \int_0^1 \int_{-\infty}^t \left|F_2^{-1}(v)\right| c\Big(F_1\left(y_1 F_2^{-1}(v)\right), v\Big) f_1\left(y_1 F_2^{-1}(v)\right) dy_1 dv. \tag{12}$$

By changing variable, $u = F_1\left(y_1 F_2^{-1}(v)\right)$, we get $du = F_2^{-1}(v) f_1\left(y_1 F_2^{-1}(v)\right) dy_1$ and we note that

$$F_2^{-1}(v) \geq 0 \Longleftrightarrow v \in [0, F_2(0)], \text{ and } F_2^{-1}(v) \leq 0 \Longleftrightarrow v \in [F_2(0), 1].$$

This yields

$$\begin{aligned} F_{Y_1}(t) &= -\int_0^{F_2(0)}\int_1^{F_1(tF_2^{-1}(v))}\frac{\partial^2}{\partial u\partial v}C(u,v)\,dudv+\int_{F_2(0)}^1\int_0^{F_2(tF_1^{-1}(v))}\frac{\partial^2}{\partial u\partial v}C(u,v)\,dvdu \\ &= -\int_0^{F_2(0)}\left[\frac{\partial}{\partial v}C\left(F_1\left(tF_2^{-1}(v)\right),v\right)-\frac{\partial}{\partial v}C(1,v)\right]dv+\int_{F_2(0)}^1\frac{\partial}{\partial v}C\left(F_1\left(tF_2^{-1}(v)\right),v\right)dv \quad (13) \\ &= F_2(0)+\int_0^1\operatorname{sgn}\left(F_2^{-1}(v)\right)\frac{\partial}{\partial v}C\left(F_1\left(tF_2^{-1}(v)\right),v\right)dv. \end{aligned}$$

Thus, the assertions of Proposition 1 hold. □

From Proposition 1 and applying Equation (8), we obtain the following corollary on both density and distribution functions for the quotient $Y=\frac{X_1}{X_2}$ of two independent random variables X_1 and X_2 by using copulas:

Corollary 1. *When X_1 and X_2 are independent, then its copula $C(u,v)=uv$ has the density $c(u,v)=1$, $\forall u,v\in\mathbb{I}$ and the density $f_Y(y)$ of the ratio $Y:=\frac{X_1}{X_2}$ of two independent random variables becomes*

$$f_Y(y)=\int_{-\infty}^{\infty}|x|f_1(xy)f_2(x)dx.$$

This result is well known in the literature.

Next, we apply Equation (9) to derive both density and distribution function for $Y=\frac{X_1}{X_2}$ in case X_1 and X_2 are normal random variables and their dependence structure is captured by Gaussian Copulas. We first obtain the following corollary:

Corollary 2. *Assume that $X_1\sim N(\mu_1,\sigma_1^2)$, $X_2\sim N(\mu_2,\sigma_2^2)$ and (X_1,X_2) follows Gaussian Copulas $C_r(u,v)$, $|r|<1$, given in (46). Then, the density $f_Y(y)$ and distribution function $F_Y(y)$ of $Y=\frac{X_1}{X_2}$ have the forms*

$$\begin{aligned} f_Y(y) &= \int_0^1 sgn\left(\sigma_2\Phi^{-1}(v)+\mu_2\right)\varphi\Big(\frac{(y\sigma_2-r\sigma_1)\Phi^{-1}(v)+y\mu_2-\mu_1}{\sigma_1\sqrt{1-r^2}}\Big)\frac{\sigma_2\Phi^{-1}(v)+\mu_2}{\sigma_1\sqrt{1-r^2}}dv, \quad (14) \\ F_Y(y) &= \Phi\Big(-\frac{\mu_2}{\sigma_2}\Big)+\int_0^1 sgn\left(\sigma_2\Phi^{-1}(v)+\mu_2\right)\Phi\Big(\frac{(y\sigma_2-r\sigma_1)\Phi^{-1}(v)+y\mu_2-\mu_1}{\sigma_1\sqrt{1-r^2}}\Big)dv, \quad (15) \end{aligned}$$

respectively, where $\varphi(x)$ and $\Phi(x)$ are PDF and CDF of the standard normal distribution, respectively, and $\Phi^{-1}(x)$ denotes for the inverse function of $\Phi(x)$.

Proof. Let $X_1\sim N(\mu_1,\sigma_1^2)$, and $X_2\sim N(\mu_2,\sigma_2^2)$; then, their CDFs and inverse functions can be expressed in the following form:

$$F_i(x)=\Phi\Big(\frac{x-\mu_i}{\sigma_i}\Big),\quad F_i^{-1}(v)=\sigma_i\Phi^{-1}(v)+\mu_i,\quad i=1,2.$$

Given Gaussian Copulas $C_r(u,v)$ with $|r|<1$, one can obtain its derivative $\frac{\partial C_r}{\partial v}(u,v)$, see Meyer (2013), as shown in the following:

$$\frac{\partial C_r}{\partial v}(u,v)=\Phi\Big(\frac{\Phi^{-1}(u)-r\Phi^{-1}(v)}{\sqrt{1-r^2}}\Big).$$

Now, applying Equation (9), we can simplify it to be

$$\begin{aligned} F_Y(y) &= \Phi\Big(-\frac{\mu_2}{\sigma_2}\Big) + \int_0^1 sgn\Big(\sigma_2\Phi^{-1}(v) + \mu_2\Big)\Phi\Big(\frac{\Phi^{-1}\Big(\Phi\Big(\frac{y\sigma_2\Phi^{-1}(v)+y\mu_2-\mu_1}{\sigma_1}\Big)\Big) - r\Phi^{-1}(v)}{\sqrt{1-r^2}}\Big) \\ &= \Phi\Big(-\frac{\mu_2}{\sigma_2}\Big) + \int_0^1 sgn\Big(\sigma_2\Phi^{-1}(v) + \mu_2\Big)\Phi\Big(\frac{(y\sigma_2 - r\sigma_1)\Phi^{-1}(v) + y\mu_2 - \mu_1}{\sigma_1\sqrt{1-r^2}}\Big)dv. \end{aligned}$$

Taking derivative of $F_y(y)$ with respect to y, one gets the density $f_Y(y)$ defined as in (14). The assertions of Corollary 2 hold. □

We note that the probability exhibited in (15) can be easily computed by using the following Monte Carlo algorithm: For each $y \in \mathbb{R}$, we generate V from the uniform distribution on the unit interval $[0,1]$ with sample size N, say $N = 10{,}000$, and then the estimated probability is given by

$$\widehat{F}_y(y) \approx \Phi\Big(-\frac{\mu_2}{\sigma_2}\Big) + \frac{1}{N}\sum_{i=1}^{N} sgn\Big(\sigma_2\Phi^{-1}(v_i) + \mu_2\Big)\Phi\Big(\frac{(y\sigma_2 - r\sigma_1)\Phi^{-1}(v_i) + y\mu_2 - \mu_1}{\sigma_1\sqrt{1-r^2}}\Big). \tag{16}$$

Using the result, we obtain the following corollary:

Corollary 3. *If $X_1 \sim N(\mu_1, \sigma_1^2)$, $X_2 \sim N(0, \sigma_2^2)$, and (X_1, X_2) follows Gaussian Copulas $C_r(u,v)$, $|r| < 1$, given in (46), then the median of $Y = \frac{X_1}{X_2}$ satisfies*

$$median(Y) = r\frac{\sigma_1}{\sigma_2}, \quad \text{for all } \mu_1 \in \mathbb{R}. \tag{17}$$

Proof. Since $X_2 \sim N(0, \sigma_2^2)$, $\Phi(-\frac{\mu_2}{\sigma_2}) = \Phi(0) = 0.5$. Hence, it is sufficient to prove that the integral term given in (15) is equal to zero. In fact, we find that

$$\begin{aligned} F_Y\Big(r\frac{\sigma_1}{\sigma_2}\Big) &= 0.5 + \int_0^1 sgn(\sigma_2\Phi^{-1}(v))\Phi\Big(\frac{-\mu_1}{\sigma_1\sqrt{1-r^2}}\Big)dv \\ &= 0.5 + \Phi\Big(\frac{-\mu_1}{\sigma_1\sqrt{1-r^2}}\Big)\Big[-\int_0^{1/2} dv + \int_{1/2}^1 dv\Big] \\ &= 0.5. \end{aligned}$$

Hence, the quantity $r\frac{\sigma_1}{\sigma_2}$ is the median of Y. The proof is complete. □

We turn to develop the proposition on density and distribution functions for the ratio of one variable over the sum of two variables $Z := \frac{X_1}{X_1+X_2}$ of two dependent random variables X_1 and X_2 by using copulas as stated in the following:

Proposition 2. *Suppose that (X_1, X_2) is a vector of two absolutely continuous random variables X_1 and X_2 with the marginal distributions F_1 and F_2, respectively, and let C be an absolutely continuous copula modeling dependence structure of the random vector (X_1, X_2), and define Z as*

$$Z := \frac{X_1}{X_1 + X_2}. \tag{18}$$

Then, the density $f_Z(z)$ and distribution function $F_Z(z)$ of Z are

$$f_Z(z) = \begin{cases} \int_0^1 \frac{|F_1^{-1}(u)|}{z^2} c\Big(u, F_2\Big(\frac{1-z}{z}F_1^{-1}(u)\Big)\Big) f_2\Big(\frac{1-z}{z}F_1^{-1}(u)\Big)\,du, & \text{if } z \neq 0, \\ f_1(0)\int_0^1 |F_2^{-1}(v)| c\Big(F_1(0), v\Big)dv, & \text{if } z = 0, \end{cases} \tag{19}$$

$$F_Z(z) = \mathbf{1}_{\{z\geq 0\}} + \int_0^1 \operatorname{sgn}\left(F_1^{-1}(u)\right)\left[\tfrac{\partial}{\partial u}C\Big(u, F_2\big(-F_1^{-1}(u)\big)\Big) - \tfrac{\partial}{\partial u}C\left(u, F_2\left(\tfrac{1-z}{z}F_1^{-1}(u)\right)\right)\right] du, \quad (20)$$

respectively, where $\mathbf{1}_{\{\cdot\}}$ *denotes an indicator function,* F_i^{-1} *denotes the inverse function of* F_i *for* $i = 1, 2$*,* c *is the density of copula* C*, and* $\operatorname{sgn}(\cdot)$ *is the sign function such that*

$$\mathbf{1}_{\{z\geq 0\}} = \begin{cases} 1, & \text{if } z \geq 0, \\ 0, & \text{if } z < 0, \end{cases} \quad \text{and} \quad \operatorname{sgn}(x) = \begin{cases} 1, & \text{if } x > 0, \\ -1, & \text{if } x < 0. \end{cases}$$

Proof. By defining

$$\begin{cases} Z_1 = \dfrac{X_1}{X_1 + X_2}, \\ Z_2 = X_1 + X_2. \end{cases}$$

Here, we note that, since X_1 and X_2 are absolutely continuous, $P(X_1 + X_2 = 0) = 0$; that is, $X_1 + X_2 \neq 0$ almost surely. Hence, the transformation $Z_1 = \frac{X_1}{X_1+X_2}$ always exists with probability 1 and we obtain the following inverse transformation:

$$\begin{cases} X_1 = Z_1 Z_2, \\ X_2 = Z_2 - Z_1 Z_2, \end{cases}$$

and the Jacobian

$$J = \begin{vmatrix} Z_2 & Z_1 \\ -Z_2 & 1 - Z_1 \end{vmatrix} = Z_2.$$

Thus, the joint density of Z_1 and Z_2 becomes

$$\begin{aligned} h(z_1, z_2) &= f(z_1 z_2, z_2 - z_1 z_2)\,|z_2| \\ &= |z_2|\, c\left(F_1(z_1 z_2), F_2\left(z_2 - z_1 z_2\right)\right) f_1(z_1 z_2) f_2\left(z_2 - z_1 z_2\right), \end{aligned}$$

which leads us to get the density of Z_1 such that

$$f_{Z_1}(z_1) = \int_{-\infty}^{\infty} |z_2|\, c\left(F_1(z_1 z_2), F_2\left(z_2 - z_1 z_2\right)\right) f_1(z_1 z_2) f_2\left(z_2 - z_1 z_2\right) dz_2. \quad (21)$$

If $z_1 = 0$, then by taking $v := F_2(z_2)$, we get

$$f_{Z_1}(0) = f_1(0) \int_0^1 \left|F_2^{-1}(v)\right| c\Big(F_1(0), v\Big) dv.$$

If $z_1 > 0$, then by taking $u := F_1(z_1 z_2)$, we obtain

$$f_{Z_1}(z_1) = \int_0^1 \frac{\left|F_1^{-1}(u)\right|}{z_1^2} c\left(u, F_2\left(\frac{1-z_1}{z_1}F_1^{-1}(u)\right)\right) f_2\left(\frac{1-z_1}{z_1}F_1^{-1}(u)\right) du.$$

If $z_1 < 0$, then also by taking $u := F_1(z_1 z_2)$, we yield

$$f_{Z_1}(z_1) = \int_1^0 \frac{\left|F_1^{-1}(u)\right|}{-z_1^2} c\left(u, F_2\left(\frac{1-z_1}{z_1}F_1^{-1}(u)\right)\right) f_2\left(\frac{1-z_1}{z_1}F_1^{-1}(u)\right) du. \quad (22)$$

Hence, for $z_1 \neq 0$, we obtain

$$f_{Z_1}(z_1) = \int_0^1 \frac{\left|F_1^{-1}(u)\right|}{z_1^2} c\Big(u, F_2\left(\frac{1-z_1}{z_1}F_1^{-1}(u)\right)\Big) f_2\left(\frac{1-z_1}{z_1}F_1^{-1}(u)\right) du \quad \text{for} \quad z_1 \neq 0.$$

As a consequence, the distribution of Z_1 becomes

$$F_{Y_1}(t) = \int_0^1 \int_{-\infty}^t \frac{\left|F_1^{-1}(u)\right|}{z_1^2} c\Big(u, F_2\left(\frac{1-z_1}{z_1}F_1^{-1}(u)\right)\Big) f_2\left(\frac{1-z_1}{z_1}F_1^{-1}(u)\right) dz_1 du. \tag{23}$$

Setting $v = F_2\left(\frac{1-z_1}{z_1}F_1^{-1}(u)\right) \implies dv = -\frac{F_1^{-1}(u)}{z_1^2} f_2\left(\frac{1-z_1}{z_1}F_1^{-1}(u)\right) dz_1$, and note that

$$F_1^{-1}(u) \geq 0 \Longleftrightarrow u \geq F_1(0), \text{ and } F_1^{-1}(u) \leq 0 \Longleftrightarrow u \leq F_1(0),$$

we consider two cases as follows:

(i) **Case 1:** For $t < 0$, we have

$$\begin{aligned}
F_{Z_1}(t) &= \int_0^{F_1(0)} \int_{F_2(-F_1^{-1}(u))}^{F_2\left(\frac{1-t}{t}F_1^{-1}(u)\right)} \frac{\partial^2}{\partial u \partial v} C(u,v)\, dv du - \int_{F_1(0)}^1 \int_{F_2(-F_1^{-1}(u))}^{F_2\left(\frac{1-t}{t}F_1^{-1}(u)\right)} \frac{\partial^2}{\partial u \partial v} C(u,v)\, dv du \\
&= \int_0^{F_1(0)} \left[\frac{\partial}{\partial u} C\left(u, F_2\left(\frac{1-t}{t}F_1^{-1}(u)\right)\right) - \frac{\partial}{\partial u} C\Big(u, F_2\big(-F_1^{-1}(u)\big)\Big)\right] du \\
&\quad - \int_{F_1(0)}^1 \left[\frac{\partial}{\partial u} C\left(u, F_2\left(\frac{1-t}{t}F_1^{-1}(u)\right)\right) - \frac{\partial}{\partial u} C\Big(u, F_2\big(-F_1^{-1}(u)\big)\Big)\right] du \\
&= \int_0^1 \operatorname{sgn}\left(F_1^{-1}(u)\right) \left[\frac{\partial}{\partial u} C\Big(u, F_2\big(-F_1^{-1}(u)\big)\Big) - \frac{\partial}{\partial u} C\left(u, F_2\left(\frac{1-t}{t}F_1^{-1}(u)\right)\right)\right] du.
\end{aligned} \tag{24}$$

(ii) **Case 2:** For $t \geq 0$, we first split the integrals

$$\begin{aligned}
F_{Z_1}(t) &= \int_0^1 \int_{-\infty}^0 \frac{\left|F_1^{-1}(u)\right|}{z_1^2} c\Big(u, F_2\left(\frac{1-z_1}{z_1}F_1^{-1}(u)\right)\Big) f_2\left(\frac{1-z_1}{z_1}F_1^{-1}(u)\right) dz_1 du \\
&\quad + \int_0^1 \int_0^t \frac{\left|F_1^{-1}(u)\right|}{z_1^2} c\Big(u, F_2\left(\frac{1-z_1}{z_1}F_1^{-1}(u)\right)\Big) f_2\left(\frac{1-z_1}{z_1}F_1^{-1}(u)\right) dz_1 du \\
&=: I_1 + I_2.
\end{aligned} \tag{25}$$

We then apply (24) to obtain the following expression for the integral I_1:

$$\begin{aligned}
I_1 = F_{Z_1}(0) &= \int_0^1 \operatorname{sgn}\left(F_1^{-1}(u)\right) \left[\frac{\partial}{\partial u} C\Big(u, F_2\big(-F_1^{-1}(u)\big)\Big) - \frac{\partial}{\partial u} C\left(u, \lim_{t \to 0^-} F_2\left(\frac{1-t}{t}F_1^{-1}(u)\right)\right)\right] du \\
&= \int_0^1 \operatorname{sgn}\left(F_1^{-1}(u)\right) \frac{\partial}{\partial u} C\Big(u, F_2\big(-F_1^{-1}(u)\big)\Big) du \\
&\quad + \int_0^{F_1(0)} \frac{\partial}{\partial u} C(u,1)\, du - \int_{F_1(0)}^1 \frac{\partial}{\partial u} C(u,0)\, du \\
&= \int_0^1 \operatorname{sgn}\left(F_1^{-1}(u)\right) \frac{\partial}{\partial u} C\Big(u, F_2\big(-F_1^{-1}(u)\big)\Big) du + F_1(0),
\end{aligned} \tag{26}$$

and obtain the following expression for the integral I_2:

$$\begin{aligned}
I_2 &= \int_0^{F_1(0)} \int_0^{F_2\left(\frac{1-t}{t}F_1^{-1}(u)\right)} \frac{\partial^2}{\partial u \partial v} C(u,v)\, dv du - \int_{F_1(0)}^1 \int_1^{F_2\left(\frac{1-t}{t}F_1^{-1}(u)\right)} \frac{\partial^2}{\partial u \partial v} C(u,v)\, dv du \\
&= \int_0^{F_1(0)} \left[\frac{\partial}{\partial u} C\left(u, F_2\left(\frac{1-t}{t}F_1^{-1}(u)\right)\right) - \frac{\partial}{\partial u} C\big(u, 0\big)\right] du \\
&\quad - \int_{F_1(0)}^1 \left[\frac{\partial}{\partial u} C\left(u, F_2\left(\frac{1-t}{t}F_1^{-1}(u)\right)\right) - \frac{\partial}{\partial u} C\big(u, 1\big)\right] du \\
&= 1 - F_1(0) - \int_0^1 \operatorname{sgn}\left(F_1^{-1}(u)\right) \frac{\partial}{\partial u} C\Big(u, F_2\big(-F_1^{-1}(u)\big)\Big) du.
\end{aligned} \tag{27}$$

From (26) and (27), we get the following for $t \geq 0$,

$$F_{Z_1}(t) = 1 + \int_0^1 \text{sgn}\left(F_1^{-1}(u)\right)\left[\frac{\partial}{\partial u}C\left(u, F_2\left(-F_1^{-1}(u)\right)\right) - \frac{\partial}{\partial u}C\left(u, F_2\left(\frac{1-t}{t}F_1^{-1}(u)\right)\right)\right]du. \quad (28)$$

Combining (24) and (28) imply (20), we complete the proof. □

In the situation X_1 and X_2 are independent, applying Proposition 2, we obtain the following corollary:

Corollary 4. *When X_1 and X_2 are independent, then its copula $C(u,v) = uv$ has the density $c(u,v) = 1$, $\forall u, v \in \mathbb{I}$ and the density $f_Z(z)$ and distribution function $F_Z(z)$ for the ratio of one variable over the sum of two variables $Z := \frac{X_1}{X_1+X_2}$ of two independent random variables X_1 and X_2 become*

$$f_Z(z) = \int_{-\infty}^{\infty} |x|\, f_1(xz) f_2\left((1-x)z\right) dx$$

and

$$\begin{aligned} F_Z(z) &= \mathbf{1}_{\{z\geq 0\}} + \int_0^1 \text{sgn}\left(F_1^{-1}(u)\right)\left[F_2\left(-F_1^{-1}(u)\right) - F_2\left(\frac{1-z}{z}F_1^{-1}(u)\right)\right]du, \\ &= \mathbf{1}_{\{z\geq 0\}} + \int_{-\infty}^{\infty} \text{sgn}(x)\left[F_2(-x) - F_2\left(\frac{1-z}{z}x\right)\right] f_1(x)dx, \end{aligned}$$

respectively.

Next, we apply Equation (20) to derive the distribution function of $Z := \frac{X_1}{X_1+X_2}$ in the situation that both X_1 and X_2 are normal distributed such that their dependence structure can be captured by Gaussian Copulas as shown in the following corollary:

Corollary 5. *Assume that $X_1 \sim N(\mu_1, \sigma_1^2)$, $X_2 \sim N(\mu_2, \sigma_2^2)$, and (X_1, X_2) follows Gaussian Copulas $C_r(u,v)$, $|r| < 1$, given in (46). Then, distribution function $F_Z(z)$ of $Z := \frac{X_1}{X_1+X_2}$ has the form*

$$\begin{aligned} F_Z(z) = &= \mathbf{1}_{\{z\geq 0\}} + 2\Phi\left(\frac{\mu_1}{\sigma_1}\right) - 1 - \int_0^1 sgn\left(\sigma_1\Phi^{-1}(u) + \mu_1\right)\Phi\left(\frac{(\sigma_1 + r\sigma_2)\Phi^{-1}(u) + \mu_1 - \mu_2}{\sigma_2\sqrt{1-r^2}}\right)du \\ &\quad - \int_0^1 sgn\left(\sigma_1\Phi^{-1}(u) + \mu_1\right)\Phi\left(\frac{[(1-z)\sigma_1 - zr\sigma_2]\Phi^{-1}(u) - z(\mu_1+\mu_2) + \mu_1}{z\sigma_2\sqrt{1-r^2}}\right)du, \end{aligned} \quad (29)$$

where $\Phi(x)$ and $\Phi^{-1}(x)$ are CDF and its inverse of the standard normal random variable, respectively.

Proof. Let $X_1 \sim N(\mu_1, \sigma_1^2)$ and $X_2 \sim N(\mu_2, \sigma_2^2)$, the CDFs and their inverse functions can be written in the form

$$F_i(x) = \Phi\left(\frac{x-\mu_i}{\sigma_i}\right), \quad F_i^{-1}(v) = \sigma_i\Phi^{-1}(v) + \mu_i, \quad i = 1, 2.$$

Given Gaussian Copulas $C_r(u,v)$, $|r| < 1$, we apply the results from Meyer (2013) to obtain its derivative $\frac{\partial C_r}{\partial u}(u,v)$ as shown in the following:

$$\frac{\partial C_r}{\partial u}(u,v) = \Phi\left(\frac{\Phi^{-1}(v) - r\Phi^{-1}(u)}{\sqrt{1-r^2}}\right).$$

Applying Equation (20), one can simplify it to be

$$\begin{aligned}
F_Z(z) &= \mathbf{1}_{\{z\geq 0\}} + \int_0^1 sgn\Big(\sigma_1\Phi^{-1}(u)+\mu_1\Big)\Phi\Big(\frac{\Phi^{-1}\Big(\Phi\Big(-\frac{\sigma_1\Phi^{-1}(u)+\mu_1-\mu_2}{\sigma_2}\Big)\Big)-r\Phi^{-1}(u)}{\sqrt{1-r^2}}\Big)du \\
&\quad -\int_0^1 sgn\Big(\sigma_1\Phi^{-1}(u)+\mu_1\Big)\Phi\Big(\frac{\Phi^{-1}\Big(\Phi\Big(\frac{\frac{1-z}{z}\big[\sigma_1\Phi^{-1}(u)+\mu_1\big]-\mu_2}{\sigma_2}\Big)\Big)-r\Phi^{-1}(u)}{\sqrt{1-r^2}}\Big)du \\
&= \mathbf{1}_{\{z\geq 0\}} + \int_0^1 sgn\Big(\sigma_1\Phi^{-1}(u)+\mu_1\Big)\Phi\Big(-\frac{(\sigma_1+r\sigma_2)\Phi^{-1}(u)+\mu_1-\mu_2}{\sigma_2\sqrt{1-r^2}}\Big)du \\
&\quad -\int_0^1 sgn\Big(\sigma_1\Phi^{-1}(u)+\mu_1\Big)\Phi\Big(\frac{[(1-z)\sigma_1-zr\sigma_2]\Phi^{-1}(u)-z(\mu_1+\mu_2)+\mu_1}{z\sigma_2\sqrt{1-r^2}}\Big)du \\
&= \mathbf{1}_{\{z\geq 0\}} + 2\Phi\Big(\frac{\mu_1}{\sigma_1}\Big)-1-\int_0^1 sgn\Big(\sigma_1\Phi^{-1}(u)+\mu_1\Big)\Phi\Big(\frac{(\sigma_1+r\sigma_2)\Phi^{-1}(u)+\mu_1-\mu_2}{\sigma_2\sqrt{1-r^2}}\Big)du \\
&\quad -\int_0^1 sgn\Big(\sigma_1\Phi^{-1}(u)+\mu_1\Big)\Phi\Big(\frac{[(1-z)\sigma_1-zr\sigma_2]\Phi^{-1}(u)-z(\mu_1+\mu_2)+\mu_1}{z\sigma_2\sqrt{1-r^2}}\Big)du.
\end{aligned}$$

In the last step of the above, we use the property $\Phi(-x) = 1-\Phi(x)$ and

$$\int_0^1 sgn\Big(\sigma_1\Phi^{-1}(u)+\mu_1\Big)du = -\int_0^{\Phi\left(-\frac{\mu_1}{\sigma_1}\right)} du + \int_{\Phi\left(-\frac{\mu_1}{\sigma_1}\right)}^1 du = 1-2\Phi\Big(-\frac{\mu_1}{\sigma_1}\Big) = 2\Phi\Big(\frac{\mu_1}{\sigma_1}\Big)-1.$$

The proof is complete. □

We note that the probability given in (29) can also be easily computed by using the following Monte Carlo algorithm: For each $z \in \mathbb{R}$, we first generate U from the uniform distribution on the unit interval $[0,1]$ with sample size N, say $N = 10,000$. Then, we obtain the following estimated probability:

$$\begin{aligned}
\widehat{F}_Z(z) &\approx \mathbf{1}_{\{z\geq 0\}} + 2\Phi\Big(\frac{\mu_1}{\sigma_1}\Big)-1-\frac{1}{N}\sum_{i=1}^N sgn\Big(\sigma_1\Phi^{-1}(u_i)+\mu_1\Big)\Phi\Big(\frac{(\sigma_1+r\sigma_2)\Phi^{-1}(u_i)+\mu_1-\mu_2}{\sigma_2\sqrt{1-r^2}}\Big) \\
&\quad -\frac{1}{N}\sum_{i=1}^N sgn\Big(\sigma_1\Phi^{-1}(u_i)+\mu_1\Big)\Phi\Big(\frac{[(1-z)\sigma_1-zr\sigma_2]\Phi^{-1}(u_i)-z(\mu_1+\mu_2)+\mu_1}{z\sigma_2\sqrt{1-r^2}}\Big). \qquad (30)
\end{aligned}$$

Using the above results, we obtain the following corollary:

Corollary 6. *Assume that $X_1 \sim N(0,\sigma^2)$, $X_2 \sim N(0,\sigma^2)$ and (X_1,X_2) follows Gaussian Copulas $C_r(u,v)$, $|r| < 1$, given in (46). Then, the median of $Z := \frac{X_1}{X_1+X_2}$ is equal to $\frac{1}{2}$.*

Proof. Because $X_1 \sim N(0,\sigma^2)$, $X_2 \sim N(0,\sigma^2)$ and $sgn\Big(\sigma\Phi^{-1}(u)\Big) = sgn\Big(\Phi^{-1}(u)\Big)$, we obtain CDF of the ratio $Z := \frac{X_1}{X_1+X_2}$ from (29) as shown in the following:

$$F_Z\Big(z\Big) = \mathbf{1}_{\{z\geq 0\}} - \int_0^1 sgn\Big(\Phi^{-1}(u)\Big)\Big[\Phi\Big(\frac{(1+r)\Phi^{-1}(u)}{\sqrt{1-r^2}}\Big)+\Phi\Big(\frac{[1-z-zr]\Phi^{-1}(u)}{z\sqrt{1-r^2}}\Big)\Big]du. \qquad (31)$$

We get

$$F_Z\Big(\frac{1}{2}\Big) = 1 - \int_0^1 sgn\Big(\Phi^{-1}(u)\Big)\Big[\Phi\Big(\frac{(1+r)\Phi^{-1}(u)}{\sqrt{1-r^2}}\Big)+\Phi\Big(\frac{[1-r]\Phi^{-1}(u)}{\sqrt{1-r^2}}\Big)\Big]du. \qquad (32)$$

Hence, it is sufficient to prove that the integral term given in (32) is equal to $\frac{1}{2}$. We let

$$I_1 := \int_0^1 sgn\Big(\Phi^{-1}(u)\Big)\Phi\Big(\frac{(1+r)\Phi^{-1}(u)}{\sqrt{1-r^2}}\Big)du, \quad I_2 := \int_0^1 sgn\Big(\Phi^{-1}(u)\Big)\Phi\Big(\frac{(1-r)\Phi^{-1}(u)}{\sqrt{1-r^2}}\Big)du,$$

and denote $\partial_i C_r(u,v), i=1,2$ to be the partial derivative of $C_r(u,v)$ with respect to the ith variable, That is, $\partial_1 C_r(u,v) = \frac{\partial}{\partial u}C_r(u,v)$ and $\partial_2 C_r(u,v) = \frac{\partial}{\partial v}C_r(u,v)$. One can observe that

$$\partial_1 C_r(u,u) = \Phi\Big(\frac{(1-r)\Phi^{-1}(u)}{\sqrt{1-r^2}}\Big).$$

Since Gaussian Copulas is symmetric; that is, $C_r(u,v) = C_r(v,u)$, we have $\partial_1 C_r(u,v) = \partial_2 C_r(v,u)$, and, thus, for $u=v$, we can derive $\partial_1 C_r(u,u) = \partial_2 C_r(u,u)$. Thereafter, the differentiation of $C_r(u,u)$ can be obtained:

$$dC_r(u,u) = \big[\partial_1 C_r(u,u) + \partial_2 C_r(u,u)\big]du = 2\partial_1 C_r(u,u),$$

and we get

$$\begin{aligned} I_2 = \frac{1}{2}\int_0^1 sgn\Big(\Phi^{-1}(u)\Big)dC_r(u,u) &= -\frac{1}{2}\int_0^{1/2} dC_r(u,u) + \frac{1}{2}\int_{1/2}^1 dC_r(u,u) \\ &= -\frac{1}{2}C_r\Big(\frac{1}{2},\frac{1}{2}\Big) + \frac{1}{2}\Big[1 - C_r\Big(\frac{1}{2},\frac{1}{2}\Big)\Big] \\ &= \frac{1}{2} - C_r\Big(\frac{1}{2},\frac{1}{2}\Big). \end{aligned} \tag{33}$$

Similarly, for I_1, since $\Phi^{-1}(1-u) = -\Phi^{-1}(u)$, we obtain

$$\partial_1 C_r(u,1-u) = \Phi\Big(\frac{\Phi^{-1}(1-u) - r\Phi^{-1}(u)}{\sqrt{1-r^2}}\Big) = \Phi\Big(-\frac{(1+r)\Phi^{-1}(u)}{\sqrt{1-r^2}}\Big) = 1 - \Phi\Big(\frac{(1+r)\Phi^{-1}(u)}{\sqrt{1-r^2}}\Big),$$

and get

$$I_1 = \int_0^1 sgn\Big(\Phi^{-1}(u)\Big)du - \int_0^1 sgn\Big(\Phi^{-1}(u)\Big)\partial_1 C_r(u,1-u)du = -\int_0^1 sgn\Big(\Phi^{-1}(u)\Big)\partial_1 C_r(u,1-u)du,$$

in which we apply $\int_0^1 sgn\Big(\Phi^{-1}(u)\Big)du = 0$. From symmetry of the Gaussian Copulas, we also have $\partial_1 C_r(1-u,u) = \partial_2 C_r(u,1-u)$, obtain the differentiation of $C_r(u,1-u)$ given by

$$dC_r(u,1-u) = \big[\partial_1 C_r(u,1-u) - \partial_2 C_r(u,1-u)\big]du = \big[\partial_1 C_r(u,1-u) - \partial_1 C_r(1-u,u)\big]du,$$

and get

$$\begin{aligned} I_1 &= \int_0^{1/2} \partial_1 C_r(u,1-u)du - \int_{1/2}^1 \partial_1 C_r(u,1-u)du, \\ &= \int_0^{1/2} \partial_1 C_r(u,1-u)du + \int_{1/2}^0 \partial_1 C_r(1-u,u)du \\ &= \int_0^{1/2} \partial_1 C_r(u,1-u)du - \int_0^{1/2} \partial_1 C_r(1-u,u)du \\ &= \int_0^{1/2} dC_r(u,1-u) \\ &= C_r\Big(\frac{1}{2},\frac{1}{2}\Big). \end{aligned} \tag{34}$$

Combining (32)–(34), we find that $F_Z\left(\frac{1}{2}\right) = \frac{1}{2}$. Hence, the quantity $\frac{1}{2}$ is the median of Z. The proof is complete. □

Applying the proof of Corollary 6, we extend the result to obtain the following corollary for a larger family of symmetric distribution and symmetric copulas:

Corollary 7. *Assume that X_1 and X_2 are identically and symmetrically distributed with distribution F that has zero median and the dependence structure of (X_1, X_2) is modelled by a family of symmetric copulas $C(u,v)$, i.e., $C(u,v) = C(v,u)$ for all $u, v \in \mathbb{I}$. Then, the median of $Z := \frac{X_1}{X_1+X_2}$ is equal to $\frac{1}{2}$.*

Proof. Since X_1 and X_2 are identically and symmetrically distributed with distribution F and zero median, we have $F(-x) = 1 - F(x)$, for all $x \in \mathbb{R}$. By applying Equation (20), we obtain the CDF of the ratio Z, which is defined by

$$\begin{aligned} F_Z(z) &= \mathbf{1}_{\{z\geq 0\}} + \int_0^1 \operatorname{sgn}\left(F^{-1}(u)\right)\left[\frac{\partial}{\partial u}C\Big(u, F\big(-F^{-1}(u)\big)\Big) - \frac{\partial}{\partial u}C\left(u, F\left(\frac{1-z}{z}F^{-1}(u)\right)\right)\right]du, \\ &= \mathbf{1}_{\{z\geq 0\}} + \int_0^1 \operatorname{sgn}\left(F^{-1}(u)\right)\left[\frac{\partial}{\partial u}C\big(u, 1-u\big) - \frac{\partial}{\partial u}C\left(u, F\left(\frac{1-z}{z}F^{-1}(u)\right)\right)\right]du. \end{aligned} \tag{35}$$

Since copulas $C(u,v)$ are symmetric, i.e., $C(v,u) = C(u,v)$, we have $\partial_1 C(v,u) = \partial_2 C(u,v)$, and, thus, for $v = 1-u$, one can easily obtain $\partial_1 C(1-u,u) = \partial_2 C(u,1-u)$ and find that the differentiation of $C(u,1-u)$ with respect to u satisfies

$$dC(u,1-u) = \left[\partial_1 C(u,1-u) - \partial_2 C(u,1-u)\right]du = \left[\partial_1 C(u,1-u) - \partial_1 C(1-u,u)\right]du.$$

In addition, because the distribution F has zero median; that is, $F(0) = 0.5$, we have

$$\begin{aligned} \int_0^1 \operatorname{sgn}\left(F^{-1}(u)\right)\frac{\partial}{\partial u}C\big(u, 1-u\big) &= -\int_0^{1/2} \partial_1 C(u,1-u)du + \int_{1/2}^1 \partial_1 C(u,1-u)du, \\ &= -\int_0^{1/2} \partial_1 C(u,1-u)du - \int_{1/2}^0 \partial_1 C(1-u,u)du \\ &= -\int_0^{1/2} \partial_1 C(u,1-u)du + \int_0^{1/2} \partial_1 C(1-u,u)du \\ &= -\int_0^{1/2} dC(u,1-u) = -C\Big(\frac{1}{2},\frac{1}{2}\Big). \end{aligned}$$

Therefore, we get

$$F_Z(z) = \mathbf{1}_{\{z\geq 0\}} - C\Big(\frac{1}{2},\frac{1}{2}\Big) - \int_0^1 \operatorname{sgn}\left(F^{-1}(u)\right)\frac{\partial}{\partial u}C\left(u, F\left(\frac{1-z}{z}F^{-1}(u)\right)\right)du, \tag{36}$$

and thus,

$$F_Z(0.5) = 1 - C\Big(\frac{1}{2},\frac{1}{2}\Big) - \int_0^1 \operatorname{sgn}\left(F^{-1}(u)\right)\frac{\partial}{\partial u}C(u,u)\,du. \tag{37}$$

Similarly, since copulas $C(u,v)$ are symmetric, i.e., $C(v,u) = C(u,v)$, we have $\partial_1 C(v,u) = \partial_2 C(u,v)$, and, thus, for $v = u$, we get $\partial_1 C(u,u) = \partial_2 C(u,u)$. Thus, the differentiation of $C(u,u)$ with respect to u satisfies

$$dC(u,u) = \left[\partial_1 C(u,u) + \partial_2 C(u,u)\right]du = 2\partial_1 C(u,u)du.$$

Applying this relation, we find

$$\int_0^1 \operatorname{sgn}\left(F^{-1}(u)\right) \frac{\partial}{\partial u} C(u,u)\, du = \frac{1}{2}\int_0^1 \operatorname{sgn}\left(F^{-1}(u)\right) dC(u,u) = \frac{1}{2} - C\left(\frac{1}{2},\frac{1}{2}\right).$$

Hence, $F_Z(0.5) = 0.5$, i.e., $\frac{1}{2}$ is median of Z. The proof is complete. □

Remark: In the literature, they are many symmetric distributions with zero median, for example, normal $N(0,\sigma)$, Student-t t_ν, Cauchy distribution with location parameter $\alpha = 0$, uniform $U(-a,a), a \in \mathbb{R}_+$, and logistic distribution with zero location. In addition, Elliptical copulas (Gaussian, Student-t copulas) and Archimedian copulas (Clayton, Gumbel, Frank, Joe,...) are classes of symmetric copulas. Thus, if we apply (X_1, X_2) with these distributions and these copulas, Proposition 7 tells us that the random variable $Z := \frac{X_1}{X_1+X_2}$ always gets the median one-half. This theoretical result is consistent with our simulation results displayed in the next section.

We note that the formulas given in (8), (9), (19) and (20) may not have closed forms. However, they are easily computed by using the Monte Carlo (MC) simulation method or any techniques of numerical integration. We provide simulation studies in the next section.

5. A Simulation Study

Since the density and the CDF formula of the ratio $Y = \frac{X_1}{X_2}$ $[Z = \frac{X_1}{X_1+X_2}]$ expressed in (8) and (9) ((19) and (20)) are in terms of integrals and are very complicated, we cannot obtain the exact forms of the density and the CDF. To circumvent the difficulty, in this paper, we propose to use the Monte Carlo algorithm, numerical analysis and graphical approach to study behavior of density and distribution and the changes of their shapes when parameters are changing.

Suppose that X_1 and X_2 are normally distributed and denoted by $X_i \sim N(\mu_i, \sigma_i^2)$ for $i = 1,2$ with PDF given by

$$f_{X_i}(x) = \frac{1}{\sqrt{2\pi}\sigma_i} \exp\left(-\frac{(x-\mu_i)^2}{2\sigma_i^2}\right).$$

Without loss of generality, we consider $\mu_1 = \mu_2 = 0$ and $\sigma_1 = \sigma_2 = 1$. We note that, if X_1 and X_2 are independent and standard normal distributed, then it is well known that $Y = \dfrac{X_1}{X_2}$ follows standard Cauchy distribution. The Cauchy distribution is a type of distribution that has no mean and also does not exist any higher moments. To circumvent this problem, one may use median to measure the central tendency and use range or interquartile range to measure the spread of the distribution. The general Cauchy distribution has the following PDF:

$$f(x) := \frac{\beta}{\pi(\beta^2 + (x-\alpha)^2)}, \quad \alpha \in \mathbb{R}, \beta > 0.$$

Thus, location and scale parameter of $Y = \dfrac{X_1}{X_2}$ are $\alpha = 0$ and $\beta = 1$, respectively, if X_1 and X_2 are identically independent standard normal distributed.

We now investigate different dependence structures of X_1 and X_2 through several families of copulas and observe the shapes of the corresponding distributions for both Y and Z as well as estimate their percentiles at different levels including $0.05, 0.25, 0.5, 0.75, 0.95$ and denote the corresponding percentiles to be $Q_{0.05}, Q_{0.25}, Q_{0.50}, Q_{0.75}$, and $Q_{0.95}$, respectively. In risk analysis, the percentile $Q_{0.05}$ is often used as the Value-at-Risk (VaR) 5% while $Q_{0.25}$ and $Q_{0.75}$ are, respectively, called the first and third quartile of the random variable and the interquartile range (IQR) is defined by their difference; that is, $IQR = Q_{0.75} - Q_{0.25}$. The median measures the center of the distribution, which is equal to $Q_{0.5}$.

For each copula $C_\theta(u,v)$, the PDF and CDF of Y and Z can be plotted on the interval $[-4,4]$ by using the following steps:

(i) For each pair of y and $z \in \{-4, -3.9, -3.8, \ldots, 4\}$, generate the uniform random variables U and V on the unit interval; that is, $U, V \sim Uniform[0,1]$ with the sample size N, say $N = 10{,}000$;
(ii) estimate the values for $f_Y(y)$, $F_Y(y)$, $f_Z(z)$, and $F_Z(z)$ by using

$$\widehat{f}_Y(y) \approx \frac{1}{N}\sum_{i=1}^{N} |F_2^{-1}(v_i)| c\Big(F_1\left(yF_2^{-1}(v_i)\right), v_i\Big) f_1\left(yF_2^{-1}(v_i)\right), \tag{38}$$

$$\widehat{F}_Y(y) \approx F_2(0) + \frac{1}{N}\sum_{i=1}^{N} \operatorname{sgn}\left(F_2^{-1}(v_i)\right) \frac{\partial}{\partial v} C\left(F_1\left(yF_2^{-1}(v_i)\right), v_i\right), \tag{39}$$

$$\widehat{f}_Z(z) \approx \begin{cases} \frac{1}{N}\sum_{i=1}^{N} \frac{|F_1^{-1}(u_i)|}{z^2} c\Big(u_i, F_2\left(\frac{1-z}{z}F_1^{-1}(u_i)\right)\Big) f_2\left(\frac{1-z}{z}F_1^{-1}(u_i)\right), & \text{if } z \neq 0, \\ \frac{f_1(0)}{N}\sum_{i=1}^{N} \left|F_2^{-1}(v_i)\right| c\Big(F_1(0), v_i\Big), & \text{if } z = 0, \end{cases} \tag{40}$$

$$\widehat{F}_Z(z) \approx \mathbf{1}_{\{z \geq 0\}} + \frac{1}{N}\sum_{i=1}^{N} \operatorname{sgn}\left(F_1^{-1}(u_i)\right) \Big[\frac{\partial C}{\partial u}(u_i, F_2(-F_1^{-1}(u_i))) - \frac{\partial C}{\partial u}(u_i, F_2(\frac{1-z}{z}F_1^{-1}(u_i)))\Big], \tag{41}$$

in which the density copula $c_\theta(u_i, v_i)$, the derivatives $\frac{\partial}{\partial u}C_\theta(u_i, v_i)$ and $\frac{\partial}{\partial v}C_\theta(u_i, v_i)$ can be obtained by using the packages of *VineCopula* in *R* language; and

(iii) plot $\widehat{f}_Y(y)$, $\widehat{F}_Y(y)$, $\widehat{f}_Z(z)$, and $\widehat{F}_Z(z)$ with $y, z \in \{-4, -3.9, -3.8, \cdots, 3.9, 4\}$.

To estimate percentiles Q_α's of Y and Z, we first construct the joint distribution of (X_1, X_2) by using Sklar's Theorem as shown in the following: For each copula $C_\theta(u, v)$, we first obtain the joint CDF of (X_1, X_2) such that

$$H_\theta(x_1, x_2) = C_\theta\Big(F_1(x_1), F_2(x_2)\Big).$$

We then repeat 5000 times, $k = 1, 2, \ldots, 5000$ for the following steps in the computation:

(1) For each repetition $k = 1, 2, \ldots, 5000$,

(i) generate (X_1, X_2) from $H_\theta(x_1, x_2)$ of sample size 10^4 by using the package *copula* in *R* language and define

$$y_i^{(k)} = \frac{x_{1i}^{(k)}}{x_{2i}^{(k)}}, \quad i = 1, 2, \cdots, 10^4; \tag{42}$$

$$z_i^{(k)} = \frac{x_{1i}^{(k)}}{x_{1i}^{(k)} + x_{2i}^{(k)}}, \quad i = 1, 2, \cdots, 10^4; \tag{43}$$

(ii) estimate the percentiles Q_α with $\alpha = 0.05, 0.25, 0.5, 0.75, 0.95$ for both Y and Z by using the following formula

$$\widehat{Q}_\alpha(Y^{(k)}) = y^{(k)}_{(\lfloor h \rfloor)} + (h - \lfloor h \rfloor)\left(y^{(k)}_{(\lfloor h \rfloor+1)} - y^{(k)}_{(\lfloor h \rfloor)}\right), \quad \textit{with } h = \left(10^4 - 1\right)\alpha + 1, \tag{44}$$

$$\widehat{Q}_\alpha(Z^{(k)}) = z^{(k)}_{(\lfloor h \rfloor)} + (h - \lfloor h \rfloor)\left(z^{(k)}_{(\lfloor h \rfloor+1)} - z^{(k)}_{(\lfloor h \rfloor)}\right), \tag{45}$$

where $y^{(k)}_{(1)} \leq y^{(k)}_{(2)} \leq \ldots \leq y^{(k)}_{(N)}$ and $z^{(k)}_{(1)} \leq z^{(k)}_{(2)} \leq \ldots \leq z^{(k)}_{(N)}$ denote the order statistics of both $y^{(k)}$ and $z^{(k)}$, respectively, and $\lfloor h \rfloor$ denotes integer part of h.

(2) Finally, we take average for each of the above quantities by using the following formula:

$$\widehat{Q}_\alpha(Y) = \frac{1}{5000}\sum_{k=1}^{5000}\widehat{Q}_\alpha(Y^{(k)}),$$
$$\widehat{Q}_\alpha(Z) = \frac{1}{5000}\sum_{k=1}^{5000}\widehat{Q}_\alpha(Z^{(k)}),$$

to obtain the estimates of the percentiles for Y and Z.

We note that the algorithms discussed in the above can be applied to any non-symmetric marginal distribution and skewed copulas family with absolutely continuous random variable that could contain negative range, for example, generalized skewed-t distribution and skewed-t Copulas. [2]

5.1. Gaussian Copulas

We first investigate dependence structures of X_1 and X_2 through Gaussian Copulas $C_r(u,v)$ and observe the shapes of the corresponding distributions for both Y and Z.

$$C_r(u,v) = \frac{1}{2\pi\sqrt{1-r^2}}\int_{-\infty}^{\Phi^{-1}(u)}\int_{-\infty}^{\Phi^{-1}(v)}\exp\left(-\frac{s^2-2rst+t^2}{2(1-r^2)}\right)dsdt, \tag{46}$$

where $\Phi^{-1}(x)$ is the inverse of standard normal CDF and r is Pearson's correlation coefficient between X_1 and X_2, $|r| < 1$. We now consider the cases with $r = -0.9, -0.5, 0, 0.5, 0.9$. When $r = 0$, it is corresponding to the independence situation, and we get PDFs and CDFs of Y and Z shown in Figures 1 and 2, respectively. As can be seen from the Figures and Tables 1 and 2, when the parameter r varies from negative to positive, the median is totally equal to the Pearson's correlation coefficient r. The more correlated, that is, the higher $|r|$ between X_1 and X_2 is, the smaller the spread, that is, IQR of Y, becomes. In contrast to Y, the center of Z is definitely unchanged (0.5), but the scale parameter of Z is smaller indicated by the higher height of the density. The shapes of both Y and Z are symmetric.

Table 1. Some percentiles of $Y = X_1/X_2$, where (X_1, X_2) follows Gaussian Copulas.

r	$\tau(C_r)$	$Q_{0.05}$	$Q_{0.25}$	Median	$Q_{0.75}$	$Q_{0.95}$
−0.9	−0.71	−3.65	−1.34	−0.9	−0.46	1.85
−0.5	−0.33	−5.97	−1.37	−0.5	0.37	4.97
0	0	−6.31	−1.00	0.0	1.00	6.31
0.5	0.33	−4.97	−0.37	0.5	1.37	5.97
0.9	0.71	−1.85	0.46	0.9	1.34	3.65

Table 2. Some percentiles of $Z = X_1/(X_1 + X_2)$, where (X_1, X_2) follows Gaussian Copulas.

r	$\tau(C_r)$	$Q_{0.05}$	$Q_{0.25}$	Median	$Q_{0.75}$	$Q_{0.95}$
−0.9	−0.71	−13.27	−1.68	0.5	2.68	14.26
−0.5	−0.33	−4.97	−0.37	0.5	1.37	5.97
0	0	−2.66	0.00	0.5	1.00	3.65
0.5	0.33	−1.33	0.21	0.5	0.79	2.32
0.9	0.71	−0.23	0.39	0.5	0.61	1.22

[2] We would like to show our appreciation to the anonymous reviewer to giving us helpful comments so that we could draw this conclusion.

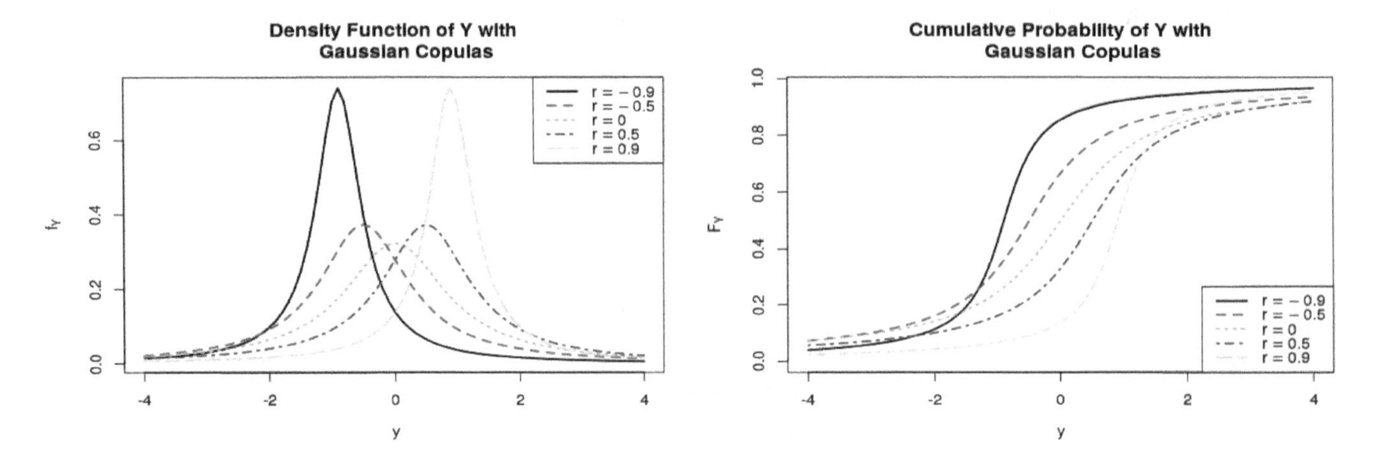

Figure 1. PDFs and CDFs of the ratio $Y = \frac{X_1}{X_2}$, where (X_1, X_2) follows Gaussian Copulas.

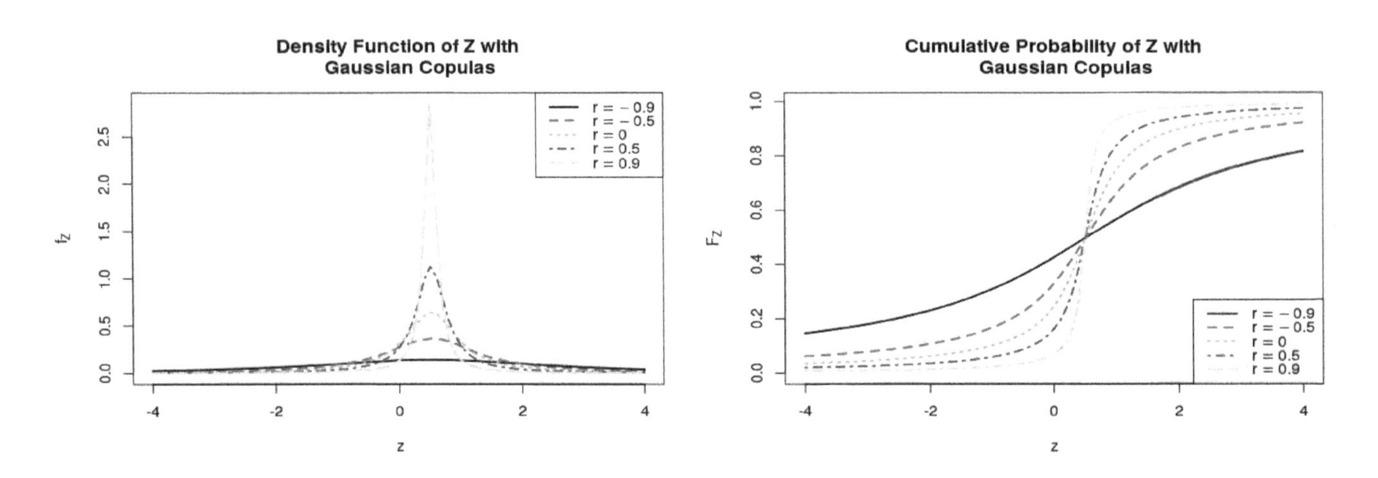

Figure 2. PDFs and CDFs of the ratio $Z = \frac{X_1}{X_1+X_2}$, where (X_1, X_2) follows Gaussian Copulas.

5.2. Student-t Copulas

We then investigate dependence structures of X_1 and X_2 through Student-t Copulas $C_{r,\nu}(u,v)$ and observe the shapes of the corresponding distributions of both Y and Z:

$$C_{r,\nu}(u,v) = \frac{1}{2\pi\sqrt{1-r^2}} \int_{-\infty}^{t_\nu^{-1}(u)} \int_{-\infty}^{t_\nu^{-1}(v)} \left(1 + \frac{s^2 - 2rst + t^2}{v(1-r^2)}\right)^{(v+2)/2} dsdt,$$

where $t_\nu^{-1}(x)$ is the inverse of Student CDF with degrees of freedom ν, r denotes Pearson's correlation coefficient between X_1 and X_2, and $|r| < 1$, and $\nu > 2$ is the degrees of freedom. We also consider $r = -0.9, -0.5, 0, 0.5, 0.9$ with three degrees of freedom ($\nu = 3$), where $r = 0$ is corresponding to no linear correlation. The PDFs and CDFs of Y of Z are, respectively, represented in Figures 3 and 4. Some percentiles are estimated and displayed in Tables 3 and 4. Similarly to Gaussian copula, in this case, the center and spread of Y and Z are also varying in the same way. However, one can see a representation of skewness and tailedness, that is, right skewed if $r < 0$ and left skewed if $r > 0$, since the fact that Student-t Copulas can capture tail dependence between X_1 and X_2.

Table 3. Some percentiles of $Y = X_1/X_2$, where (X_1, X_2) follows Student-t Copulas, $\nu = 3$.

r	$\tau(C_r)$	$Q_{0.05}$	$Q_{0.25}$	Median	$Q_{0.75}$	$Q_{0.95}$
−0.9	−0.71	−3.42	−1.26	−0.92	−0.50	1.82
−0.5	−0.33	−5.18	−1.28	−0.56	0.40	4.42
0	0	−5.40	−1.00	0.00	1.00	5.41
0.5	0.33	−4.42	−0.40	0.56	1.28	5.18
0.9	0.71	−1.81	0.50	0.92	1.26	3.42

Table 4. Some percentiles of $Y = X_1/(X_1 + X_2)$, where (X_1, X_2) follows Student-t Copulas, $\nu = 3$.

r	$\tau(C_r)$	$Q_{0.05}$	$Q_{0.25}$	Median	$Q_{0.75}$	$Q_{0.95}$
−0.9	−0.71	−18.20	−2.05	0.5	3.05	19.21
−0.5	−0.33	−6.62	−0.45	0.5	1.45	7.61
0	0	−3.35	0.00	0.5	1.00	4.35
0.5	0.33	−1.54	0.24	0.5	0.76	2.54
0.9	0.71	−0.24	0.40	0.5	0.60	1.24

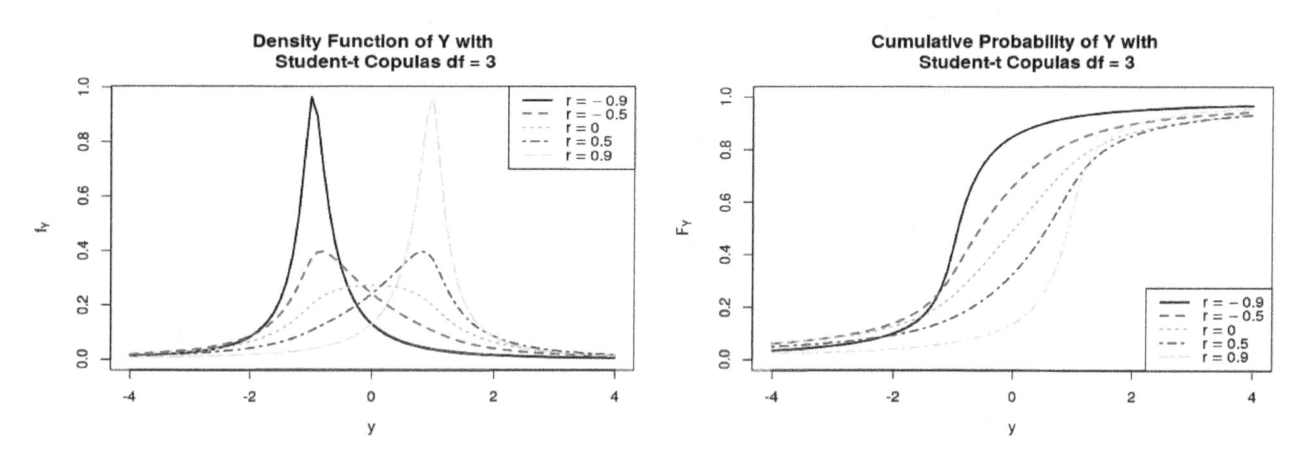

Figure 3. PDFs and CDFs of the ratio $Y = \frac{X_1}{X_2}$, where (X_1, X_2) follows Student-t Copulas, $\nu = 3$.

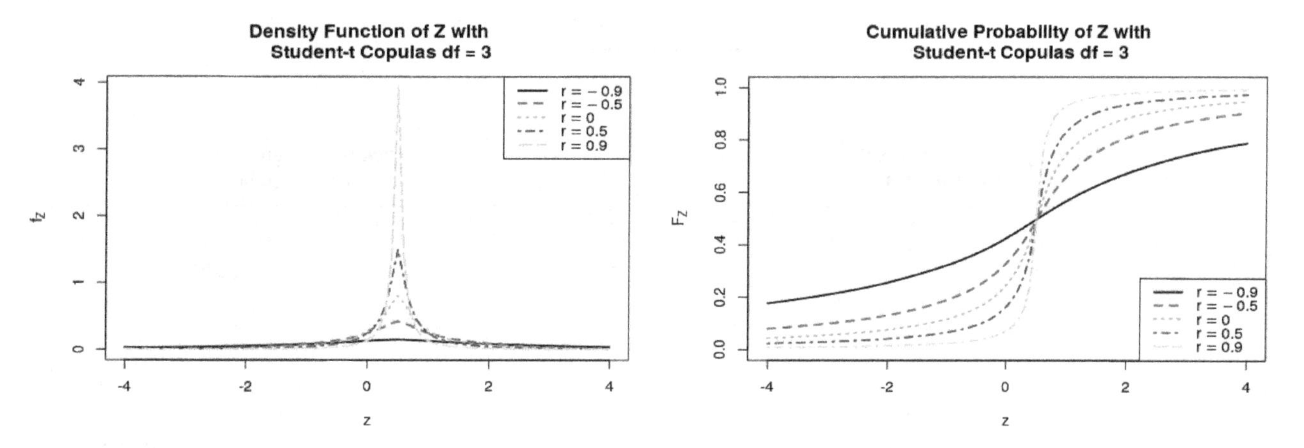

Figure 4. PDFs and CDFs of the ratio $Z = \frac{X_1}{X_1+X_2}$, where (X_1, X_2) follows Student-t Copulas, $\nu = 3$.

5.3. Clayton Copulas

We turn to investigate dependence structures of X_1 and X_2 through the following Clayton Copulas $C_\theta(u, v)$ and observe the shapes of the corresponding distributions of both Y and Z:

$$C_\theta(u,v) = \max\left\{u^{-\theta} + v^{-\theta} - 1\right\}^{-\frac{1}{\theta}}, \ \theta \in [-1; +\infty) \setminus 0.$$

In practice, we use $\theta > 0$ that leads to

$$C_\theta(u,v) = \left(u^{-\theta} + v^{-\theta} - 1\right)^{-\frac{1}{\theta}}, \ \theta > 0.$$

For $\theta = 1,2,3,4$, we obtain CDFs and PDFs of Y and Z and exhibit them in Figures 5 and 6, respectively, and their percentiles as shown in Tables 5 and 6. Clearly, Clayton Copulas affect Y to get heavier left tail and the more positive dependence is; that is, $\theta \to \infty$, the greater the median and the smaller the IQR of Y tend to be. On the other hand, the shape of distribution of Z is still symmetric with unchanged median, less spread, and more spike.

Table 5. Some percentiles of $Y = X_1/X_2$, where (X_1, X_2) follows Clayton Copulas.

θ	$\tau(C_\theta)$	$Q_{0.05}$	$Q_{0.25}$	**Median**	$Q_{0.75}$	$Q_{0.95}$
1	0.33	−4.97	−0.36	0.53	1.31	5.83
2	0.5	−3.83	0.03	0.76	1.34	5.24
3	0.60	−2.87	0.25	0.86	1.34	4.71
4	0.67	−2.12	0.39	0.91	1.32	4.25

Table 6. Some percentiles of $Y = X_1/(X_1 + X_2)$, where (X_1, X_2) follows Clayton Copulas.

θ	$\tau(C_\theta)$	$Q_{0.05}$	$Q_{0.25}$	**Median**	$Q_{0.75}$	$Q_{0.95}$
1	0.33	−1.31	0.22	0.5	0.78	2.31
2	0.5	−0.73	0.30	0.5	0.70	1.73
3	0.60	−0.42	0.35	0.5	0.65	1.42
4	0.67	−0.24	0.37	0.5	0.63	1.24

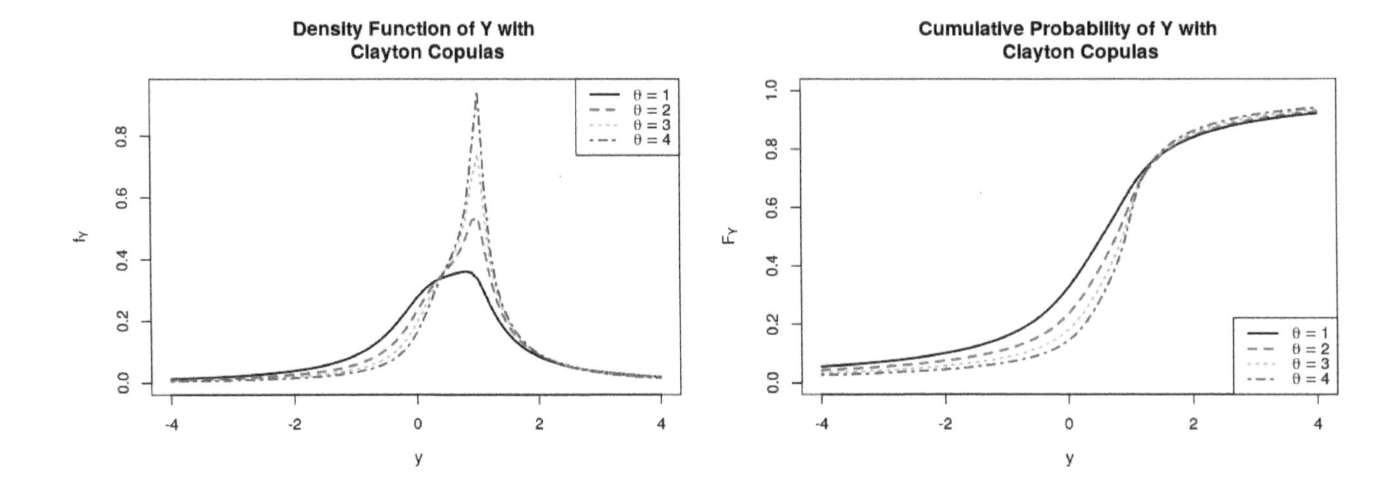

Figure 5. PDFs and CDFs of the ratio $Y = \frac{X_1}{X_2}$, where (X_1, X_2) follows Clayton Copulas.

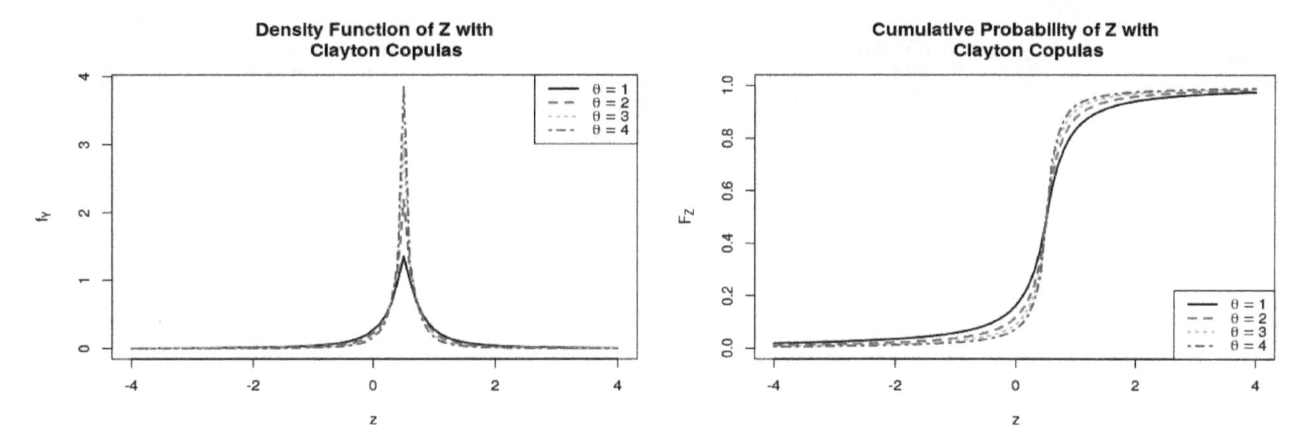

Figure 6. PDFs and CDFs of the ratio $Z = \frac{X_1}{X_1+X_2}$, where (X_1, X_2) follows Clayton Copulas.

5.4. Gumbel Copulas

We now investigate dependence structures of X_1 and X_2 through the following Gumbel Copulas $C_\theta(u, v)$ and observe the shapes of the corresponding distributions of both Y and Z:

$$C_\theta(u, v) = \exp\left(-\left[(-\ln u)^\theta + (-\ln v)^\theta\right]^{\frac{1}{\theta}}\right), \quad \theta > 0.$$

The parameter $\theta = 1$ implies uncorrelated (X_1, X_2). Figures 7 and 8 show the behavior of Y and Z via the copulas. Tables 7 and 8 represent some estimated percentiles for both Y and Z. In the comparison with Clayton, Gumbel Copula gets left skewness, higher median, and less spread for Y, but gets a symmetric shape, unchanged median, less spread, and smaller scale (higher spike) for Z. However, the shape of Y tends to be more symmetric when one increases the parameter θ.

Table 7. Some percentiles of $Y = X_1/X_2$, where (X_1, X_2) follows Gumbel Copulas.

θ	$\tau(C_\theta)$	$Q_{0.05}$	$Q_{0.25}$	Median	$Q_{0.75}$	$Q_{0.95}$
1	0	−6.31	−1.00	0.00	1.00	6.32
2	0.5	−3.70	0.00	0.74	1.36	5.01
3	0.67	−2.26	0.38	0.89	1.32	3.95
4	0.75	−1.45	0.56	0.94	1.27	3.30

Table 8. Some percentiles of $Y = X_1/(X_1 + X_2)$, where (X_1, X_2) follows Gumbel Copulas.

θ	$\tau(C_\theta)$	$Q_{0.05}$	$Q_{0.25}$	Median	$Q_{0.75}$	$Q_{0.95}$
1	0	−2.66	0.00	0.5	1.00	3.66
2	0.5	−0.82	0.30	0.5	0.70	1.82
3	0.67	−0.34	0.37	0.5	0.63	1.34
4	0.75	−0.12	0.41	0.5	0.59	1.12

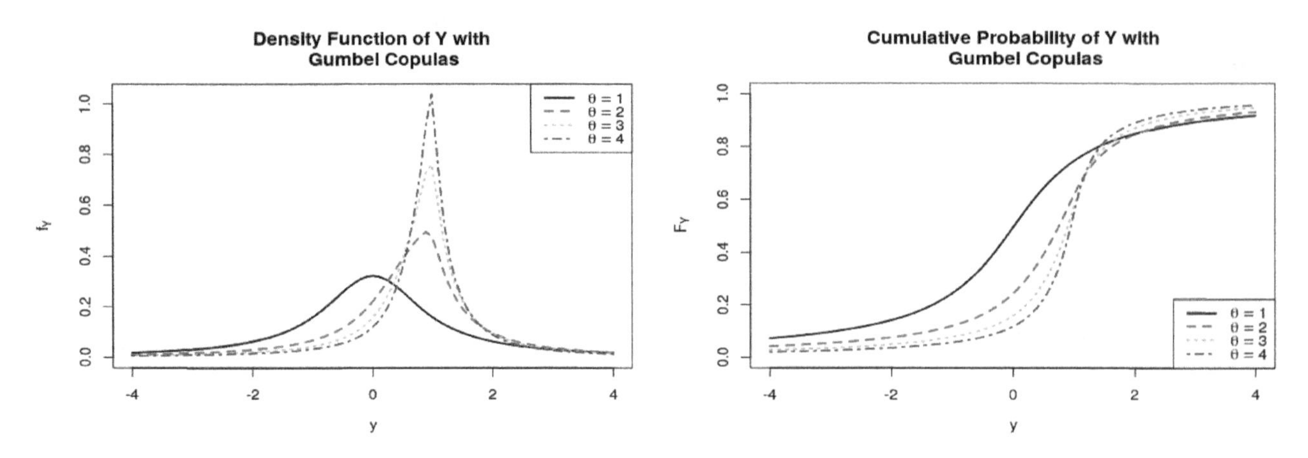

Figure 7. PDFs and CDFs of the ratio $Y = \frac{X_1}{X_2}$, where (X_1, X_2) follows Gumbel Copulas.

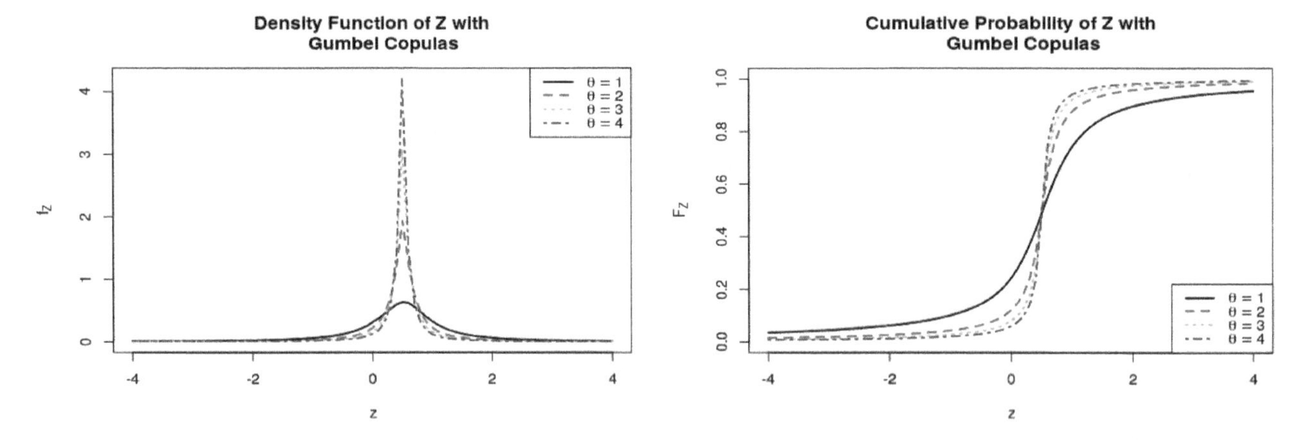

Figure 8. PDFs and CDFs of the ratio $Z = \frac{X_1}{X_1+X_2}$, where (X_1, X_2) follows Gumbel Copulas.

5.5. Frank Copulas

In addition, we investigate dependence structures of X_1 and X_2 through the following Frank Copulas $C_\theta(u, v)$ and observe the shapes of the corresponding distributions of both Y and Z:

$$C_\theta(u,v) = -\frac{1}{\theta} \ln \left(1 + \frac{\left(e^{-\theta u} - 1\right)\left(e^{-\theta v} - 1\right)}{e^{-\theta} - 1} \right), \quad \theta \in \mathbb{R} \backslash \{0\}.$$

For Frank Copulas, the parameter θ represents two independent random variables when it tends to zero, becomes more monotonic structure when $\theta \to \infty$, and becomes more counter-monotonic when $\theta \to -\infty$. For $\theta = 1, 2, 3, 4$, we have Tables 9 and 10, and Figures 9 and 10. In contrast to both Clayton and Gumbel Copulas, the density of Y via Frank Copulas is more symmetric and the density of Z is not scaling too much, but for the median and spread of Y, it behaves like the Gumbel Copulas and Clayton Copulas; that is, if we increase the parameter θ, the median also increases, whereas the spread (via IQR) decreases.

Table 9. Some percentiles of $Y = X_1/X_2$, where (X_1, X_2) follows Frank Copulas.

θ	$\tau(C_\theta)$	$Q_{0.05}$	$Q_{0.25}$	Median	$Q_{0.75}$	$Q_{0.95}$
1	0.11	−6.03	−0.79	0.19	1.18	6.42
2	0.21	−5.59	−0.56	0.36	1.31	6.37
3	0.31	−5.05	−0.33	0.50	1.40	6.19
4	0.39	−4.46	−0.14	0.60	1.46	5.90

Table 10. Some percentiles of $Y = X_1/(X_1 + X_2)$, where (X_1, X_2) follows Frank Copulas.

θ	$\tau(C_\theta)$	$Q_{0.05}$	$Q_{0.25}$	Median	$Q_{0.75}$	$Q_{0.95}$
1	0.11	−2.09	0.09	0.5	0.91	3.09
2	0.21	−1.62	0.16	0.5	0.84	2.62
3	0.31	−1.25	0.21	0.5	0.79	2.25
4	0.39	−0.96	0.25	0.5	0.75	1.96

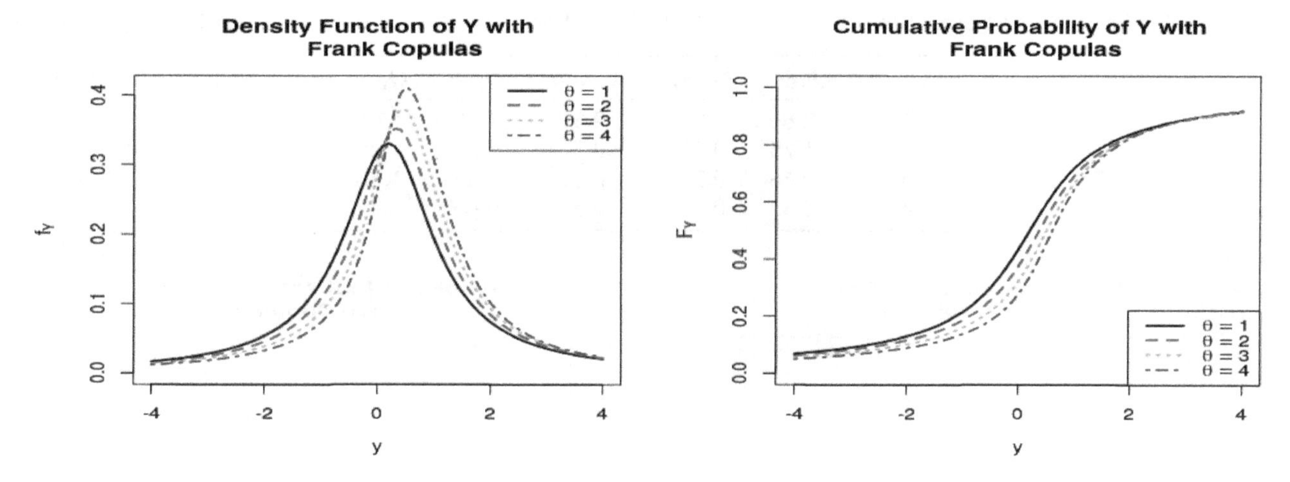

Figure 9. PDF and CDF of quotient of the ratio $Y = \frac{X_1}{X_2}$, where (X_1, X_2) follows Frank Copulas.

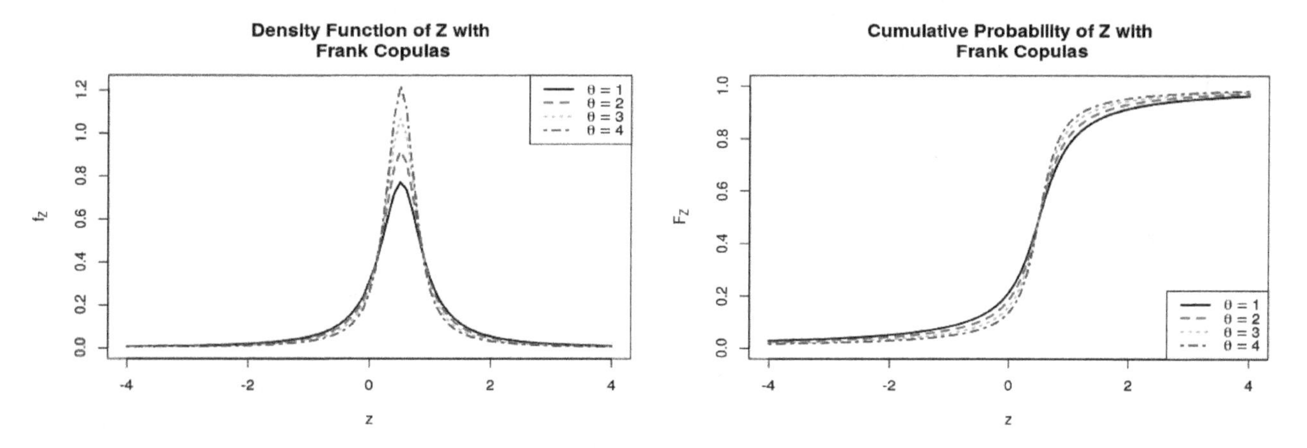

Figure 10. PDF and CDF of the ratio $Z = \frac{X_1}{X_1+X_2}$, where (X_1, X_2) follows Frank Copulas.

5.6. Joe Copulas

Finally, we investigate dependence structures of X_1 and X_2 through the following Joe Copulas $C_\theta(u, v)$ and observe the shapes of the corresponding distributions of both Y and Z:

$$C_\theta(u,v) = 1 - [(1-u)^\theta + (1-v)^\theta - (1-u)^\theta(1-v)^\theta]^{1/\theta}, \quad \theta \in [1, \infty).$$

Similar to Gumbel Copulas, Joe Copula shows independence when $\theta = 1$ and becomes more monotonicity if $\theta \to \infty$. With assistance of the tables and graphs with $\theta = 1, 2, 3, 4$, Tables 11 and 12 and Figures 11 and 12 tell us that the distribution behaviors of Y are also affected with higher median, less IQR, smaller scale (higher spike), and the shape is more asymmetric if one increases the parameter θ. On the other hand, Z still gets median unchanged with *median* $= 0.5$ and less spread with the sum of the first quartile and third quartile is always equal to 1, i.e., $(Q_{0.25} + Q_{0.75} = 1)$.

Table 11. Some percentiles of $Y = X_1/X_2$, where (X_1, X_2) follows Joe Copulas.

θ	$\tau(C_\theta)$	$Q_{0.05}$	$Q_{0.25}$	Median	$Q_{0.75}$	$Q_{0.95}$
1	0	−6.31	−1.00	0.00	1.00	6.32
2	0.36	−4.82	−0.32	0.57	1.30	5.67
3	0.52	−3.66	0.07	0.79	1.33	5.13
4	0.61	−2.71	0.29	0.88	1.33	4.62

Table 12. Some percentiles of $Y = X_1/(X_1 + X_2)$, where (X_1, X_2) follows Joe Copulas.

θ	$\tau(C_\theta)$	$Q_{0.05}$	$Q_{0.25}$	Median	$Q_{0.75}$	$Q_{0.95}$
1	0	−2.66	0.00	0.5	1.00	3.66
2	0.36	−1.25	0.23	0.5	0.77	2.25
3	0.52	−0.66	0.31	0.5	0.69	1.66
4	0.61	−0.37	0.35	0.5	0.65	1.37

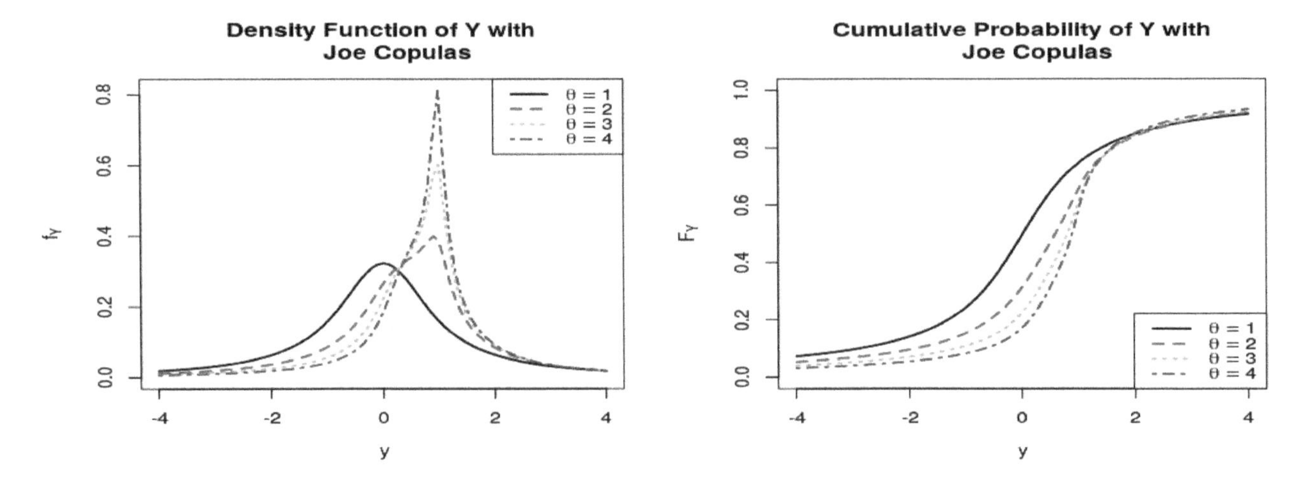

Figure 11. PDFs and CDFs of the ratio $Y = \frac{X_1}{X_2}$, where (X_1, X_2) follows Joe Copulas.

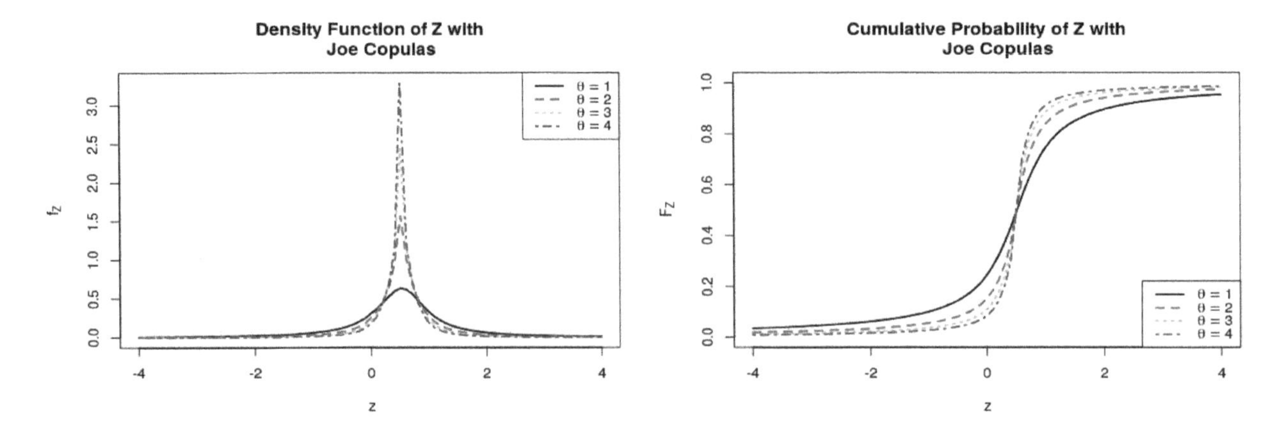

Figure 12. PDFs and CDFs of the ratio $Z = \frac{X_1}{X_1+X_2}$, where (X_1, X_2) follows Joe Copulas.

5.7. Comparison of Copulas with the Same Measure of Dependence

In this section, we investigate the effects of the six copulas families as discussed above on the shapes of different distributions for the random variables $Y := X_1/X_2$ and $Z := X_1/(X_1 + X_2)$ when they have the same measure of dependence—the Kendall's coefficient τ. Here, the parameters are chosen to each copula to correspond to Kendall $\tau = 0.49$. We exhibit the corresponding CDFs and PDFs of both Y and Z in Figures 13 and 14, estimate the percentiles Q_α for some $\alpha = 0.05, 0.25, 0.5, 0.75, 0.95$, and display the values in Tables 13 and 14. As can be seen from the tables and the figures, Y attains the greatest median (0.76) and the smallest IQR (1.32) with left skewed shape when both X_1 and X_2 follow Joe Copula and Student-t Copula. In contrast, Gaussian Copula produces the smallest median (0.70) and the largest IQR (1.42) with symmetric shape. Using Clayton Copula, Y gets the second biggest median (0.74). The ratio random variable Y has the same median (0.72) for both Gumbel and Frank Copula, but the Frank makes Y get higher IQR than the Gumbel ($1.41 > 1.33$). On the other hand, the random variable Z has symmetric shape with unchanged median (0.5) among all investigated copulas. It only changes the scale, where the Joe Copula affects Z with the smallest scale (i.e., the tallest height of density) whilst the Frank Copula causes Z to get the greatest scale (i.e., the shortest height of density).

Table 13. Some percentiles of $Y = X_1/X_2$, where (X_1, X_2) modelled with six copulas having the same Kendall coefficient $\tau = 0.49$.

Copulas	Parameters	$\tau(C)$	$Q_{0.05}$	$Q_{0.25}$	Median	$Q_{0.75}$	$Q_{0.95}$
Gaussian	0.7	0.49	−3.81	−0.01	0.70	1.41	5.21
Student-t	$0.7, \nu = 3$	0.49	−3.52	−0.02	0.76	1.31	4.63
Clayton	1.90	0.49	−3.93	0.00	0.74	1.34	5.30
Gumbel	1.95	0.49	−3.80	−0.03	0.72	1.36	5.07
Frank	5.5	0.49	−3.60	0.08	0.72	1.49	5.41
Joe	2.8	0.49	−3.88	0.01	0.76	1.33	5.24

Table 14. Some percentiles of $Y = X_1/(X_1 + X_2)$, where (X_1, X_2) modelled with six copulas having the same Kendall coefficient $\tau = 0.49$.

Copulas	Parameters	$\tau(C)$	$Q_{0.05}$	$Q_{0.25}$	Median	$Q_{0.75}$	$Q_{0.95}$
Gaussian	0.7	0.49	−0.83	0.29	0.5	0.71	1.83
Student-t	$0.7, \nu = 3$	0.49	−0.92	0.31	0.5	0.69	1.92
Clayton	1.90	0.49	−0.77	0.30	0.5	0.70	1.77
Gumbel	1.95	0.49	−0.86	0.30	0.5	0.70	1.86
Frank	5.5	0.49	−0.65	0.29	0.5	0.71	1.65
Joe	2.8	0.49	−0.74	0.30	0.5	0.70	1.74

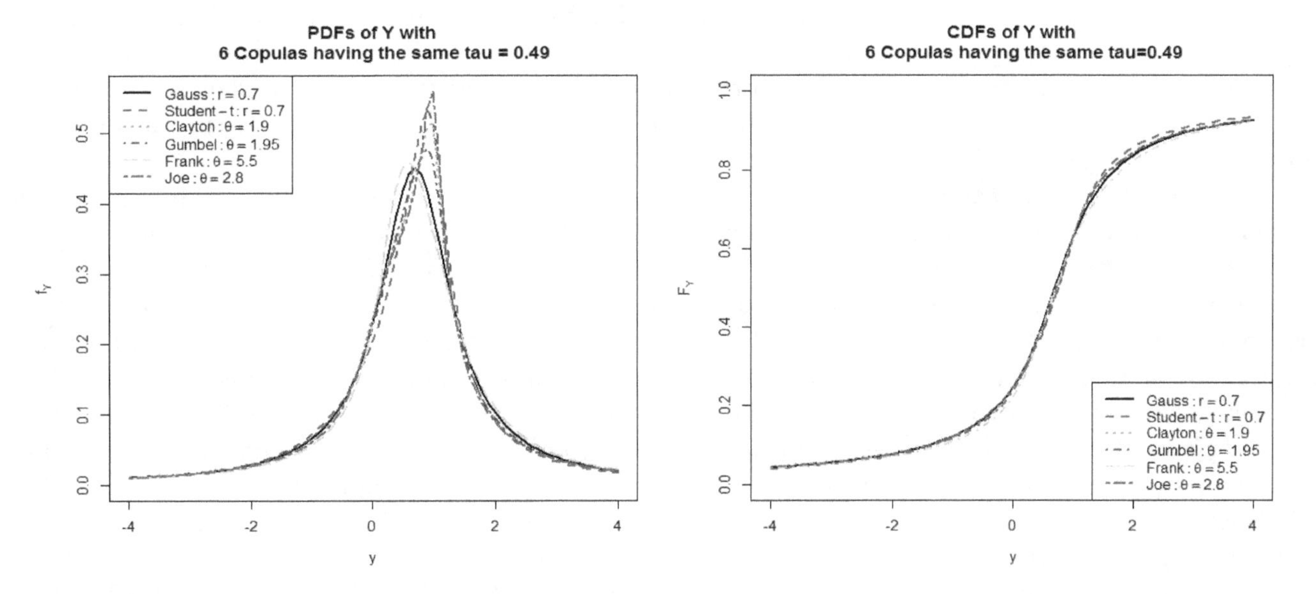

Figure 13. PDFs and CDFs of the ratio $Y = \frac{X_1}{X_2}$, where (X_1, X_2) modeled with six Copulas having the same $\tau = 0.49$.

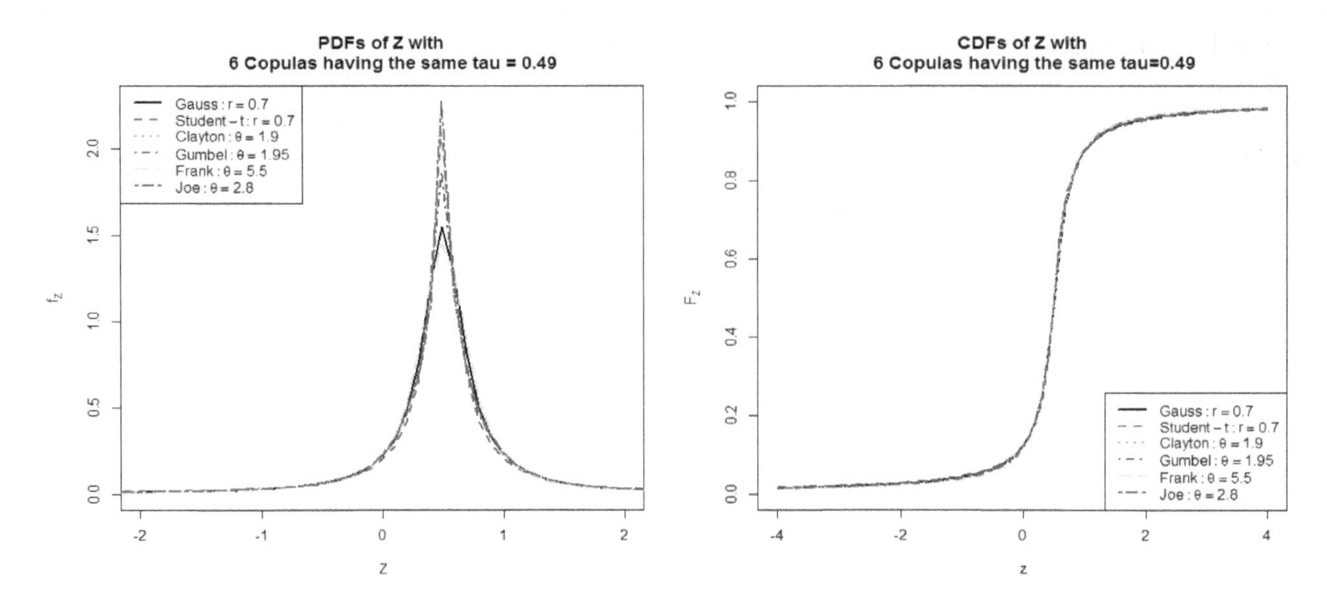

Figure 14. PDF and CDF of the ratio $Z = \frac{X_1}{X_1+X_2}$, where (X_1, X_2) modeled with six Copulas having the same $\tau = 0.49$.

6. Conclusions

Determining distributions of the functions of random variables is a very crucial task and this problem has attracted a number of researchers because there are numerous applications in Economics, Science, and many other areas, especially in the areas of finance including risk management and option pricing. However, to the best of our knowledge, the problem of determining distribution functions of quotient of dependent random variables using copulas has not been widely studied and, as far as we know, no published paper or working paper has done the work we are doing in this paper. Thus, to bridge the gap in the literature, in this paper, we first develop two general propositions on both density and distribution functions for the quotient $Y = \frac{X_1}{X_2}$ and the ratio of one variable over the sum of two variables $Z := \frac{X_1}{X_1+X_2}$ of two dependent random variables X_1 and X_2 by using copulas. We then derive two corollaries on both density and distribution functions for the two quotients $Y = \frac{X_1}{X_2}$ and $Z = \frac{X_1}{X_1+X_2}$ of two dependent normal random variables X_1 and X_2 in case of Gaussian Copulas by applying the two main general propositions developed in our paper. From the results, we derive the corollaries on the median for the ratios of both Y and Z of two normal random variables X_1 and X_2. Furthermore, the result of median for Z is also extended to a larger family of symmetric distributions and symmetric copulas of X_1 and X_2.

Since the density and the CDF formula of the ratios of both Y and Z are in terms of integrals and are very complicated, we cannot obtain the exact forms of the density and the CDF. To circumvent the difficulty, in this paper, we propose to use the Monte Carlo algorithm, numerical analysis, and graphical approach that can efficiently compute complicated integrals and study the behaviors of both density and distribution and the changes of their shapes when parameters are changing. We illustrate our proposed approaches by using a simulation study with ratios of normal random variables on several different copulas, including Gaussian, Student-t, Clayton, Gumbel, Frank, and Joe Copulas. We find that copulas make big impacts on behavior of distributions, and since Gaussian and Student-t Copulas belong to an elliptical family, they similarly act on shapes of Y and Z in the same fashion. We also document the effects when using Archimedean copulas including Clayton, Gumbel, Frank, and Joe Copulas. However, there are also some differences, especially on location and scale effects. For example, distribution of Z does not change the median and its shape is always symmetric for all investigated copulas while the random variable Y is affected in skewness, median and spread. Our findings are useful for academics in their study of the shapes, center and spread of both density and distribution functions for the ratios by using different copulas. Since the ratios in different copulas are widely used in many important empirical studies in Finance, Economics, and many other areas, our findings are useful to practitioners in Finance, Economics, and many other areas who need to study the shapes, center and spread of the ratios by using different copulas in their analysis and useful to policy makers if they need to consider the shapes, center and spread of both density and distribution functions for the ratios by using different copulas in their policy decision-making. Finally, we note that, although all of the propositions and corollaries developed in our paper are relatively easy to derive, all the results developed in our paper are useful to both academics and practitioners because there is a wide range of applications with variables that have a negative range. Readers may refer to (Chang et al. 2016, 2018a, 2018b, 2018c; Chang et al. 2015; Wong 2016) for more information on the applications in different areas.

Author Contributions: Writing—original draft preparation, S.L. (Sel Ly) and K.-H.P.; writing—review and editing,W.-K.W.; visualization, S.L. (Sel Ly) and S.L. (Sal Ly).

Acknowledgments: The authors are grateful to Michael McAleer, the Editor-in-Chief, and anonymous referees for substantive comments that have significantly improved this manuscript. The fourth author would like to thank Robert B. Miller and Howard E. Thompson for their continuous guidance and encouragement.

References

Arnold, Barry C., and Patrick L. Brockett. 1992. On distributions whose component ratios are cauchy. *The American Statistician* 46: 25–26.

Bithas, Petros S., Nikos C. Sagias, Theodoros A. Tsiftsis, and George K. Karagiannidis. 2007. Products and ratios of two gaussian class correlated weibull random variables. Paper presented at the 12th International Conference on Applied Stochastic Models and Data Analysis (ASMDA 2007), Chania, Crete, May 29–June 1.

Cedilnik, Anton, Katarina Kosmelj, and Andrej Blejec. 2004. The distribution of the ratio of jointly normal variables. *Metodoloski zvezki* 1: 99.

Chang, Chia-Lin, Michael McAleer, and Wing-Keung Wong. 2015. Informatics, Data Mining, Econometrics and Financial Economics: A Connection. Technical Report 1. Available online: https://repub.eur.nl/pub/79219/ (accessed on 8 March 2019).

Chang, Chia-Lin, Michael McAleer, and Wing-Keung Wong. 2016. Management Science, Economics and Finance: A Connection. Available online: http://eprints.ucm.es/37904/ (accessed on 8 March 2019).

Chang, Chia-Lin, Michael McAleer, and Wing-Keung Wong. 2018a. Big data, computational science, economics, finance, marketing, management, and psychology: connections. *Journal of Risk and Financial Management* 11: 15. [CrossRef]

Chang, Chia-Lin, Michael McAleer, and Wing-Keung Wong. 2018b. Decision Sciences, Economics, Finance, Business, Computing, and Big Data: Connections. Available online: https://papers.ssrn.com/sol3/papers.cfm?abstract_id=3140371 (accessed on 8 March 2019).

Chang, Chia-Lin, Michael McAleer, and Wing-Keung Wong. 2018c. Management Information, Decision Sciences, and Financial Economics: A Connection. *Decision Sciences, and Financial Economics: A Connection (January 17, 2018). Tinbergen Institute Discussion Paper*, p. 4. Available online: https://papers.ssrn.com/sol3/papers.cfm?abstract_id=3103807 (accessed on 8 March 2019).

Cherubini, Umberto, Elisa Luciano, and Walter Vecchiato. 2004. *Copula Methods in Finance*. New York: John Wiley & Sons.

Dolati, Ali, Rasool Roozegar, Najmeh Ahmadi, and Zohreh Shishebor. 2017. The effect of dependence on distribution of the functions of random variables. *Communications in Statistics-Theory and Methods* 46: 10704–717. [CrossRef]

Donahue, James D. 1964. *Products and Quotients of Random Variables and Their Applications*. Technical Repor. Fort Collins: MARTIN CO DENVER CO.

Hinkley, David V. 1969. On the ratio of two correlated normal random variables. *Biometrika* 56: 635–39. [CrossRef]

Joe, Harry. 1997. *Multivariate Models and Multivariate Dependence Concepts*. Boca Raton: Chapman and Hall/CRC.

Ly, Sel, H. Uyen Pham, and Radim Bris. 2016. On the distortion risk measure using copulas. In *Applied Mathematics in Engineering and Reliability: Proceedings of the 1st International Conference on Applied Mathematics in Engineering and Reliability (Ho Chi Minh City, Vietnam, 4–6 May 2016)*. Boca Raton: CRC Press, p. 309.

Ly, Sel, Kim-Hung Pho, Sal Ly, and Wing-Keung Wong. 2019. Determining distribution for the product of random variables by using copulas. *Risks* 7: 23. [CrossRef]

Macalos, Milburn O., and Jayrold P Arcede. 2015. On The Distribution of the Sums, Products and Quotient Of Singh-Maddala Distributed Random Variables Based on Fgm Copula. Available online: https://www.researchgate.net/profile/Jayrold_Arcede/publication/310773588_On_the_Distribution_of_the_Sums_Products_and_Quotient_of_Singh-Maddala_Distributed_Random_Variables_Based_on_FGM_Copula/links/5836ba3a08ae503ddbb54a46/On-the-Distribution-of-the-Sums-Products-and-Quotient-of-Singh-Maddala-Distributed-Random-Variables-Based-on-FGM-Copula.pdf (accessed on 8 March 2019).

Marsaglia, George. 1965. Ratios of normal variables and ratios of sums of uniform variables. *Journal of the American Statistical Association* 60: 193–204. [CrossRef]

Matović, Ana, Edis Mekić, Nikola Sekulović, Mihajlo Stefanović, Marija Matović, and Časlav Stefanović. 2013. The distribution of the ratio of the products of two independent-variates and its application in the performance analysis of relaying communication systems. *Mathematical Problems in Engineering* 2013. [CrossRef]

Mekić, Edis, Mihajlo Stefanović, Petar Spalević, Nikola Sekulović, and Ana Stanković. 2012. Statistical analysis of ratio of random variables and its application in performance analysis of multihop wireless transmissions. *Mathematical Problems in Engineering* 2012. [CrossRef]

Meyer, Christian. 2013. The bivariate normal copula. *Communications in Statistics-Theory and Methods* 42: 2402–22. [CrossRef]

Nadarajah, Saralees, and Mariano Ruiz Espejo. 2006. Sums, products, and ratios for the generalized bivariate pareto distribution. *Kodai Mathematical Journal* 29: 72–83. [CrossRef]

Nadarajah, Saralees, and Samuel Kotz. 2006a. On the product and ratio of gamma and weibull random variables. *Econometric Theory* 22: 338–44. [CrossRef]

Nadarajah, Saralees, and Samuel Kotz. 2006b. Sums, products, and ratios for downton's bivariate exponential distribution. *Stochastic Environmental Research and Risk Assessment* 20: 164–70. [CrossRef]

Nelsen, Roger B. 2007. *An Introduction to Copulas*. New York: Springer Science & Business Media.

Pham-Gia, Thu, Noyan Turkkan, and E. Marchand. 2006. Density of the ratio of two normal random variables and applications. *Communications in Statistics—Theory and Methods* 35: 1569–91. [CrossRef]

Pham-Gia, Thu. 2000. Distributions of the ratios of independent beta variables and applications. *Communications in Statistics-Theory and Methods* 29: 2693–715. [CrossRef]

Rathie, Pushpa Narayan, Luan Carlos de SM Ozelim, and Cira E. Guevara Otiniano. 2016. Exact distribution of the product and the quotient of two stable lévy random variables. *Communications in Nonlinear Science and Numerical Simulation* 36: 204–18. [CrossRef]

Sakamoto, H. 1943. On the distributions of the product and the quotient of the independent and uniformly distributed random variables. *Tohoku Mathematical Journal, First Series* 49: 243–60.

Springer, Melvin Dale. 1979. *The Algebra of Random Variables (1. print. in the USA)*. Hoboken: Wiley.

Tran, D. Hien, Uyen H. Pham, Sel Ly, and Trung Vo-Duy. 2015. A new measure of monotone dependence by using sobolev norms for copula. In *International Symposium on Integrated Uncertainty in Knowledge Modelling and Decision Making*. New York: Springer, pp. 126–37.

Tran, Hien D., Uyen H. Pham, Sel Ly, and Trung Vo-Duy. 2017. Extraction dependence structure of distorted copulas via a measure of dependence. *Annals of Operations Research* 256: 221–36. [CrossRef]

Wong, Wing-Keung. 2016. Behavioural, financial, and health and medical economics: A connection. *Journal of Health & Medical Economics* 2: 1.

Examination and Modification of Multi-Factor Model in Explaining Stock Excess Return with Hybrid Approach in Empirical Study of Chinese Stock Market

Jian Huang * and Huazhang Liu

Division of Business Management, Beijing Normal University-HongKong Baptist University United International College, Zhuhai 519087, China; spot_light@outlook.com or k530002087@mail.uic.edu.hk
* Correspondence: k530002046@mail.uic.edu.hk or jianhuang.951111@gmail.com

Abstract: To search significant variables which can illustrate the abnormal return of stock price, this research is generally based on the Fama-French five-factor model to develop a multi-factor model. We evaluated the existing factors in the empirical study of Chinese stock market and examined for new factors to extend the model by OLS and ridge regression model. With data from 2007 to 2018, the regression analysis was conducted on 1097 stocks separately in the market with computer simulation based on Python. Moreover, we conducted research on factor cyclical pattern via chi-square test and developed a corresponding trading strategy with trend analysis. For the results, we found that except market risk premium, each industry corresponds differently to the rest of six risk factors. The factor cyclical pattern can be used to predict the direction of seven risk factors and a simple moving average approach based on the relationships between risk factors and each industry was conducted in back-test which suggested that SMB (size premium), CMA (investment growth premium), CRMHL (momentum premium), and AMLH (asset turnover premium) can gain positive return.

Keywords: multi-factor model; risk factors; OLS and ridge regression model; python; chi-square test

1. Introduction

Financial markets are rife with uncertainties which make it difficult to forecast future market trends. Especially in the stock market, while providing investors with remarkable return, at the same time, it entails tremendous risk. As people pursue higher returns, they must bear the corresponding risks.

A pricing model with multiple factors is a promising approach to predict future stock prices. If the relationship between the risk and return can be expressed by multiple factors in a mathematical model, the standards of stock selection and corresponding investment strategies can be established. To be specific, each investor has their own risk preference, for instance, risk averse investors tend to bear lower risk and receive lower return. This specific type of investors has their preference in stock selection. In order to find suitable selection criteria, we need to quantify the relationship of risk premium factor and excess return. Investors can refer to the selection criteria to establish a corresponding trading strategy which can achieve their target excess return.

In the process of model examination, regression was conducted on single stocks which differed from the previous research which used portfolios. Especially, ridge regression was conducted instead of OLS regression. As for the process of model modification, two new factors were added to achieve higher explanatory power. Furthermore, we discussed the endogeneity and exogeneity for the risk premium factor. On the basis of economic objectives, we established a trading strategy for Chinese stock market.

Although previous research has worked well in the American stock market, these findings may be less practically applied to the Chinese market due the investor component. Since individual investors contribute nearly 80% of the trading volume, investment behavior and preference can largely impact the market average return. However, the asymmetric information and investment concepts may lead to irrational behaviors. Therefore, we use risk premium factors to explain the excess return and the coefficients to measure the sensitivity of investor reactions to the risk premium.

To begin, we conducted single stock regression to examine the effectiveness of a five-factor model in Chinese stock market. Then we compare the coefficients of different factors under specific company types to discover the leading factor. We use the t-statistic to evaluate the significance of each factor. Regarding previous articles and research, we will add new factors in the model for better performance. Moreover, we conducted inter-factor cyclical research to study the pattern of factors. We can predict the rise and fall of the coefficient on a quarterly basis.

2. Literature Review

The multi-factor risk model has undergone a series of developments which can be divided into four major steps. At the beginning, in order to figure out the leading factors for security return, several researchers have contributed to the development of an asset pricing model. Initially, Markowitz (1952) proposed a mean-variance model to illustrate the statistical relationship between security risk and return in terms of standard deviation and expected rerun. It has established the foundation of modern finance theory. However, he had not specified the factors that explain expected security returns.

Based on the modern portfolio theory, the capital asset pricing model (CAPM) was developed by Sharpe (1964) and Lintner (1965) to illustrate the linear relationship between expected return and market risk premium.

$$\mathrm{Ri} = \mathrm{Rf} + \beta_{market}(\mathrm{Rm} - \mathrm{Rf}) \quad (1)$$

With the empirical tests in the stock market, they managed to find out the pattern of stock returns in line with the general stock market index. To be specific, the model estimates the relationships between the return on market index (the explanatory variable) and the return on the stock (the dependent variable). The regression coefficient of the single index model is referred to as beta which is a measure of the sensitivity of a stock to general movement in the market index. This empirical research symbolized the transition from qualitative analysis to quantitative analysis which also laid the foundation for the subsequent asset pricing models.

Since then, the CAPM model has been widely applied in research and empirical testing. However, with increasing abnormal returns which cannot be explained by existing factors, the market beta was no longer sufficient to describe expected return (Fama 1996). Therefore, Fama and French proposed a multifactor model consisting of three factors for market risk (Rm − Rf), market value, and book-to-market ratio (Fama and French 1993). In this model, Ri, Rf, and Rm stands for security expected return, risk free rate, and market return. SMB and HML are the risk premium factors. They illustrated that small stocks can generate higher returns than large stocks while value stocks can generate higher returns than growth stocks.

$$R_i - R_f = \mathrm{a} + \beta_{market}\left(R_m - R_f\right) + \beta_{size}\mathrm{SMB} + \beta_{BM}\mathrm{HML} \quad (2)$$

It is also a supplement of arbitrage pricing theory (Ross 1976) which emphasizes that the expected return is not only affected by market risk but also a series of other factors. The generalized model (Bodie et al. 2014) illustrated the linear relationship of expected return and different factors (Bodie et al. 2017). In Equation (3), F_j represents the factors and b represents the coefficients.

$$r_i = a_i + \sum\nolimits_{j=1}^{k} b_{ij}F_j + \epsilon_i,\ i = 1,\ 2\ldots,\ N \quad (3)$$

Moreover, Banz (1981) illustrated that return of securities are affected by several index including B/M ratio and E/P ratio which represent a series of risk premium. These articles largely contribute to the development of the multi-factor model.

In latter decades, the three-factor model was faced with a series of challenges. As the three-factor model was applied to stock trading and empirical testing, it appeared that some of the phenomenon cannot be explained by the model which can lead to unpredictable abnormal return. Therefore, more specific factors need to be added into the model to improve the accuracy. Novy-marx (2013) proved that profitability, measured by gross profits-to-assets, has roughly the same power as book-to-market value in explaining the average return. The following equation is based on the dividend discount model, $Y_{t+\tau}$ represents the earning for period $t + \tau$, $dB_{t+\tau}$ is the change in book equity, r represents the expected return. It implied that higher market value leads to lower expected return, while higher earnings imply higher expected return.

$$M_t = \sum_{\tau=1}^{\infty} E(Y_{t+\tau} - dB_{t+\tau})/(1+r)^{\tau} \tag{4}$$

Aharoni et al. (2013) recorded an insignificant but statistically reliable relationship between investment pattern and average return. Other evidence also illustrated that the profitability and investment factors can explain some of the variation in average return.

Carhart (1997) conducted a study on the common factors in stock return and investment expense by adding momentum factor in three-factor model. It measures the tendency of price changes with a portfolio of long previous-12-month return winners and short previous-12-month loser stocks, which had an 8% accumulated yield. The momentum factor can explain 6.4% of the excess return.

Therefore, Fama and French (2015) added the two new factors, investment (CMA) and profitability (RMW), to build the five-factor model for measuring the effects of company size, valuation, profitability, and investment pattern in average stock returns. In this equation, they divide both sides by book value at time t to create book-to-market ratio and present the relationship between return r and valuation factor.

$$R_i - R_f = \mathrm{a} + \beta_{market}\left(R_m - R_f\right) + \beta_{size}\mathrm{SMB} + \beta_{BM}\mathrm{HML} + \beta_{profitability}\mathrm{RMW} + \beta_{investment}\mathrm{CMA} + \varepsilon \tag{5}$$

Moreover, they use the RMW factor to represent the profitability and the CMA factor to represent investment. According to empirical study, the five-factor model has a higher effectiveness than the three-factor model, which can explain 71–94% of the variation of average return. In the test, researchers divided all the stocks with three sets of factors. The result also implied that book-to-market factor is redundant for describing average return using the American stock data from 1963 to 2013. Also, the model fails to capture the low average return for small firms with low profitability and high investment.

Sehgal and Vasishth (2015) tested the model in various emerging markets and discovered that the change of price and trading volume are partly risk based and partly behavioral. The research indicated that behavioral factors are necessary in the study of the multi-factor model. Except for the existing risk premium factors in Fama and French model, Peng et al. (2014) studied the effect of different investment sentiments on the market return from which they found out customer satisfaction is a significant factor for abnormal return. Moreover, a human capital factor was considered in terms of compensation level (Moinak and Balakrishnan 2018). Zahedi and Rounaghi (2015) applied an artificial neural network to assess the components of a multi-factor model and predict future stock prices.

Considering the features of data and the time-varying factors, a range of studies applied various methods to address problems via multi-factor models. Akter and Nobi (2018) examined the distribution and frequency distribution for both daily stock returns and volatility. Furthermore, Chen and Kawaguchi (2018) distinguish two significant regimes (a persistent bear market and a bull market) to examine market time-varying risk factors to achieve Markov regime-switching. These studies examined the model specifically in certain periods which enables the researchers to compare the model performances in different situations which can improve the applicability of the model.

Some researchers conducted modification on the basic model with specific structure. Ronzani et al. (2017) suggested that β (systematic risk) evolves over time and the model with time-varying β provide less conservative VaR measures than the static β. While Cisse et al. (2019) examined the dynamics of the model with Kalman filter and Markov switching (MS) model and proved that the former method fits better in the model. Bhattacharjee and Roy (2019) proposed a social network dependence structure to address such misspecifications. For the investment aspect, Frazzini et al. (2013) suggesting that Buffett's returns are more due to stock selection than to his effect on management.

3. Methodology

3.1. Hypotheses

Hypothesis 1. *Market premium has a positive relationship with excess return.*

Hypothesis 2. *Size premium has a positive relationship with excess return.*

Hypothesis 3. *Book-to-market premium has a positive relationship with excess return.*

Hypothesis 4. *Profitability premium has a positive relationship with excess return.*

Hypothesis 5. *Investment growth premium has a positive relationship with excess return.*

Hypothesis 6. *Momentum premium has positive a relationship with excess return.*

Hypothesis 7. *Asset turnover premium has a positive relationship with excess return.*

Hypotheses are conducted on the relationship between dependent variable and independent variables. To be specific, the positive relationship stated that risk premium factor can generate higher excess return. For instance, a company with higher profitability can achieve higher returns than a low profitability company. Since previous researchers have not examined the effect of factors on single stocks, we decided to retest the effect of seven factors on excess return for each security and summary by industries.

3.2. Research Design

The data were collected from the Choice, Tongdaxin, and Resset database. Choice and Tongdaxin can provide historical values of stock financial ratios and Resset provides the basic information of companies and interest rates. The samples are 1097 stocks in the Chinese stock market, including Shanghai securities exchange market and Shenzhen securities exchange market from, 2007 to 2018 because large quantity of firms transformed their non-tradable shares into tradable shares in the process of Chinese Reformation of National Owned Stock from 2005 to 2006 and the time range after 2007 provides the research with more companies' data. The type of data is quarterly data which includes 47 quarterly data points for each stock. In the process of data cleaning, those stocks with missing data will be deleted in the samples.

Even if Fama and French (2015) had delivered the five-factor model, yet they only consider the factor performance in different groups of portfolio but not the factors' effect on each security in the stock market. Thus, by applying variable-intercept models mentioned in the book of Hsiao (2003), we ran the panel data for each stock to test the effect of these five premiums in the Model 1. The study bridged the relationships between the excess return of each security and different premium factors via OLS (ordinary least squares regression), which means each stock has their individual coefficient and there are more than 1000 regression functions (Johnson and Wichern 2008). During the research process, this paper classified the securities into different groups to calculate risk premium factors based on market value, book-to-market ratio, ROE, and the growth of investment. Then, this research considered the average coefficients' level, average significant level, and their distribution.

In Model 2, the momentum factor and turnover factor are added into the model. We developed an innovative method to study each single stock in the market, in order to find out the general pattern of the stock market. In contrast to the former ways that examine the model with the diversified portfolio, we conducted ridge regression on single stocks and divided the stocks in industries. The 28-industry classification standards are from ShenWan industry index.

Figure 1 illustrates the research structure and their corresponding functions for the results. The research includes two major parts, which are multi-factor examination with single stock regression and time-series analysis for risk factors.

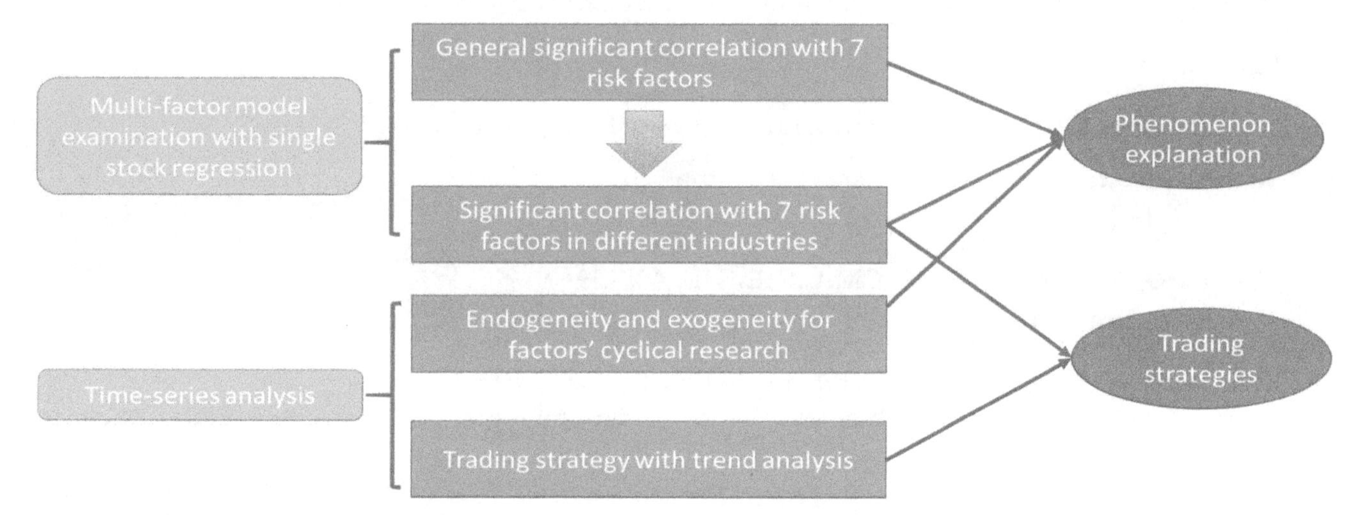

Figure 1. Multi-factor model examination with single stock regression and time-series analysis.

The first part involved stability test, OLS regression, ridge regression, and robustness test to find the significant correlations between stock excess return and seven risk factors. In addition, a chi-square test was conducted to examine whether the effect of factors various in different industries. Moreover, by measuring the percentage of positive and negative correlations in each industry, the significant relationships between factors and industry was discovered.

The second part of research covered chi-square test to find out the endogeneity and exogeneity (or called pattern of fluctuation) for seven risk factors and a back test of trading strategy with trend analysis. For details, for chi-square testing, we firstly recorded the rise and fall pattern in the neighboring quarters. There were four kinds of pattern (rise after rise, rise after fall, fall after rise, fall after fall). We calculated each patterns' amount and conducted chi-square testing on the pattern of each factor. If a factor passes the test, it can be inferred that the current pattern is in accordance with the previous period. Therefore, with the endogeneity and exogeneity of the factors, we can predict the future rise and fall based on the current pattern. Moreover, a trading strategy was established based on the trend analysis of factors. As for the investment portfolio, stocks were selected according to the effect of seven factors on industries. A simple back test was also conducted to examine the performance of the trading strategy.

The result of general significant correlations and the factors' effect in various industry can be used in some economic phenomenon explanation. Furthermore, combing the significant relationships between risk factors and corresponding industries with trend analysis, trading strategies can be built.

Factor Selection

We modified the multi-factor model to adapt for the Chinese stock market. Since some of the existing factors in the model have poor performance in the regression analysis and empirical test, it is significant to add or delete some of the factors to improve the model.

In accordance with previous studies, we selected two new factors (momentum factor and turnover factor). Firstly, the momentum effect is remarkable in the medium term of stock market. Momentum effect was proposed by Jegadeesh and Titman (1993) to illustrate the phenomenon that stock performed well in the past is more likely to achieve higher return. Regarding the composition of investor and investment behavior in stock market, we suggested that the momentum effect can explain the part of the excess return. Secondly, we added a turnover factor into the model. Asset turnover rate is a component of DuPont analysis, which is used to analyze the return of shareholder's equity (Wild 2016). Asset turnover rate is a vital measure of a company's operation capacity. A higher asset turnover rate indicates that a company has better operation capacity and higher efficiency. When conducting value investing, this financial ratio is primarily regarded before investment, as it can directly reflect the condition of a company, for example profitability, operation capacity, and so on.

3.3. Research Process

3.3.1. Multi-Factor Model Examination with Single Stock Regression Analysis

- **Stability Test**

DV: Ri-Rf
IV: Rm-Rf, SMB, RMW, HML, CMA, CRMHL, AMLH
CV: Time, Time^2, Season

If the three control variables which refers to time, the square of time and season can pass the *t*-test with high p-level, these three variables should be put into the regression models, otherwise they can be deleted.

- **OLS Regression**

We conducted OLS regression on 47 quarters for each single stock, in order to examine the multi-factor model. According to OLS, to calculate the beta (coefficient) of each independent variable, the matrix operation combines X and y. In this formula, X is a 47×7 matrix, in which 47 is the 47 time-series of data and 7 represents seven proposed risk factors. The y is a 47×1 matrix and 1 indicates the excess return of a single stock. Since there are 1097 stock, X matrix is fixed yet y matrix is the data of different stocks.

$$\beta = (X^T X)^{-1} X^T y \tag{6}$$

$$\begin{bmatrix} \beta_1 \\ \beta_2 \\ \vdots \\ \vdots \\ \beta_7 \end{bmatrix} = \left(\begin{bmatrix} SMB_1 & RMW_1 & \cdots & AMLH_1 \\ \vdots & \vdots & & \vdots \\ \vdots & \vdots & & \vdots \\ SMB_{47} & RMW_{47} & \cdots & AMLH_{47} \end{bmatrix}^T \begin{bmatrix} SMB_1 & RMW_1 & \cdots & AMLH_1 \\ \vdots & \vdots & & \vdots \\ \vdots & \vdots & & \vdots \\ SMB_{47} & RMW_{47} & \cdots & AMLH_{47} \end{bmatrix} \right)^{-1} \begin{bmatrix} SMB_1 & RMW_1 & \cdots & AMLH_1 \\ \vdots & \vdots & & \vdots \\ \vdots & \vdots & & \vdots \\ SMB_{47} & RMW_{47} & \cdots & AMLH_{47} \end{bmatrix}^T \begin{bmatrix} y_1 \\ y_2 \\ \vdots \\ \vdots \\ y_7 \end{bmatrix} \tag{7}$$

We used computer simulation in Python, which can help us automatically run regression 1097 times. The python program would print out the result of regression and calculate the number of mean value of coefficients and *t*-value, which were collected in a 7×1097 matrix. We also drew the frequency histogram for each factor to study the distribution of the coefficients. Based on these assumptions, the majority of the coefficients should be positive. If most of them come out negative, we may infer that the five-factor model is not suitable for the Chinese stock market.

- **Ridge Regression**

Since we added two more factors in the model, we applied ridge regression instead of OLS regression. On one hand, ridge regression can prevent an over-fitting result from extra factors. On the other, it can prevent multi-collinearity and increase the significance of the factors. Ridge regression is developed based on OLS regression by adding regularization term λI. The regularization term can discover the factor with collinearity and force the coefficient to approach zero to ensure that the effect of multi-collinearity is minimized. To calculate the beta (coefficient) of each independent variable, the matrix operation combines X, y, λ, and I. X and y are the same as the matrix in the OLS model. λ is an optimization hyperparameter and I is a 7 × 7 identity matrix.

$$\beta = (X^T X + \lambda I)^{-1} X^T y \tag{8}$$

$$\begin{bmatrix} \beta_1 \\ \beta_2 \\ \vdots \\ \vdots \\ \beta_7 \end{bmatrix} = \left(\begin{bmatrix} SMB_1 & RMW_1 & \cdots & AMLH_1 \\ \vdots & \vdots & & \vdots \\ \vdots & \vdots & & \vdots \\ SMB_{47} & RMW_{47} & \cdots & AMLH_{47} \end{bmatrix}^T \begin{bmatrix} SMB_1 & RMW_1 & \cdots & AMLH_1 \\ \vdots & \vdots & & \vdots \\ \vdots & \vdots & & \vdots \\ SMB_{47} & RMW_{47} & \cdots & AMLH_{47} \end{bmatrix} + \begin{bmatrix} \lambda & 0 & 0 & 0 \\ 0 & \lambda & 0 & 0 \\ \vdots & \vdots & \ddots & \vdots \\ 0 & 0 & 0 & \lambda \end{bmatrix} \right)^{-1} \begin{bmatrix} SMB_1 & RMW_1 & \cdots & AMLH_1 \\ \vdots & \vdots & & \vdots \\ \vdots & \vdots & & \vdots \\ SMB_{47} & RMW_{47} & \cdots & AMLH_{47} \end{bmatrix}^T \begin{bmatrix} y_1 \\ y_2 \\ \vdots \\ \vdots \\ y_7 \end{bmatrix} \tag{9}$$

When we applied the ridge regression, we need to find a λ that can provide the largest sum of R-square which are the number of positive factors and percentage of positive coefficient. We set the summary of these three measures as our target function to find out the optimal λ with iteration.

$$\lambda = \operatorname{argmax}(Q) \tag{10}$$

Q = P average value + number of positive coefficients + R square mean + number of variables whose p value over 0.9 (9).

- **Robustness Test**

In this part, a zero-mean test for OLS and ridge regression model is adopted to check the robustness of models. Both OLS and ridge regression models are fixed-coefficient models, which means that they have an assumption that the covariance of errors between different units (stocks) are equal to zero. Thereafter, a residual covariance matrix between 1097 stocks' regression can be drawn by Python and calculate the mean and standard deviation of all numbers in covariance matrix. Then a t-test is conducted to test whether $E(u_{i,j})$ is equal to zero.

Covariance matrix of residual

$$\begin{bmatrix} u_{1,1} & u_{1,2} & \cdots & u_{1,1097} \\ u_{2,1} & \ddots & \cdots & \vdots \\ \vdots & \vdots & \ddots & \vdots \\ u_{1097,1} & \cdots & \cdots & u_{1097,1097} \end{bmatrix} \tag{11}$$

$u_{1,2}$ means the covariance between stock 1 and 2.

$$\text{T-value} = \frac{E(u_{i,j})}{STDE(u_{i,j})} \tag{12}$$

3.3.2. Time-Series Analysis for Risk Factors

- **Chi-Square Test**

We use chi-square in two steps respectively. Firstly, it was used to examine the different effect of factors in different industries. In the test we divided the significance of factors based on industries and calculated the total value χ^2.

Secondly, it was applied to find out the pattern in the inter-factor direction prediction analysis. We divided the pattern into four types with a combination of increase and decrease. Then we used chi-square to find out whether there is a pattern for factor change direction. With the increase and decrease pattern, we can predict the next quarter movement based on the current situation. As we can see in Figure 2, suppose "0" represents the increase of risk factor in one term (or one season) and "1" represents the decrease of risk factors.

$$R = \text{Max}(F1, F2) + \text{Max}(F3, F4) \tag{13}$$

$$W = \text{Min}(\text{F1}, \text{F2}) + \text{Min}(\text{F3}, \text{F4}) \quad (14)$$

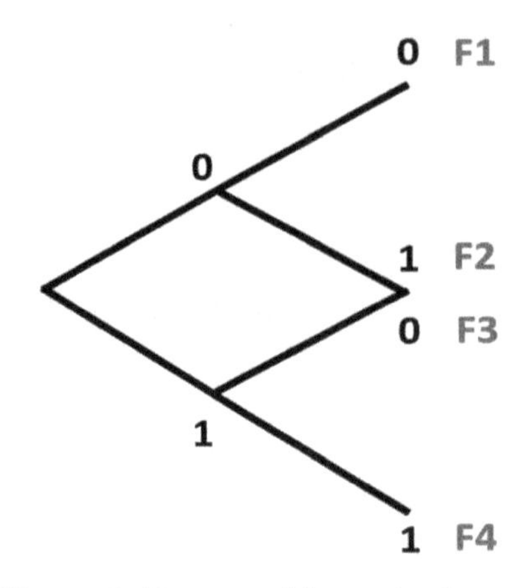

Figure 2. Pattern of factor's change.

In Figure 3, if investors spontaneously make investment decision, the frequencies of right and wrong decisions are equitable. Both are T/2. (T is number of total transaction time)

H0: *There is no relationship between right or wrong investment decisions and direction prediction rules.*

H1: *There is a relationship between right or wrong investment decisions and direction prediction rules.*

	Follow the rule	Randomly select	Total
Right investment decision	R	$T/2$	$R+T/2$
Wrong investment decision	W	$T/2$	$W+T/2$
Total	$R+W$	T	$R+W+T$

Figure 3. Observed frequencies: f_o.

In Figure 4, the expected frequency of right decisions (R_e) and wrong decisions (W_e) made by the rule can be calculated by Equations (15) and (16). Also, the expected frequency of right or wrong decision made by random selection ($(T/2)_e$) can be calculated by Equation (17).

$$R_e = (R + W) \times (R + T/2)/(R + W + T) \quad (15)$$

$$W_e = (R + W) \times (W + T/2)/(R + W + T) \quad (16)$$

$$(T/2)_e = T \times (R + T/2)/(R + W + T) \quad (17)$$

	Follow the rule	Randomly select
Right investment decision	R_e	$(T/2)_e$
Wrong investment decision	W_e	$(T/2)_e$

Figure 4. Expected frequencies: f_e.

According to the significant table (Figure 5) of chi-square test, if total chi-square level is >2.706, we have 90% confidence to reject the null hypothesis (H0). If total chi-square level is >3.841, we have 95% confidence to reject the null hypothesis (H0). If total chi-square level is >6.635, we have 99% confidence to reject the null hypothesis (H0).

$$\text{Total chi-square level} = (R - R_e)^2/R_e + (W - W_e)^2/W_e + 2[T/2 - (T/2)_e{}^2]/(T/2)_e \quad (18)$$

$$\text{The degree of freedom} = (\text{column} - 1)(\text{row} - 1) = (2-1)(2-1) = 1 \quad (19)$$

	Follow the rule	Randomly select
Right investment decision	$(R - R_e)^2 / R_e$	$[T/2 - (T/2)_e]^2 / (T/2)_e$
Wrong investment decision	$(W - W_e)^2 / W_e$	$[T/2 - (T/2)_e]^2 / (T/2)_e$

Figure 5. Chi-square level: $\chi^2 = (f_o - f_e)^2 f$.

- **Back-Test for Trading Strategy**

After factors' cyclical research and trend analysis for fluctuation of risk factors, the trading strategy based on moving average approach and correlations between factors and industries can be formed. By deciding the time to do the transactions and stocks which should be bought or sold, the float return from 2007S4 to 2018S3 and annual expected return can then be calculated.

3.4. Assumptions for Multi-Factor Examination

The following assumptions are made in applying the multi-factor model:

1. Perfect market: there are no tax and transaction costs.
2. People are risk averse or rational.
3. People can lend or borrow money at a risk-free rate freely.
4. There is a trade-off between risk and return for all securities.
5. Everyone can obtain market information equally and freely.
6. Investors have the same expectations for this market.

The intuition and assumption behind the hypotheses are presented in the Appendix A with measurements of factors and graphs.

3.5. Models and Variable Definitions

Model 1:

$$R_i - R_f = \mathrm{a} + \beta_{market}\left(R_m - R_f\right) + \beta_{size}\mathrm{SMB} + \beta_{BM}\mathrm{HML} + \beta_{profitability}\mathrm{RMW} + \beta_{investment}\mathrm{CMA} + \varepsilon$$

Model 2:

$$R_i - R_f = \mathrm{a} + \beta_{market}\left(R_m - R_f\right) + \beta_{size}\mathrm{SMB} + \beta_{BM}\mathrm{HML} + \beta_{profitability}\mathrm{RMW} + \beta_{investment}\mathrm{CMA} + \beta_{momentum}\mathrm{CRMHL} + \beta_{turnover}\mathrm{AMHL} + \varepsilon$$

The details of the factors (Rm-Rf, SMB, HML, RMW, CMA, CRMHL, AMHL) including explanation and graphs are presented in Appendix A.

Table 1 displays the dependent variable, independent variables and the corresponding labels and explanations.

Table 1. Dependent variable and independent variables.

Name	Label	Note
	DV	
Excess return	$R_i - R_f$	The return of security mines risk-free rate of return

Table 1. *Cont.*

Name	Label	Note
	IV	
Abnormal return	a	The constant term of formula
Market premium	$R_m - R_f$	The return of market index (in this model, market index is Shanghai stock exchange market index) mines risk-free rate of return
Size premium (Small minus Big)	SMB	The return on a diversified set of small stocks minus the return on a diversified set of big stocks.
Book-to-market premium (High minus low)	HML	The difference between the returns on diversified portfolios of high and low B/M stocks.
Profitability premium (Robust minus weak)	RMW	The difference between the returns on diversified portfolios of stocks with robust and weak profitability.
Investment growth premium (Conservative minus aggressive)	CMA	The difference between the returns on diversified portfolios of the stocks of low and high investment firms, which we label as conservative and aggressive.
Momentum premium (High momentum minus low momentum)	CRMHL	The difference between higher momentum (higher accumulated return) companies' average return and lower momentum (lower accumulated return) companies' average return in a diversified portfolio or in the market.
Asset turnover premium (Low turnover rate minus high turnover rate)	AMLH	The difference between low asset turnover companies' average return and higher turnover companies' average return or in the market.

4. Results

4.1. Multi-Factor Model Examination with Single Stock Regression Analysis

4.1.1. Stability Test

DV: Ri-Rf
IV: Rm-Rf, SMB, RMW, HML, CMA, CRMHL, AMLH
CV: Time, Time^2, Season

According to the result of stability test (presented in Appendix E), the control variable including Time, Time^2, Season did not pass the *t*-test, the significance level is far less than 0.9. Thus, they are removed from the model. It means that excess return is not affected by time.

4.1.2. Regression Models

- **OLS Regression**

Model 1: Five-factor model

DV: Ri-Rf
IV: Rm-Rf, SMB, RMW, HML, CMA

Model 2: Seven-factor model

DV: Ri-Rf
IV: Rm-Rf, SMB, RMW, HML, CMA, CRMHL, AMLH

Model 3: Modified optimal model

DV: Ri-Rf
IV: Rm-Rf, RMW, HML, CMA, CRMHL, AMLH

- **Ridge Regression**

Model 4: Ridge regression model

DV: Ri-Rf
IV: Rm-Rf, SMB, RMW, HML, CMA, CRMHL, AMLH

We conducted OLS regression on Model 1, 2, and 3, and ridge regression on Model 4. After conducting the regression analysis, we recorded each factors' coefficients and mean of R square. We created frequency histograms to observe the distribution of coefficients. The graphs provide a general view of the coefficients.

In Table 2, above each histogram, there are two numbers, mean of coefficients and the proportion of coefficients that passed the *t*-test (in the parenthesis).

To be specific, we divided the coefficients into uncorrelated, positively correlated, and uncorrected, based on 95% confidence interval. Table 3 shows the percentage of coefficient. SMB, CMA, Rm-Rf, CRMHL, and AMHL have more positive coefficients. Whereas RMW and HML have more negative coefficients.

In ridge regression, we obtain the λ by calculating the maximum of the objective function Q. When λ equals to 0.008, the result of ridge regression is maximized. In Figure 6, we can find out the largest Q value at 0.008.

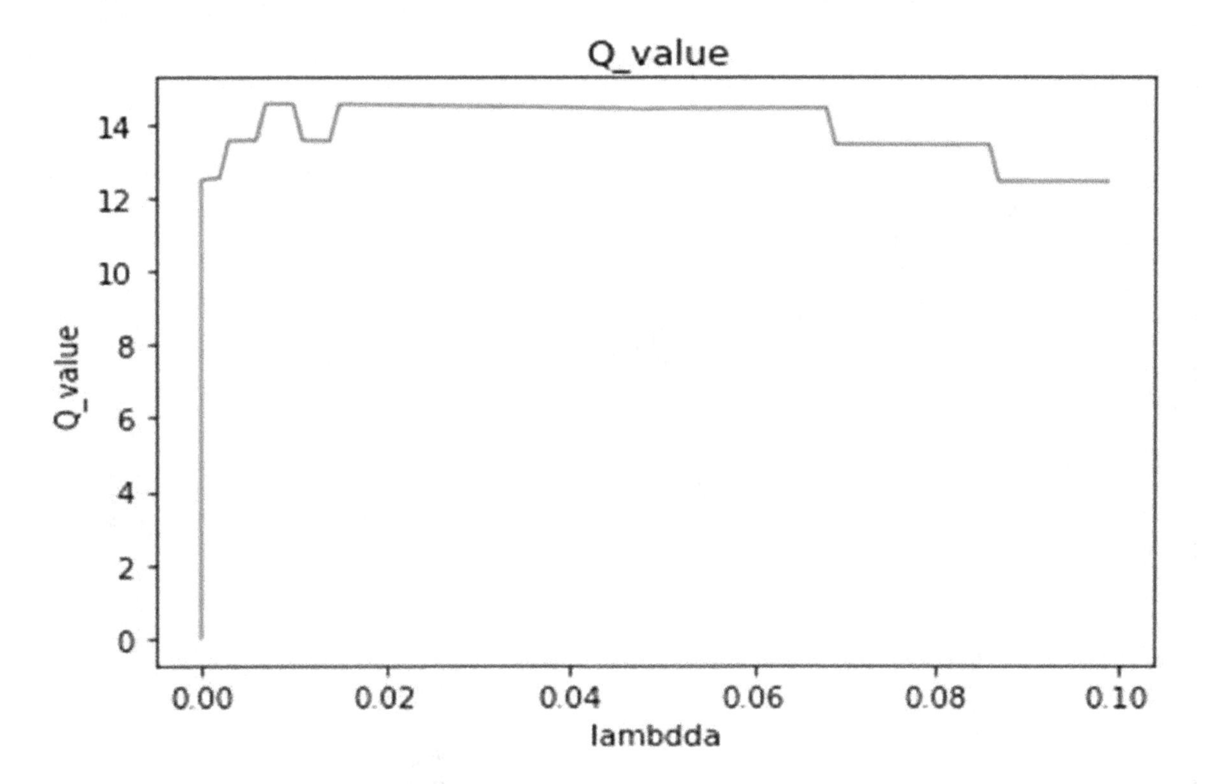

Figure 6. Target function optimization.

Some factors have negative effects on the excess return, which deviates from our assumption. Moreover, regarding the factors that have an even distribution. We need to classify them in industry groups to discover a further pattern.

Table 2. R square mean, coefficients' mean and p-level for four models.

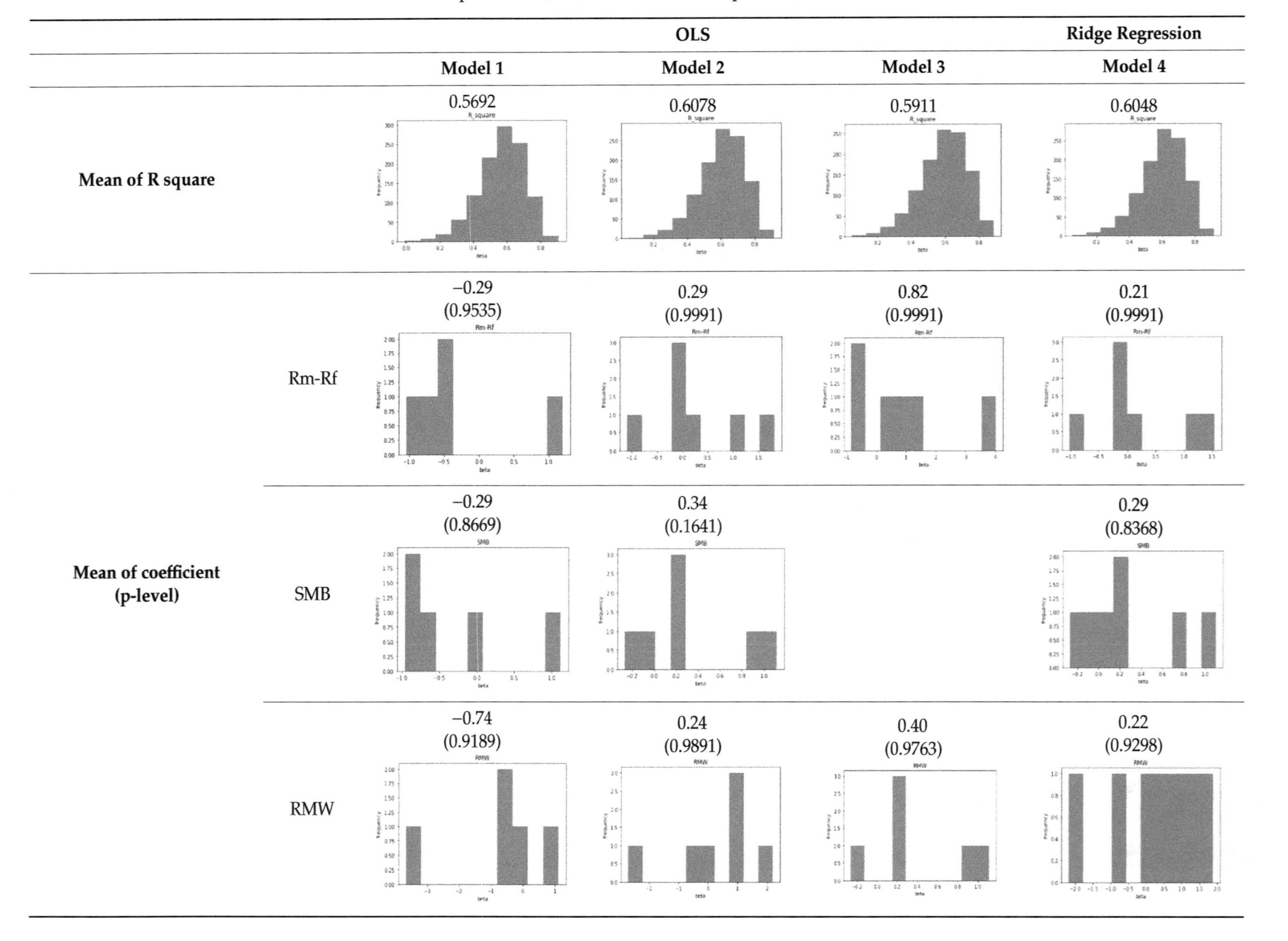

		OLS			Ridge Regression
		Model 1	**Model 2**	**Model 3**	**Model 4**
Mean of R square		0.5692	0.6078	0.5911	0.6048
Mean of coefficient (p-level)	Rm-Rf	−0.29 (0.9535)	0.29 (0.9991)	0.82 (0.9991)	0.21 (0.9991)
	SMB	−0.29 (0.8669)	0.34 (0.1641)		0.29 (0.8368)
	RMW	−0.74 (0.9189)	0.24 (0.9891)	0.40 (0.9763)	0.22 (0.9298)

Table 2. *Cont.*

		OLS			Ridge Regression
		Model 1	**Model 2**	**Model 3**	**Model 4**
Mean of coefficient (p-level)	HML	0.13 (0.3874)	0.30 (0.9690)	−0.27 (0.9444)	0.22 (0.9681)
	CMA	−0.11 (0.9335)	0.58 (0.9900)	−0.50 (0.9763)	0.38 (0.9909)
	CRMHL		−0.14 (0.9617)	0.33 (0.9243)	0.02 (0.9535)
	AMLH		−0.04 (0.9991)	−0.72 (0.9991)	0.06 (0.9991)

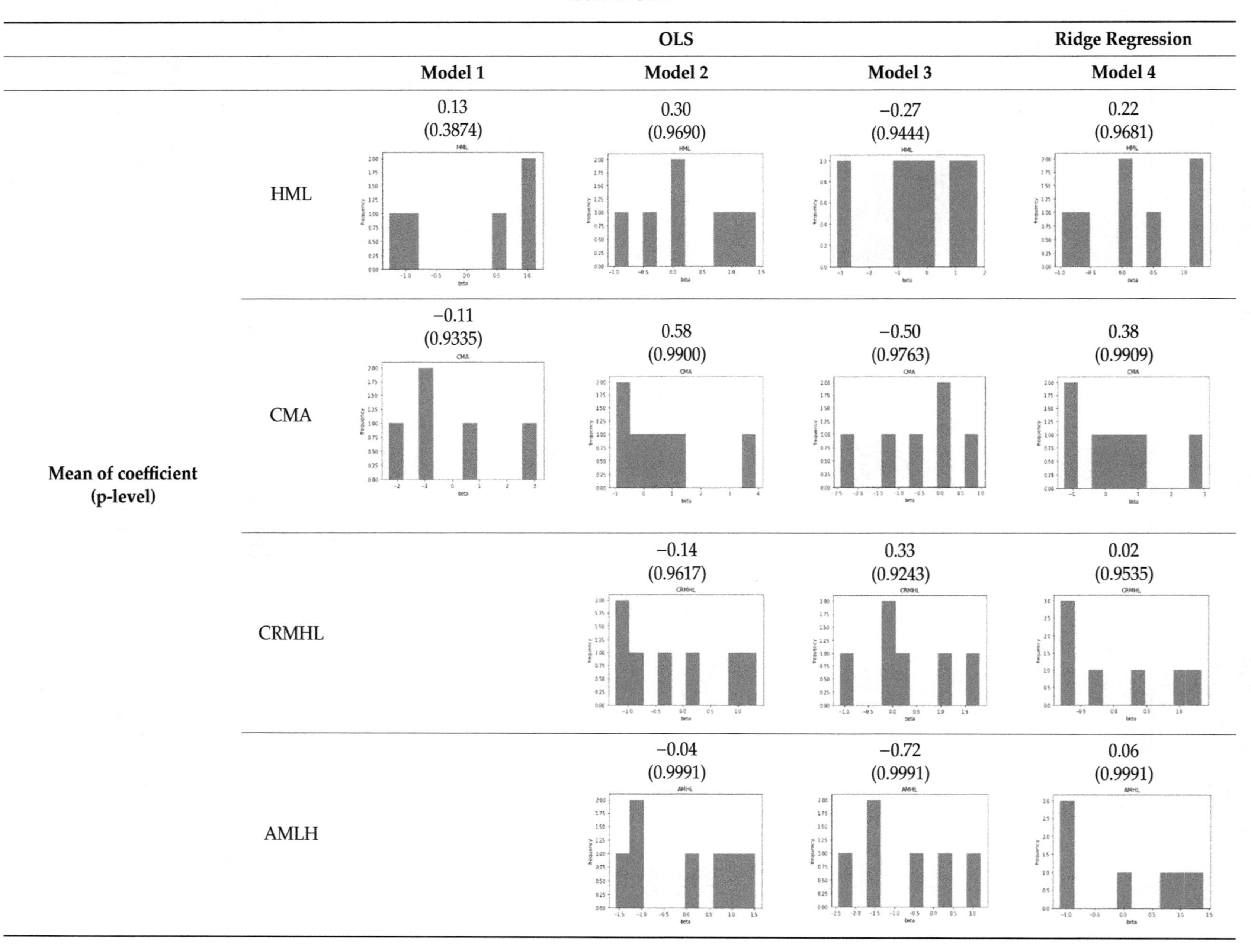

Table 3. Correlations between seven factors and excess return of stocks.

	SMB	RMW	HML	CMA	AMLH	CRMHL	Rm-Rf
Positive	61.0%	24.2%	23.0%	56.2%	62.4%	77.8%	99.8%
Negative	22.7%	68.7%	73.8%	42.8%	37.5%	17.6%	0.1%
Uncorrelated	16.3%	7.0%	3.2%	0.9%	0.1%	0.1%	0.1%

4.1.3. Robustness Test: Zero Mean Residual Testing

According to Table 4, since t-values in four models are less than 2.021, we have 95% confidence not to reject the null hypothesis, which means the residuals are randomly independent for all stocks, both the OLS and ridge regression model pass the zero mean residual test.

Table 4. Zero mean residual testing.

OLS t-Value			Ridge Regression t-Value
Model 1	Model 2	Model 3	Model 4
0.5811	0.5393	0.5416	0.5387

4.1.4. Industry Analysis

According to the result of chi-square test, each industry corresponds differently to the factors. Therefore, effect of factors needs to be discussed based on different industries respectively. The details of the test results are presented in the Appendix B.

We classify the stocks in 28 industries. Also, we divided the coefficients in positive correlation and negative correlation. As for data cleaning, we only preserved the factors that contained over 70% coefficients are of the same signs. The specific table is presented in the Appendix C.

In the Table 5, the industries are classified in manifold groups based on the significance level and the relationship with each factor.

According to the table, there is no industry that can be explained by seven factors simultaneously. The banking and steel industries are best fitted in the seven-factors model, since they have six significant factors and the highest sum of percentage of significant coefficients.

Each factor has different relationships with different industries. For market premium (Rm-Rf), it can apply to all industries. For SMB, it is positively related with seven industries (Group A), yet SMB is negatively correlated with Steel industry. With respect to RMW, it has negative relationships with most industries (Group D). However, RMW has significant positive relationships with bank industry. HML is negative related to multiple industries (Group F). Also, HML has significant positive relationships with the banking industry. CMA is positively correlated with five industries (Group G). Meanwhile, CMA is negatively correlated to three industries (Group H) including extractive, banking, and nonferrous metal. CRMHL is positive correlated with most of industries (Group I) except media, electrical equipment, non-bank finance, animal husbandry and fishery, commercial, and comprehensive industries. AMLH has a negative correlation with Group K. At the same time, AMLH is positively related with three other industries (Group J) including housing, non-bank finance, and banking.

Moreover, the industries with the same pattern are regarded as coordinated industry. Additionally, the industries with different pattern are regarded as deviated industries. To be specific, the banking industry is considered as a deviated industry since it has a positive RMW and HML factor, while other industries have negative factors. The graph shows four groups of coordinated industry—including telecommunication and electronics, chemistry and automobile, light manufacturing and defense, and nonferrous metal and extractive—while banking is regarded as a deviated industry.

Considering the trend of the factors, we can formulate corresponding strategies. To be specific, based on the condition of each factor and the classification from Group A to L which consist of different industries presented in Table 5, we formulated manifold strategies in Table 6.

Table 5. Relationships of industries and seven factors.

	Positive Relationships	Negative Relationships
SMB	(Group A: 7) Electrical Equipment, Electronics, Computer, Construction Material, Food, Telecommunication, and Leisure Service	(Group B: 1) Steel
RMW	(Group C: 1) Banking	(Group D: 16) Extractive, Media, Electrical Equipment, Textiles & Garments, Steel, Defense, Chemistry, Mechanical Equipment, Computer, Construction Material, Transportation, Automobile, Light Manufacturing, Leisure Service, Nonferrous Metal, and Comprehensive Industries.
HML	(Group E: 1) Banking	(Group F: 19) Extractive, Media, Electrical Equipment, Electronics, Housing, Textiles & Garments, Steel, Defense, Chemistry, Mechanical equipment, Computer, Animal Husbandry and Fishery, Automobile, Light Manufacturing, Commercial, Telecommunication, Leisure Service, Medical, Nonferrous Metal, and Comprehensive Industries.
CMA	(Group G: 5) Textiles & Garments, Utilities, Defense, Light Manufacturing, and Commercial	(Group H: 3) Extractive, Banking, and Nonferrous Metal
CRMHL	(Group I: 20) Extractive, Electronics, Housing, Textiles & Garments, Steel, Utilities, Defense, Chemistry, Mechanical Equipment, Computer, Domestic Appliance, Construction material, Transportation, Automobile, Light Manufacturing, Telecommunication, Leisure Service, Medical, Banking, and Nonferrous Metal.	
AMLH	(Group J: 3) Housing, Non-Bank Finance, and Banking.	(Group K: 14) Extractive, Electrical Equipment, Electronics, Steel, Defense, Chemistry, Domestic appliance, Automobile, Light Manufacturing, Food, Telecommunication, Leisure Service, Medical and Nonferrous Metal.
Rm-Rf	(Group L: 28) All industries	

Table 6. Trading strategies based on condition of factors.

Condition	When Factor Is Positive		When Factor Is Negative	
Strategies	BUY	SELL	BUY	SELL
SMB	Small MV companies in Group A Big MV companies in Group B	Big MV companies in Group A Small MV companies in Group B	Big MV companies in Group A Small MV companies in Group B	Small MV companies in Group A Big MV companies in Group B
RMW	Robust ROE companies in Group C Weak ROE companies in Group D	Weak ROE companies in Group C Robust ROE companies in Group D	Weak ROE companies in Group C Robust ROE companies in Group D	Robust ROE companies in Group C Weak ROE companies in Group D
HML	High B/M companies in Group E Low B/M companies in Group F	Low B/M companies in Group E High B/M companies in Group F	Low B/M companies in Group E High B/M companies in Group F	High B/M companies in Group E Low B/M companies in Group F
CMA	Conservative (low growth rate of assets) companies in Group G Aggressive (High growth rate of assets) companies in Group H	Aggressive companies in Group G Conservative companies in Group H	Aggressive companies in Group G Conservative companies in Group H	Conservative companies in Group G Aggressive companies in Group H
CRMHL	High CR companies in Group I			High CR companies in Group I
AMLH	Low Asset turnover companies in Group J High Asset turnover companies in Group K	High Asset turnover companies in Group J Low Asset turnover companies in Group K	High Asset turnover companies in Group J Low Asset turnover companies in Group K	Low Asset turnover companies in Group J High Asset turnover companies in Group K
Rm-Rf	All companies (Group L)			All companies (Group L)

Here we can take an example to illustrate the above table, when SMB is positive, it means in this market, small market value (MV) companies outperform big companies, thus for those industries which is significantly positive correlated with SMB, we should buy small MV stocks and sell large MV stocks. However, for those industries which is negatively correlated with SMB, the strategy should purchase large MV stocks and sell small MV stocks. When SMB is negative, vice versa.

According to the significant level and correlation effect between 7 factors and 28 industries, the trading portfolio and strategy can be conducted. Since the strategic making depends on the positive or negative condition of seven factors, if we can forecast the conditions of seven factors, the trading strategies can be easily made. Thus, for the next two parts, we explored the fluctuation of seven factors to answer two questions:

1. From analysis of factors' changing pattern, can we find the reasons or elements to illustrate the fluctuation of seven factors? (if we find out the driving factor which stimulate the moving of other factors, we can explain the most essential ratio that investor may focus on.)
2. Can we find an approach to forecast the moving of each factor and apply it to do the back test for trading strategies?

4.2. Time-Series Analysis for Risk Factors

4.2.1. Endogeneity and Exogeneity for Factors' Cyclical Research

- **Endogeneity**

According to Pearson Correlation matrix, last term data of SMB, RMW, HML, CMA, CRMHL, and AMLH has significant correlation with this term at the level of 0.01. Nevertheless, for the market factor (Rm-Rf), its last term data are nearly independent with this term, which means except market risk factor, other factors are possible to forecast themselves with last term data.

However, to test the endogeneity of these factors, chi-square test are necessary for examination and application of these relationships. After the test, the results show that only SMB, RMW pass the test with 99% confidence level and CMA also has a significance level of 0.05, which means these three factors exist endogeneity and can be truly used to forecast their direction and investment cycle.

- **Exogeneity**

Based on Pearson correlation matrix, the fluctuation of HML, CRHML, and AMLH depend upon the history data of other factors. For details, HML is related to the AMLH of last term. CRHML has a small correlation with HML AMLH bridge its relationships with RMW of last term. Finally, with respect to Rm-Rf, its volatility is partly connected to last term's RMW and AMLH.

Figure 7 summarizes the endogeneity and exogeneity for predicting direction of seven risk factors. In the chi-square test, RMW, SMB, and CMA factor can be only predicted by the last term direction of themselves. With confidence level of 95%, CRHML's direction can forecasted by its history data with HML and the increase and decrease of HML can be predicted by last term direction and AMLH at 90% confidence level. For market factor (Rm-Rf)—even if single stock history data cannot predict next term—with RMW and AMLH, market factor can estimate the direction of its next term with 95% confidence. With a confidence level of 90%, AMLH's fluctuation direction can forecast by RMW. The details of prediction are presented in Appendix D.

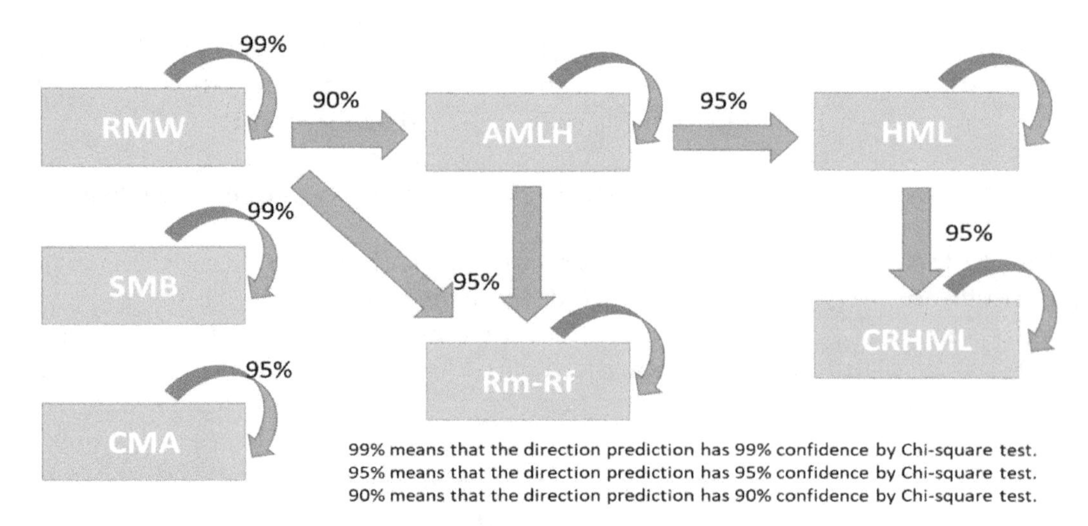

Figure 7. Summary of endogeneity and exogeneity for predicting direction of seven risk factors.

4.2.2. Trading Strategy with Trend Analysis

In the last two parts, with the examination of chi-square test, the directions of fluctuation of seven factors are predictable, yet the range of fluctuation is still unknown.

Therefore, in order to forecast the positive and negative levels of each risk factor, trend analysis is conducted.

The above seven graphs in Figure 8 describe the trend of each factor. The orange line is the four-season moving average. SMB, HML, and CMA are mostly below 0. By contrast, RMW and CRMHL are above 0. While AMLH and Rm-Rf fluctuate around 0.4.

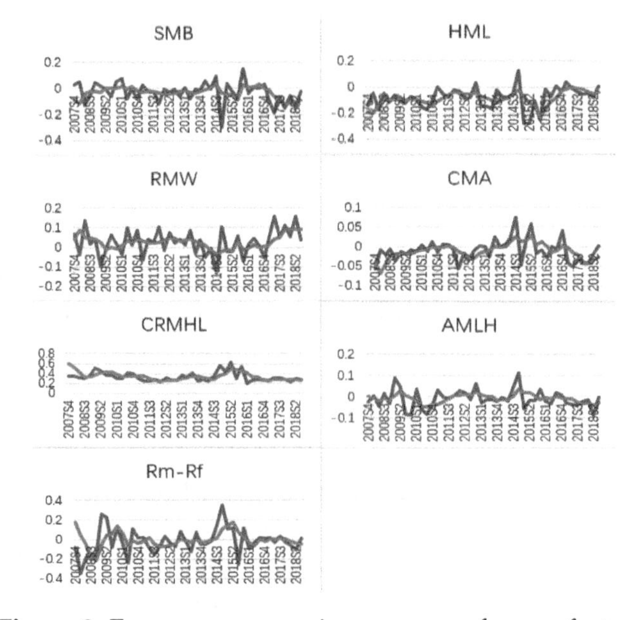

Figure 8. Four seasons moving average of seven factors.

4.2.3. Back-Testing

Table 7 illustrates the results of the back-testing from 2007S4 to 2018S3 for the trading system. With the strategy, we can achieve positive annual return in SMB, CMA, CRMHL, and AMLH for 15.43%, 3.23%, 29.13%, and 8.4% respectively. To some extent, the result proved the practicability of the seven-factor model, especially for the factor CRMHL in terms of trading strategy. Consequently, with alternative trading strategies which can predict the factor fluctuation more precisely, we can witness a greater margin of improvement.

Table 7. Back-testing float return from 2007S4 to 2018S3 and expected annual return for factors.

Factors	Float Return	Expected Annual Return
SMB	165.89%	15.43%
RMW	−6.525%	−0.61%
HML	−22.29%	−2.07%
CMA	34.67%	3.23%
CRMHL	313.09%	29.13%
AMLH	90.26%	8.40%

5. Discussion

5.1. Analysis of Multi-Factor Model

We compared three multi-factor models in OLS regression and proceed to ridge regression for higher significance and explanatory power. In the seven-factor model, only RMW and HML factor have negative effects on the stock excess return, which is deviated from the assumption. To be specific, negative RMW means that a company with lower ROE can achieve higher excess return than higher ROE company. Also, negative HML means lower book-to-market ratio result in higher excess return.

According to the previous empirical studies in Chinese stock market, the low level of ROE often corresponds to low stock price. On the contrary, relatively high ROE means that the stock price has reached a periodic top. We can infer that stocks with relatively low ROE can attract investment which can boost the stock price. As for the CMA, a company with lower growth of investments has higher excess return. We can infer that investors prefer purchase stock with lower growth of investment.

5.2. Industry Analysis

We merely discussed the industries with more than four significant factors. Since they are consistent with the model to the higher extent. As we can see in Figure A10 in Appendix C, among 12 industries, there are four groups of coordinated industries and one group of deviated industry. We found out that the coordinated industries are usually cyclical industries which are highly related to the economic wave. Cyclical industry primarily consists of two categories, resources and industrial raw material. They are closely related to macro economy cycle and supply–demand relationship such as manufacturing, automobile, metal and chemistry industry. Moreover, the coordinated industry is distributed at upstream and downstream; for example, electronics is the upstream industry for telecommunication. Thereby, the chain effect is transmitted from the upstream industry through a series of companies all the way to the downstream industry.

To be specific, regarding different factors, industries may react differently in terms of positive and negative relationship. For SMB, it has a negative effect on steel industry, while six industries stand in the opposite position including electrical equipment, electronics, computers, construction material, telecommunication, etc. There are two reasons for the adverse effect in steel industry. Firstly, steel industry is widely regarded as traditional industry. Secondly, it is dominated by a few giant companies which are supported by the government. Whereas the six companies with positive effects are in opposite condition. They are relatively small in terms of market value. Most companies in electrical equipment and computer industry are in early stage which have potential to achieve higher excess return.

As for RMW, over half of the industry are in a negative relationship, while banking is positive. Since banking is closely related to the macro-economic policy. Therefore, it responds to market situations like ROE more swiftly. Moreover, we ran the regression on the same period which may reveal the lag of the information dissemination in some of the industry. Since the market is imperfect, the companies may not respond to the information in the exact same period.

For HML, over 70% of the companies have a negative effect, while banking has a positive relationship. A higher book-to-market ratio reflects that a bank is increasing the amount of loan issue. Since the major source of incomes of banking come from loan issue, thus a bank with higher book-to-market ratio can achieve a higher excess return. While other industries with low book-to-market ratio can achieve higher excess return.

As for CMA, five companies—including textiles & garments, utilities, and light manufacturing—consist of manufacturing processes. They have constant cash flow to maintain daily operation and dividend payment. They are also relatively conservative in investment. However, banking and non-bank finance is negative related to CMA factor. It can be inferred that they are more aggressive in investment to achieve higher excess return.

For CRMHL, none of the companies are in negative relationship, while over 70% of the industries are positive to the momentum factor. This means that the momentum effect widely applied to industries, which can result in the similar movement in the next period.

When it comes to AMLH, the housing and finance industries are in a positive relationship. In these industries, assets in terms of land and money reserves are placed in the foremost position, which means that lower asset turnover rates can result in higher excess return. However, for industries with a negative relationship, since asset turnover rate is the reflection of operating capacity, a higher asset turnover rate can lead to higher excess return.

5.3. Factor Cyclical Research

According to the result of chi-square test, the future direction of SMB and CMA can be merely predicted based on the previous period. While RMW has the initial effect to the rest of the factors, this chain effect conveys through AMLH and HML to Rm-Rf and CRMHL respectively. Although the size of this effect cannot be predicted, we can find the pattern of investment in this chain effect which can reflect investor behavior in the stock market. Firstly, investors primarily focus on the ROE of the companies. Together with the asset turnover rate, we can predict the future direction of the market premium (Rm-Rf). Secondly, the future direction of CRMHL can also be predicted by HML and AMLH. It can suggest that investors usually refer to the profitability and operation capacity of companies before making investment. These two factors can affect the company evaluation and eventually reflected on the change of momentum factor. According to the investment pattern, we can establish a corresponding investment strategy.

Since the direction of market premium factor (Rm-Rf) can be predicted, to some extent, we can find out the pattern of the index. Therefore, it can be also be applied in the investment of index futures and options through call and put.

5.4. Trading Strategy and Back Test

According to the trend analysis, a moving average model is applied in the model (Hanke and Wichern 2014). However, the span of moving average is not specifically decided, which can affect the result of back-testing. We only conducted a four-season moving average as an example. In order to find out the optimal moving average, iteration can be applied to figure out the optimal trading strategies.

5.5. Significance and Limitations of Research

With the comparison of previous work and our research, academic significance can be illustrated in two aspects.

For one thing, prior studies mainly focus on the effect of factors on excess return from a portfolio aspect. Stocks are classified into different groups based on their characteristics. For example, the companies are divided into two groups based on size, then each group is divided into two groups based on the book-to-market ratio. By comparing the average return of the four groups, they can find out the relationship between factors and return. Thus, their methods can only be applied to investments of a specific portfolio, which may lack of practicability and explanatory power in application to a single stock or specific industries. For this research, nevertheless, we focus on discovering different effects of risk premium factors to the excess return of single stock via OLS and ridge regression and then summarize the significant correlations between factors and different industries, which means our research is more practical. For example, based on our model, investors can judge the factors which significantly affect a specific stock or industry.

Additionally, previous works have never discussed the risk premium factors from the aspect of time-series analysis. While forecasting the investment cycle, prior research only investigated from a macro-economic level or technical analysis. For one example, in *Business Cycles* (Lars 2006), the author merely analyzed the cycle of housing, credit, and inventory, which belonged to macro-economic level. For another example, technical analyses, like Elliott wave principle and candlestick charts, aim to describe the market price wave pattern on time series. Thereafter, there is an academic gap of discussing investment cycle of fundamental analysis. In this paper, however, the fluctuations of seven factors are studied by chi-square test of endogeneity and exogeneity of factors. Moreover, it is a new type of cycle analysis because it explains the companies' value behind the investment decision.

The contribution consists of four points. Firstly, we applied a novel method to find out the correlations between seven risk factors and each single stock. Secondly, we find out and explain the correlations between seven risk factors and each industry and specify the situations (positive or negative) of risk factors to buy or sell the industries' stocks. Thirdly, the result of cyclical analysis can be applied in forecasting the direction of risk factors, especially the market risk factor (Rm-Rf) which can be used in transaction of options and futures of market index. Lastly, a back-test was conducted in a simple trading system which suggested that SMB (size premium), CMA (investment growth premium), CRMHL (momentum premium), and AMLH (asset turnover premium) can gain positive returns.

As for the limitations, this research conducted a four-seasons-moving average method to forecast the level of seven factors. Even if it proved that SMB, CMA, CRMHL, and AMLH can be applied in trading system. However, more forecasting methods like ARIMA, VAR, and ANN can be constructed to form better trading strategies. More limitations are presented with our direction for further study.

6. Conclusions and Further Study

In the research of relationship between risk premium and excess return, the hybrid approach takes a primary position. In the examination and modification of multi-factor model, OLS and ridge regression are conducted on Models 1 to 4. The seven-factor model has an optimal combination of p level and R square. In chi-square test, the effect of factor was responded differently in each industry. Therefore, p level was calculated based on the industry in order to find out the well-fitted coordinated and deviated industry. Bank and steel industry are well-fitted in the model, while industries within the same stream—e.g., telecommunication and electronics—are found to be coordinated industries. Moreover, cyclical industry is usually coordinated except for banking.

In the factor cyclical research, chi-square and correlation tests are applied to find out the endogeneity and exogeneity. SMB, RMW, and CMA have endogeneity while the remaining four factors have both endogeneity and exogeneity. With the pattern of direction in factors, an investment strategy was established. To be specific, when moving average SMB is positive, small market value (MV) companies outperform big companies, thus for those industries which are significantly positively correlated with SMB, it would be better to buy small MV stocks and sell large MV stocks. With the strategy, we can achieve a positive return in SMB, CMA, CRMHL, and AMLH respectively in the back test.

As for the further study, the proportion of individual investors in the Chinese stock market contribute nearly 80% of the trading volume. Therefore, investor preference and the irrational behavior should be considered. The investor sentiment factor may be added to improve the explanatory power.

Previous studies also illustrated that the reversal effect exists in the long-term stock market. If our period is set to be longer than one year, we can add the reversal factor to explain the reversal effect.

Dynamic analysis can be applied to data processing, since the economic environment changes over time. We can also monitor the stock market and provide suggestion in stock selection which can fit for the target return.

Moreover, we can further study the inter-factor drive relationship, in order to establish more investment strategies. To be specific, by optimizing the back-test ratio and Sharpe ratio, we can construct a better investment portfolio.

Appendix A. Intuition and Assumption Behind the Hypotheses

There is a trade-off between return and risk. In order to find the corresponding risk for the excess return, according to CAPM model, there is a market risk. To illustrate the abnormal return, Fama used three-factor model which added size premium and book-to-market premium in 1993. In 2013, in their five-factor model, profitability and investment growth are also considered to be significant coefficients. Next, the intuition behind these premiums will be explain one by one.

1. Market Premium

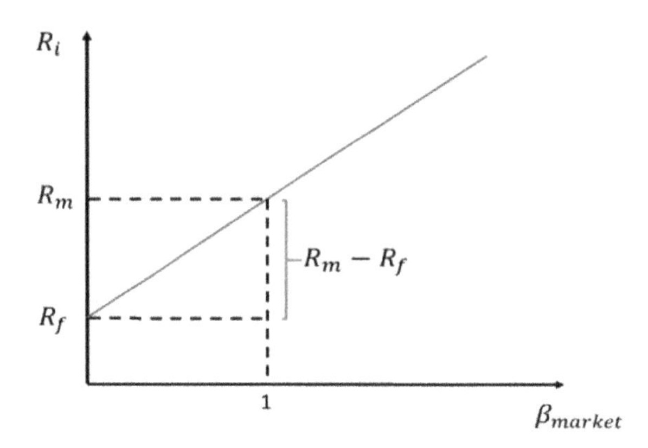

Figure A1. Positive relationship between expected return and market premium.

Market premium is represented by the difference between market return and risk-free rate. Since the fluctuation of stock market's expected return is higher than the risk-free rate, namely stock market has higher risk, the expected return of stock market should higher than risk-free rate.

2. Size Premium

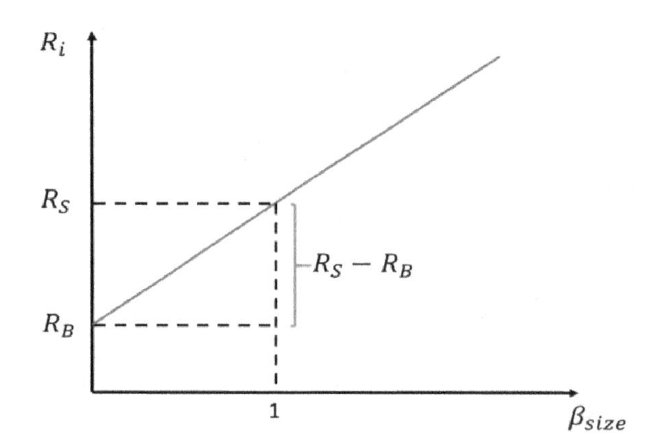

Figure A2. Positive relationship between expected return and size premium.

Size premium is the difference between small companies' average return and big companies' average return in a diversified portfolio or in the market. Normally, small companies have higher returns than the bigger ones. Because at the same period of time, small companies' profit is easy to grow faster than big companies and the growth rate of their dividend also is higher.

According to the DDM model, the price of security depends on the discounted present value of future dividend. In this formula, 'P' is the expected present price of one stock. '*D*' is the dividend of this year and 'g' is the constant growth rate of the dividend (it also is the growth rate of profit). For the last, 'i' is the dividend interest rate, which may refer to risk-free rate.

$$\begin{aligned} \mathrm{P} &= \lim_{n\to\infty}\left[\frac{D(1+g)}{1+i} + \frac{D(1+g)^2}{(1+i)^2} + \frac{D(1+g)^3}{(1+i)^3} + \ldots + \frac{D(1+g)^n}{(1+i)^n}\right] \\ &= \lim_{n\to\infty}\left[\frac{D(1+g)}{1+i} * \frac{\left(\frac{1+g}{1+i}\right)^n - 1}{\frac{1+g}{1+i} - 1}\right] \end{aligned}$$

Given by $i > g$,

$$\left(\frac{1+g}{1+i}\right)^n \to 0$$

So, $\mathrm{P} = \frac{D(1+g)}{i-g}$

$$\because P_1 = \frac{D_1(1+g_1)}{i-g_1},\ P_2 = \frac{D_2(1+g_1)}{i-g_1}$$

If we assume $D_1 = D_2$

Hence, the return rate of the stock can be represented by

$$R_i = \frac{P_2 - P_1}{P_1} = \frac{(g_2 - g_1)(i+1)}{(1+g_1)(i-g2)}$$

In terms of small companies and big companies, $g_1(the\ growth\ rate\ of\ first\ term)\ and\ i\ (risk-free\ rate)$ *are the same.*

However, in the second term, with the control of other effects, growth rate (g_2) of small companies is higher than the big one, so the expected return (R_i) of small companies is generally higher than the large one.

In other words, high returns of small companies indicate they also carry higher risk. If investors can suffer the risk brought by small companies, they can gain the risk premium which is called 'size premium'.

3. Book-to-Market Premium

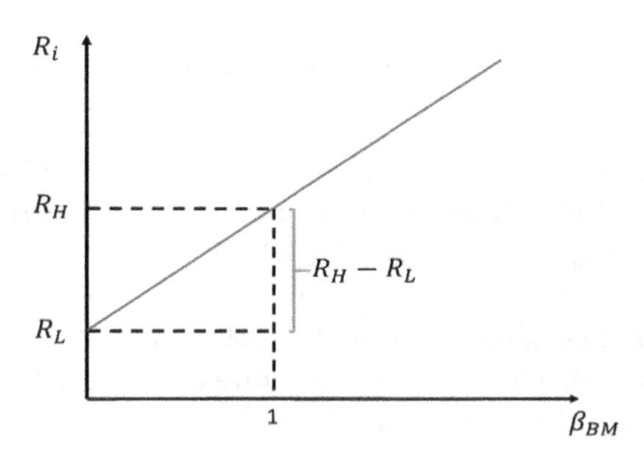

Figure A3. Positive relationship between expected return and book-to-market premium.

Book-to-market premium is the difference between high B/M ratio companies' average return and low B/M ratio companies' average return in a diversified portfolio or in the market.

There is an effect named B/M effect which indicates that higher B/M ratio companies has higher excess return. It can be illustrated by prospect theory easily. However, in this case, the x-axis is MV and the y-axis is the value.

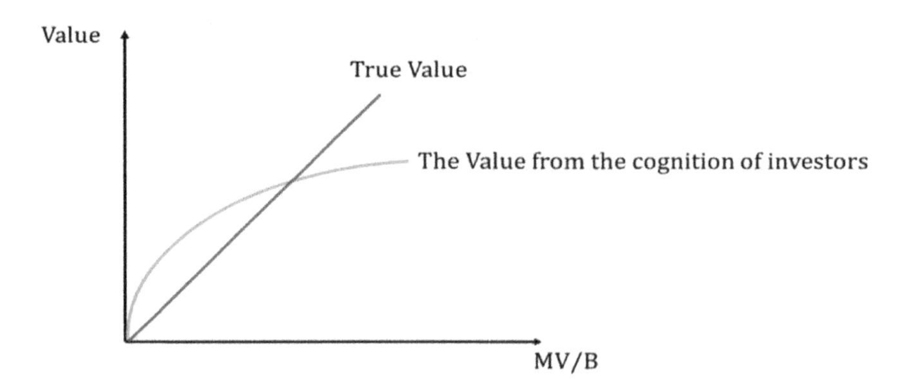

Figure A4. Relationship between true and cognitive value of stocks with the growth of MV/B.

For the higher B/M ratio companies, the MV/B is relative lower and people always overprice the true value of a stock, it leads to the demand for those companies is higher. With the growth of demand and stock price, the return of those stocks also is higher.

Vice versa, higher B/M ratio companies has lower expected return and excess return (expected return minus risk of free rate).

Therefore, if the investor prefers higher B/M ratio companies, they take the risk of B/M effect on one hand, they gain the B/M premium on the other hand.

4. Profitability Premium

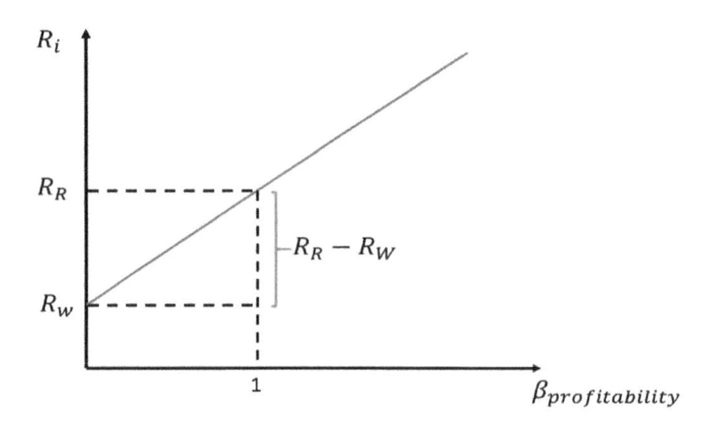

Figure A5. Positive relationship between expected return and profitability premium.

Profitability premium is the difference between robust profitability (higher ROE) companies' average return and weak profitability (lower ROE) companies' average return in a diversified portfolio or in the market.

With the control of other effects, companies with robust profitability—measured by the level of ROE (return of equity)—outperform in their expected rate of return and take greater variance. This is because companies with high profits also distribute high dividends.

$$\mathrm{P} = \frac{D(1+g)}{i-g}$$

According DDM (dividend discount model) formula, higher dividend means higher price and demand which will enhance the level of expected return. If an investor purchases companies with robust profit, they may get higher excess return and fluctuation at the same time. Vice versa, weak profitability companies bring people low excess return and risk.

5. Investment Growth Premium

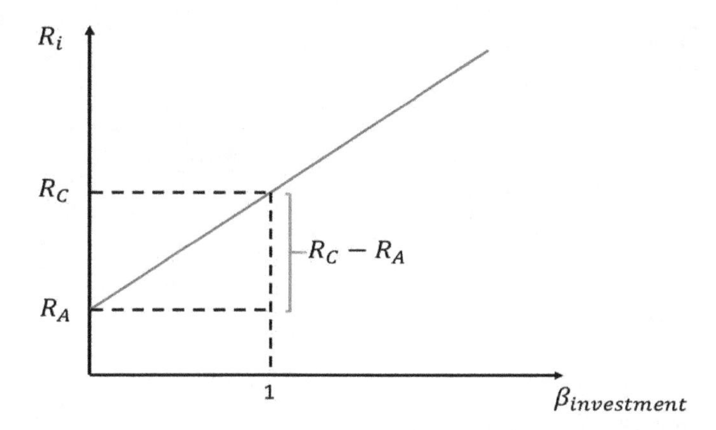

Figure A6. Positive relationship between expected return and investment growth premium.

Investment growth premium is the difference between conservative (lower growth rate of investment or lower growth rate of assets) companies' average return and aggressive (higher growth rate of investment or higher growth rate of assets) companies average return in a diversified portfolio or in the market.

The reason why aggressive companies may have low excess return and risk is that these kinds of firms allocate more profit into reinvestment rather than dividends, thus it decreases the expected price and return, which leads to low risk. Vice versa, conservative companies bring higher excess return and risk because of larger amount of dividend rather investment.

6. Momentum Premium

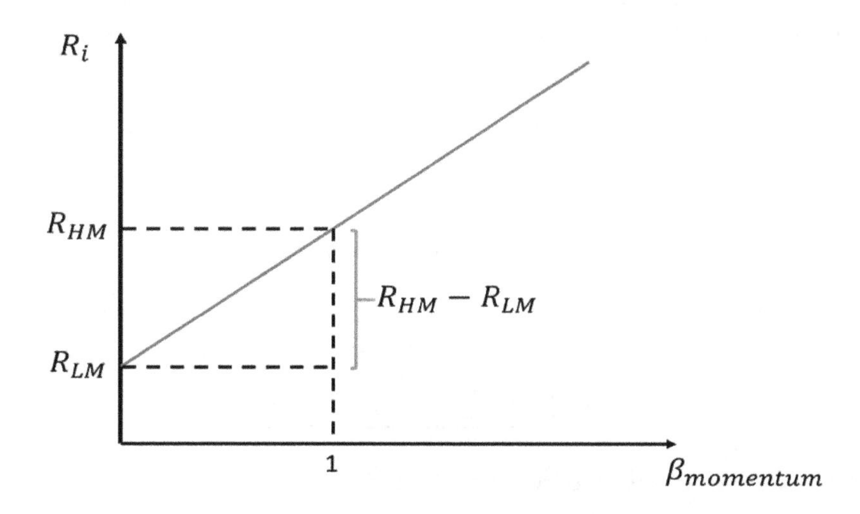

Figure A7. Positive relationship between expected return and momentum premium.

Momentum premium is the difference between higher momentum (higher accumulated return) companies' average return and lower momentum (lower accumulated return) companies' average return in a diversified portfolio or in the market.

Momentum is the accumulated return in one quarter. The higher one means the stock is popular with high return and risk. Vice versa, people invest in low momentum companies with low premium and risk.

7. Asset Turnover Premium

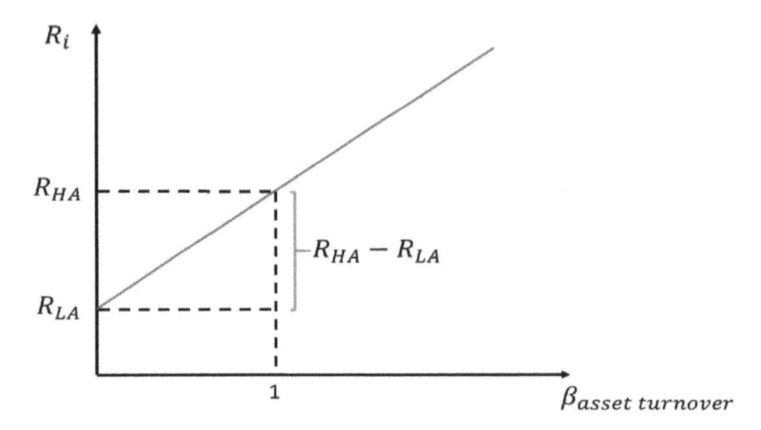

Figure A8. Positive relationship between expected return and asset turnover premium.

Asset turnover premium is the difference between higher asset turnover companies' average return and lower asset turnover companies' average return in a diversified portfolio in the market.

Asset turnover is the total revenue divided by total asset. The higher one means the stock is popular with high return and risk. Vice versa, people invest in low momentum companies with low premium and risk. The regression analysis was conducted with a opposite direction.

Appendix B. Chi-Square Test of Industry in Different Factors

Table A1 recorded the total chi-square value of 28 industries for seven factors. By conducting the chi-square test, we can find out the different effect of factors to various industries. When the total chi-square value is larger than the critical value (when df = 27), it means that each industry is responding differently to the specific factors. According to the result, six factors passed the test, while Rm-Rf is insignificant. Therefore, we need to discuss the effect of factors based on different industry.

Table A1. Chi-square test of industry in different factors.

Factors	Chi-Square (χ^2)
SMB	97.58155
RMW	102.9963
HML	134.683
CMA	85.7257
Rm-Rf	38.17769
CRMHL	66.58794
AMLH	179.6194

Appendix C. Significance Level and Correlation Effect

Table A2. Significance level of factors in different industries.

Title	SMB	RMW	HML	CMA	CRMHL	AMLH	Rm-Rf
Extractive	0	−0.76	−0.8	−0.72	0.72	−0.72	1
Media	0	−0.78571	−1	0	0	0	1
Electrical equipment	0.741935	−0.80645	−0.93548	0	0	0	1
Electronics	0.833333	0	−0.875	0	0.729167	−0.70833	1
Housing	0	0	−0.72527	0	0.714286	0.791209	1
Textiles & garments	0	−0.70833	−0.875	0.708333	0.916667	0	1
Non-bank finance	0	0	0	0	0	0.714286	0.964286
Steel	−0.73684	−0.94737	0.789474	0	1	−0.89474	1

Table A2. *Cont.*

Title	SMB	RMW	HML	CMA	CRMHL	AMLH	Rm-Rf
Utilities	0	0	0	0.710526	0.828947	0	1
Defense	0	−0.8	−0.72	0.8	0.8	−0.84	1
Chemistry	0	−0.74699	−0.78313	0	0.795181	−0.84337	1
Mechanical equipment	0	−0.76667	−0.76667	0	0.833333	0	1
Computer	0.777778	0	−0.92593	0	0.777778	0	1
Domestic appliance	0	0	0	0	0.857143	−0.7619	1
Construction material	0.714286	−0.7619	0	0	0.904762	0	1
Construction ornament	0	0	0	0	0.933333	0	1
Transportation	0	−0.81633	0	0	0.857143	0	1
Animal husbandry and fishery	0	0	−0.71429	0	0	0	1
Automobile	0	−0.86364	−0.75	0	0.818182	−0.75	1
Light manufacturing	0	−0.73077	−0.76923	0.730769	0.730769	−0.80769	1
Commercial	0	0	−0.88525	0.754098	0	0	1
Food	0.794118	0	0	0	0.794118	−0.76471	1
Telecommunication	0.727273	0	−0.77273	0	0.772727	−0.77273	1
Leisure service	0.8125	−0.875	−0.8125	0	0.8125	−0.8125	1
Medical	0	0	−0.80412	0	0.835052	−0.74227	0.989691
Bank	0	0.714286	1	−0.71429	1	1	1
Nonferrous metal	0	−0.88372	−0.74419	−0.83721	0.790698	−0.86047	1
Comprehensive	0	−0.92308	−0.80769	0	0	0	1

In this table, 0 represents that the factor is insignificant to the industry. The absolute value represents the percentage of the significant coefficients. A positive number represents that the factor has positive effect on the industry, while negative number represents that the factor has a negative effect on the industry.

Appendix D. Forecasting the Direction of Factors

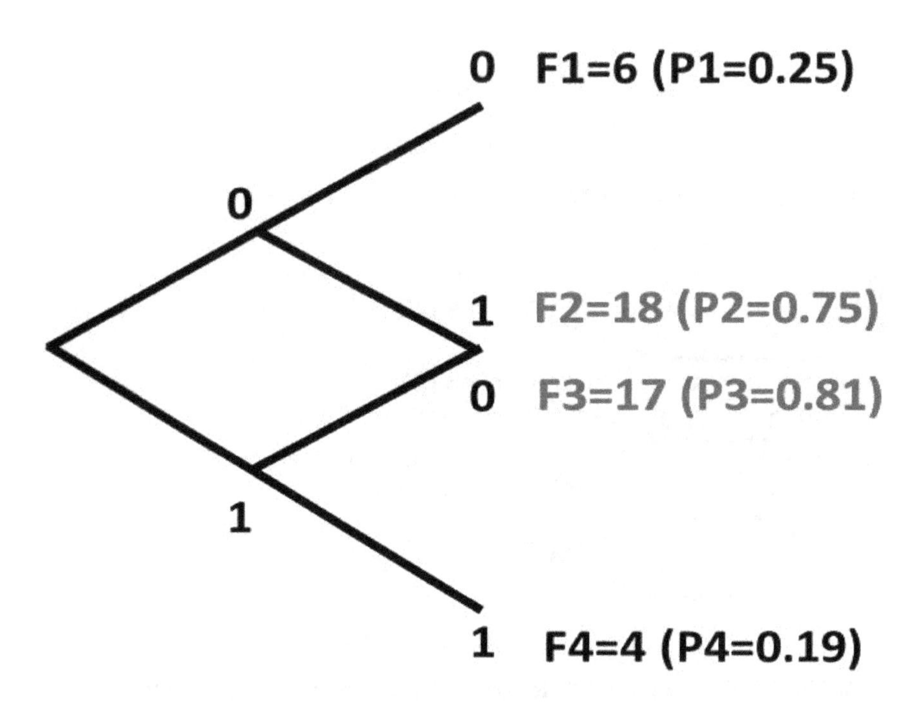

Figure A9. Use SMB to forecast the direction of SMB.

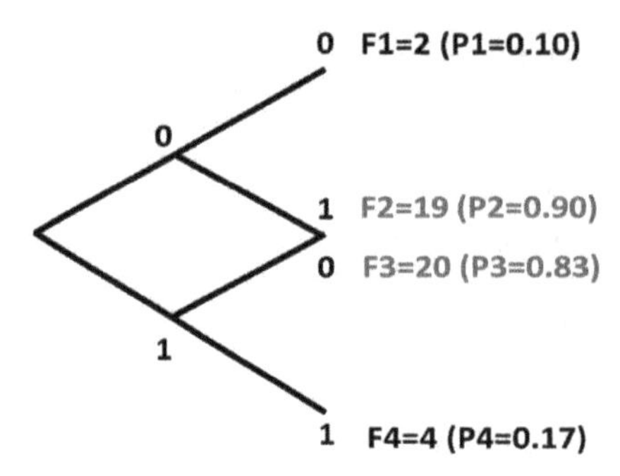

Figure A10. Use RMW to forecast the direction of RMW.

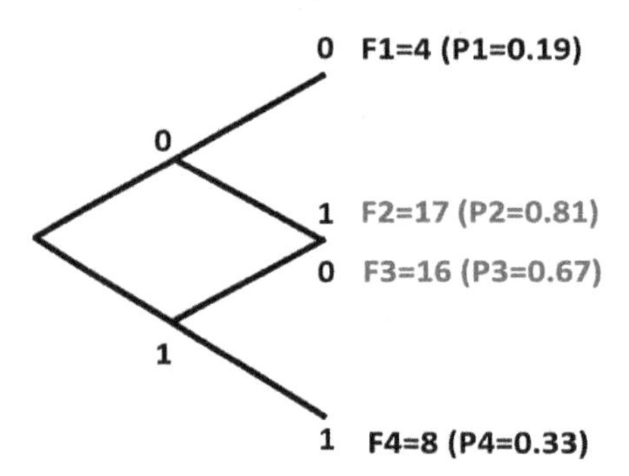

Figure A11. Use CMA to forecast the direction of CMA.

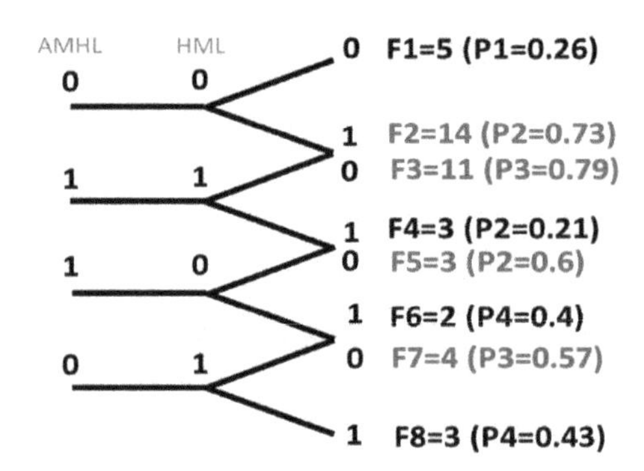

Figure A12. Use AMLH and HML to forecast the direction of HML.

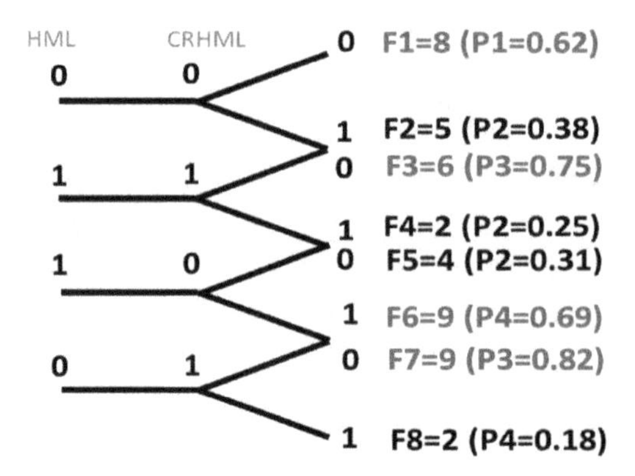

Figure A13. Use HML and CRMHL to forecast the direction of CRMHL.

Use RMW,AMLH to forecast the direction of AMLH
RMW
AMLH
0 0
1 1
1 0
0 1
0 F1=3 (P1=0.3)
1 F2=7 (P2=0.7)
0 F3=3 (P3=0.375)
1 F4=5 (P2=0.625)
0 F5=10 (P2=0.91)
1 F6=1 (P4=0.09)
0 F7=9 (P3=0.56)
1 F8=7 (P4=0.44)

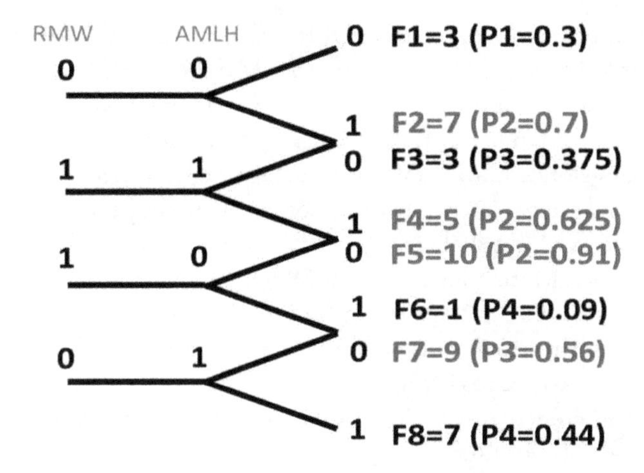

Figure A14. Use RMW and AMLH to forecast the direction of AMLH.

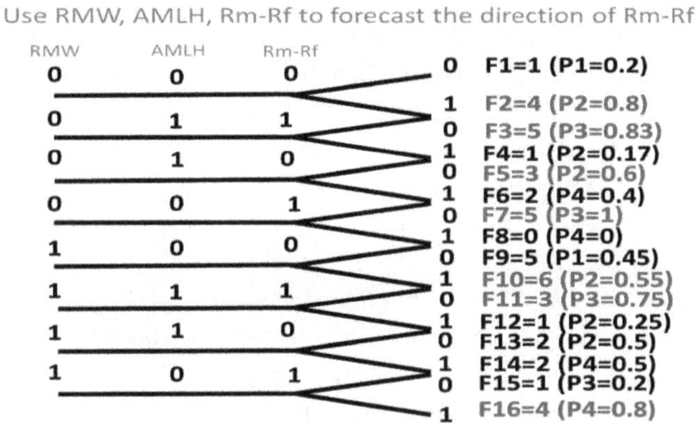

Use RMW, AMLH, Rm-Rf to forecast the direction of Rm-Rf
RMW
AMLH
Rm-Rf
0 0 0
0 1 1
0 1 0
0 0 1
1 0 0
1 1 1
1 1 0
1 0 1
0 F1=1 (P1=0.2)
1 F2=4 (P2=0.8)
0 F3=5 (P3=0.83)
1 F4=1 (P2=0.17)
0 F5=3 (P2=0.6)
1 F6=2 (P4=0.4)
0 F7=5 (P3=1)
1 F8=0 (P4=0)
0 F9=5 (P1=0.45)
1 F10=6 (P2=0.55)
0 F11=3 (P3=0.75)
1 F12=1 (P2=0.25)
0 F13=2 (P2=0.5)
1 F14=2 (P4=0.5)
0 F15=1 (P3=0.2)
1 F16=4 (P4=0.8)

Figure A15. Use RMW, AMLH, and Rm-Rf to forecast the direction of Rm-Rf.

Appendix E. Result of Stability Test

```
p_level= 0.7848678213309025
p_level= 1.0
p_level= 1.0
p_level= 1.0
p_level= 1.0
p_level= 1.0
p_level= 1.0
p_level= 0.6244302643573382
p_level= 0.11030082041932543
p_level= 0.2506836827711942
```

Figure A16. Significance level of variables.

The figures in the red frame are the significance level of time, time^2 and season.

References

Aharoni, Gil, Bruce Grundy, and Qi Zeng. 2013. Stock returns and the miller modigliani valuation formula: Revisiting the fama French analysis. *Social Science Electronic Publishing* 110: 347–57. [CrossRef]

Akter, Nahida, and Ashadun Nobi. 2018. Investigation of the financial stability of s&p 500 using realized volatility and stock returns distribution. *Journal of Risk and Financial Management* 11: 22.

Banz, Rolf W. 1981. The relationship between return and market value of common stocks. *Journal of Financial Economics* 9: 3–18. [CrossRef]

Bhattacharjee, Arnab, and Sudipto Roy. 2019. Abnormal Returns or Mismeasured Risk? Network Effects and Risk Spillover in Stock Returns. *Journal of Risk and Financial Management* 12: 50. [CrossRef]

Bodie, Zvi, Alex Kane, Alan J. Marcus, and Ravi Jain. 2014. *Investments*. New York: Mc Graw Hill Education.

Bodie, Zvi, Alex Kane, and Alan J. Marcus. 2017. *Essentials of Investments*. New York: McGraw-Hill Education.

Carhart, Mark M. 1997. On persistence in mutual fund performance. *The Journal of Finance* 52: 26. [CrossRef]

Chen, Jieting, and Yuichiro Kawaguchi. 2018. Multi-factor asset-pricing models under markov regime switches: Evidence from the chinese stock market. *International Journal of Financial Studies* 6: 54. [CrossRef]

Cisse, Mamadou, Mamadou Konte, Mohamed Toure, and Smael A. Assani. 2019. Contribution to the Valuation of BRVM's Assets: A Conditional CAPM Approach. *Journal of Risk and Financial Management* 12: 27. [CrossRef]

Fama, Eugene F. 1996. Multifactor portfolio efficiency and multifactor asset pricing. *The Journal of Financial and Quantitative Analysis* 31: 441–65. [CrossRef]

Fama, Eugene F., and Kenneth R. French. 1993. Common risk factors in the returns on stocks and bonds. *Journal of Financial Economics* 33: 3–56. [CrossRef]

Fama, Eugene F., and Kenneth R. French. 2015. A five-factor asset pricing model. *Journal of Financial Economics* 116: 1–22. [CrossRef]

Frazzini, Andrea, David Kabiller, and Lasse H. Pedersen. 2013. Buffett's alpha. *CEPR Discussion Papers* 3: 583–90.

Hanke, John E., and Dean W. Wichern. 2014. *Business Forecasting*. Zug: Pearson Schweiz Ag.

Hsiao, Cheng. 2003. *Analysis of Panel Data*. Cambridge: Cambridge University Press.

Jegadeesh, Narasimhan, and Sheridan Titman. 1993. Returns to buying winners and selling losers: Implications for stock market efficiency. *The Journal of Finance* 48: 65–91. [CrossRef]

Johnson, Richard Arnold, and Dean W. Wichern. 2008. *Applied Multivariate Statistical Analysis*. Beijing: Tsinghua University Press.

Lars, Tvede. 2006. *Business Cycles: History, Theory and Investment Reality*. Hoboken: Wiley.

Lintner, John. 1965. The valuation of risk assets and the Selection of Risky Investments in Stock Portfolios and Capital Budgets. *The Review of Economics and Statistics* 47: 13–37. [CrossRef]

Markowitz, Harry. 1952. Portfolio selection. *Journal of Finance* 7: 77–91.

Moinak, Maiti, and A. Balakrishnan. 2018. Is human capital the sixth factor? *Journal of Economic Studies* 45: 710–37.

Novy-marx, Robert. 2013. The other side of value: The gross profitability premium. *Journal of Financial Economics* 108: 1–28. [CrossRef]

Peng, Chi-Lu, Kuan-Ling Lai, Maio-Ling Chen, and An-Pin Wei. 2014. Investor sentiment, customer satisfaction and stock returns. *Social Science Electronic Publishing* 49: 827–50. [CrossRef]

Ronzani, André, Osvaldo Candido, and Wilfredo Maldonado. 2017. Goodness-of-fit versus significance: A capm selection with dynamic betas applied to the brazilian stock market. *International Journal of Financial Studies* 5: 33. [CrossRef]

Ross, Stephen A. 1976. The arbitrage theory of capital asset pricing. *Journal of Economic Theory* 13: 341–60. [CrossRef]

Sehgal, Sanjay, and Vibhuti Vasishth. 2015. Past price changes, trading volume and prediction of portfolio returns. *Journal of Advances in Management Research* 12: 330–56. [CrossRef]

Sharpe, William F. 1964. Capital asset prices: A theory of market equilibrium under conditions of risk. *The Journal of Finance* 19: 18.

Wild, John J. 2016. *Fundamental Accounting Principles*. New York: McGraw-Hill Education.

Zahedi, Javad, and Mohammad Mahdi Rounaghi. 2015. Application of artificial neural network models and principal component analysis method in predicting stock prices on tehran stock exchange. *Physica A: Statistical Mechanics and its Applications* 438: 178–87. [CrossRef]

9

Friendship of Stock Market Indices: A Cluster-Based Investigation of Stock Markets

László Nagy [1] and Mihály Ormos [2,*]

[1] Department of Finance, Budapest University of Technology and Economics, Magyar tudosok krt. 2., 1117 Budapest, Hungary; nagyl@finance.bme.hu

[2] Department of Economics, Janos Selye University, Hradná ul. 21., 94501 Komarno, Slovakia

[*] Correspondence: ormosm@ujs.sk

Abstract: This paper introduces a spectral clustering-based method to show that stock prices contain not only firm but also network-level information. We cluster different stock indices and reconstruct the equity index graph from historical daily closing prices. We show that tail events have a minor effect on the equity index structure. Moreover, covariance and Shannon entropy do not provide enough information about the network. However, Gaussian clusters can explain a substantial part of the total variance. In addition, cluster-wise regressions provide significant and stationer results.

Keywords: cluster analysis; equity index networks; machine learning

---

1. Introduction

The global stock market structure has to be well understood to diversify risk and manage cross-border equity portfolios. Appropriate portfolio construction is rather complicated. The linear dependence structure of the network is not stable (Erdős et al. 2011; Song et al. 2011; Maldonado and Anthony 1981). Moreover, exogenous shocks have major impact on the correlation structure; hence, uncorrelated assets could start moving together (Heiberger 2014). Therefore, correlation-based techniques could cause unwanted variance peaks.

Institutional economic surveys (like MSCI 2018) provide qualitatively identified network structures e.g., emerging markets and developed markets to stabilize their classification.

The main goal of this study is to provide more suitable quantitative techniques, generalize the widely used correlation-based portfolio construction framework, discover the equity index network and make diversification reliable.

The baseline concept follows the Sharpe (1964) Capital Asset Pricing Model (CAPM), in which similarity measures are calculated from correlations between logarithmic returns (Yalamova 2009). The anomalies of CAPM indicate a two-dimensional mean-beta framework that gives only a simplified picture of the real market structure. In order to explain the residuals, financial variables appeared in the famous regression (Fama and French 1996).

In this paper, we carry out a graph theory-based approach to unveil embedded network level information (Shi and Malik 2000). We propose non-linear similarity kernels that are able to deal with higher-order terms. We introduce novel jump-based similarity to investigate the effect of shocks. In addition, we test whether relative entropy of the distribution functions, that captures non-Gaussian behavior, conveys network level information. We also investigate the widely used Gaussian smoothing and correlation (Von Luxburg 2007). We compare different spectral clustering techniques and introduce the usage of the normalized Newman–Girvan cut (Bolla 2011).

Analyzing historical data supports the *a priori* assumption that clusters are homogenously connected. Thus, normalized Laplacian based techniques (Takumasa et al. 2015) are not applicable.

However, the proposed Newman-Girvan cut brings suitable, stationary clustering results. We calculate correlation, jump, relative entropy and Gaussian-based similarities. The figures show that Newman–Girvan cut outperforms normalized Laplacian technique. Analyzing the spectral property of the jump-based similarity matrix unveils that exogenous shocks have minor effect on the network. Thus, our novel results imply that shocks do not convey sufficient information about the equity index graph. Regression analysis demonstrates the stationarity and explanatory power of the clusters. Moreover, we shed some light on the node level equity index structure. We unveil that the index network has scale free properties. Nevertheless, we show that geographical and qualitative categorizations are in line with clusters.

The article structured as follows: in Section 2 we introduce our spectral clustering-based concept. In Section 3 we analyze the equity index graph, compare different similarity matrices and clustering techniques. Section 4 summarizes the article.

2. Materials and Methods

2.1. Data

The current study presents a detailed analysis of 59 stock indices. We apply USD denominated stock splits and dividend-adjusted daily closing prices between 26 September 1990 and 21 September 2015; data is provided by Thomson Reuters.

Our selection criteria for covered stock indices is based on their classification in the International Monetary Fund (IMF) Economic Outlook 2015, and the MSCI WORLD Index composition in 2015. In our analysis we allocate approximately the same weight to each region, despite an unequal number of countries and market capitalization. We rebalance the sample by choosing approximately ten indices from each IMF group. We are also interested in the role of well diversified indices e.g., MSCI WORLD and EURO STOXX600, which have also been analyzed.

In order to underline the highly different characteristics of individual stock indices, we present some monthly descriptive statistics in Table 1.

Table 1. Descriptive statistics of monthly returns.

Index	Mean	Variance	Skewness
.CSI300	0.018	0.056	−0.336
.XU100	0	0.026	−0.809
.DJI	0.012	0.009	−0.819
.UAX	−0.034	0.037	−0.721
.WORLD	0.004	0.002	−1.889

Notes: Table 1 shows the descriptive statistics of the monthly returns, where CSI300, XU100, DJI, UAX, and WORLD represent the Shanghai Composite 300, Brose Istanbul 100, Dow Jones, Ukraine UX index, and the MSCI World index respectively.

2.2. Methodology

In the 20th Century, the appearance of large, complex data sets brought new challenges to developing methods which could be used to understand complicated structures. The key concept is to classify data points according to various similarity functions. The problem is computationally extremely challenging. However, spectral clustering techniques provide optimal, lower dimensional representation of multidimensional data sets. The idea is twofold: on the one hand, similarly to principal component analysis we could calculate lower dimensional representation of the data points from the eigenvalues and eigenvectors of the similarity matrix. On the other hand, we could represent the data structure as a weighted graph and cut the graph along the different clusters. This approach leads to penalized cut optimization problems. Linear algebra and cluster analysis provide powerful methods to find the optimal representations and minimized cuts.

2.2.1. Similarity Matrix

If we would like to cluster different items, first the measurement of similarity has to be decided. In this study similarity of two stock indices (i, j) will be denoted by $W_{i,j}$. The goal is to penalize differences and reward similarities. Logarithmic returns are easy to handle and maintain all price process information.

$$r_i(t) = \ln\left(S_i(t)/S_i(t-1)\right), \tag{1}$$

where $S_i(t)$ represents the price of index i. The current study analyses multiple similarity approaches.

First, the Markowitz-based squared correlation is considered a similarity metric.

$$W_{i,j} = \mathrm{Corr}^2(r_i, r_j), \tag{2}$$

We argue this approach because logarithmic returns are not normally distributed, hence non-linear effects may also be important. However, as correlation is linear, squared correlation similarities only take into account linear dependences.

The problem of higher-order moments can be easily solved by using symmetric and positive-definite kernel functions. The idea comes from the functional analysis. Data can be transformed into a reproducing kernel Hilbert space (RKHS), where applying the usual statistics provides the same outcomes as can be attained by using non-linear statistics in the original Hilbert space (Berlinet and Christine 2011); and, in practice, the Gaussian-kernel is widely used (Gregory et al. 2008).

$$W_{i,j} = \exp\left(- \parallel r_i - r_j \parallel^2\right), \tag{3}$$

We notice that, if the sets of the relevant information and sensitivities are similar, then the relative entropy of the distribution of return processes is small. Otherwise, we can say stock indices are sensitive to different sets of information in a different manner (Ormos and Zibriczky 2014). This means that the similarity function has to be monotonically decreasing in symmetric Kullback–Leibler distance, and so we can construct a similarity measure such that:

$$W_{i,j} = 2/\left(2 + \left[\mathrm{KL}(p(r_i) \parallel p(r_j)) + \mathrm{KL}(p(r_j) \parallel p(r_i))\right]\right), \tag{4}$$

where $p(r_i)$ denotes the probability distribution function of logarithmic returns of index i and $\mathrm{KL}(p(r_i) \parallel p(r_j)) \stackrel{\text{def}}{=} \sum p(r_i = x) \ln\left(p(r_i = x)/p(r_j = x)\right)$ the relative entropy of indices i and j.

Another perspective argues that large deviations are riskier, hence similarities should be defined with tail distributions. We calculate the differences of the return series and count the number of at least two standard deviation peaks. This logic implies that indices are similar if their price processes jump together. Similarity function has to be decreasing in the number of large deviations, hence we propose the following metric:

$$W_{i,j} = 1/\left(1 + \sum_{t=1}^{T} \delta(|\, z_i(t) - z_j(t)| > 2)\right), \tag{5}$$

where z_i represents the normalized return of index i.

In the current study we compare each approach.

2.2.2. Normalized Modularity

The equity index structure is strongly connected. We cannot say that events in Africa do not have any effect on European markets, hence we have to find methods which can be used to cluster dense graphs.

Let $G(V_{N\times 1}, W_{N\times N})$ be a weighted graph, where V denotes the set of vertices and W represents the weights of the edges. A k-partition of graph $G(V, W)$ can be defined as the partition of vertices such that $\cup_{a=1}^{k} V_a = V$ and $V_i \cap V_j = \delta_{i,j} V_i$, $\forall i, j \in \{1, \ldots, k\}$.

The $W_{i,j}$ value represents the strength of the connection between nodes (i,j). If we assume that nodes are independently connected, then the guess of weight $W_{i,j}$ will be the product of the average connection strength of i and j. The average connection strength d_i and d_j are given by W,

$$d_i = \frac{1}{N}\sum_{u=1}^{N} W_{i,u},$$

Thus, $W_{i,j} - d_i d_j$ captures the information of the network structure (Bolla 2011). If we want to maximize the sum of information in each cluster, we get:

$$\max_{P_k \in \mathcal{P}_k} \sum_{a=1}^{k} \sum_{i,j \in V_a} (W_{i,j} - d_i d_j), \tag{6}$$

where P_k stands for specific k-partition in $\mathcal{P}_k$, which represents the set of all possible k-partitions.

Let $M := W - dd^T$ denotes the modularity matrix of $G(V,W)$. If we would like to get clusters with similar volumes, then we have to add a penalty to Equation (6), hence we get the normalized Newman–Girvan cut.

$$\max_{P_k \in \mathcal{P}_k} \sum_{a=1}^{k} \frac{1}{\mathrm{Vol}(V_a)} \sum_{i,j \in V_a} (W_{i,j} - d_i d_j), \tag{7}$$

where $\mathrm{Vol}(V_a) = \sum_{u \in V_a} d_u$.

Let us define the so called normalized modularity matrix:

$$M_D := D^{-1/2} M D^{-1/2}, \tag{8}$$

If we would like to cluster a weighted graph $G(V,W)$, then eigenvectors of its modularity (M) and normalized modularity matrices (M_D) can be used. Modularity and normalized modularity matrices are symmetric and 0 is always in the spectrum of M_D:

$$M_D = \sum_{i=1}^{N} \lambda_i u_i = \sum_{i=1}^{N-1} \lambda_i u_i, \tag{9}$$

where $1 > \lambda_1 \geq \lambda_2 \geq \ldots \geq \lambda_N \geq -1$ denote the eigenvalues of M_D.

If we would like to maximize Equation (7), then we can use the k-means clustering algorithm on the optimal k-dimensional representation of vertices,

$$\left(D^{-\frac{1}{2}} u_1, \ldots, D^{-\frac{1}{2}} u_k\right)^T,$$

where $u_1, \ldots, u_k$ denote the corresponding eigenvalues of $|\lambda_1(M_D)| \geq \ldots \geq |\lambda_k(M_D)|$. Moreover, if the normalized modularity matrix has large positive eigenvalues, then the graph has well-separated clusters, otherwise clusters are strongly connected.

Another natural approach is to minimize the normalized cut (Von Luxburg 2007).

$$\min_{P_k \in \mathcal{P}_k} \sum_{a=1,b=a+1}^{k-1,k} \left(\frac{1}{\mathrm{Vol}(V_a)} + \frac{1}{\mathrm{Vol}(V_b)}\right) W_{i,j}, \tag{10}$$

The optimization problem is similar to Equation (7). However, instead of the normalized-modularity matrix the normalized Laplace matrix provides the solution (Shi and Malik 2000).

$$L_D := D^{-\frac{1}{2}} (D - W) D^{-\frac{1}{2}}, \tag{11}$$

This technique works when clusters are well separated, otherwise normalized modularity gives better results.

2.2.3. Algorithm

In empirical analysis, the following steps are the backbone of the calculation (Maurizio et al. 2007).

1. Constructing the similarity matrix (W).
2. Calculating the normalized modularity matrix (M_D).
3. Based on the spectral gap, determining the number of clusters and optimal k-dimensional representation.
4. Appling k-means clustering.

2.2.4. Assessment of Clustering Methods

The relevance of different clustering techniques can be tested in multiple ways. The most common metrics follow a regression-based logic. In this framework we suppose that variance has two components: the within, and the between cluster components. Therefore, the explanatory power of given clusters can be described as:

$$\frac{\sum_{j=1}^{k}\sum_{j=1}^{N_i}\left(X_{i,j}-\overline{X}\right)^2-\sum_{i=1}^{k}\sum_{j=1}^{N_i}\left(X_{i,j}-\overline{X_i}\right)^2}{\sum_{i,j=1}^{N_i,\,N_j}\left(X_{i,j}-\overline{X}\right)^2}, \tag{12}$$

where k represents the number of clusters, N_i shows the size of clusters and $\overline{X}$, $\overline{X_i}$ stands for the total and cluster wise average (Zhao 2012). The formula penalizes dispersions within clusters, hence dense clusters would give a number close to 1. Moreover, calculating the ratios with a different number of clusters highlights the optimal number of clusters.

3. Results

This study presents a broad analysis of the equity index network structure. Logarithmic returns of 59 stock indices are clustered in different ways. Our investigations reveal stock indices are homogenously connected, and large price changes have limited effect on the network structure.

3.1. Similarity Metrics

Defining similarity is a key aspect in clustering. In general, it is not usually possible to find an optimal kernel, but different approaches can be tested and compared to specific data sets.

This study analyzes correlation, jump, entropy, and Gaussian-based similarity kernels. When calculating the similarity matrices, we expect strongly connected indices have coefficients close to one, whereas loosely connected close to zero. Level plots (Figure 1) give a feeling about the network structure which seems to be homogeneous; thus, clusters could not be well separated.

Figure 1 displays the correlation, Gaussian-kernel, relative entropy and jump-based similarity structure of the equity index graph, in which the whiter the color the stronger the connection between the indices. Indices are sorted alphabetically and (i,j) represents the similarity between index i and j.

Different similarity measures imply similar patterns, which is in line with our *a priori* intuition. However, the spectra of normalized Laplace and normalized modularity matrices help us to find the most adequate kernel function: the wider the spectral gap, the better the clustering property. This means, we have to find similarity metrics, which in turn implies large gaps in the spectrum of normalized Laplacian and modularity matrix (Chung 1997).

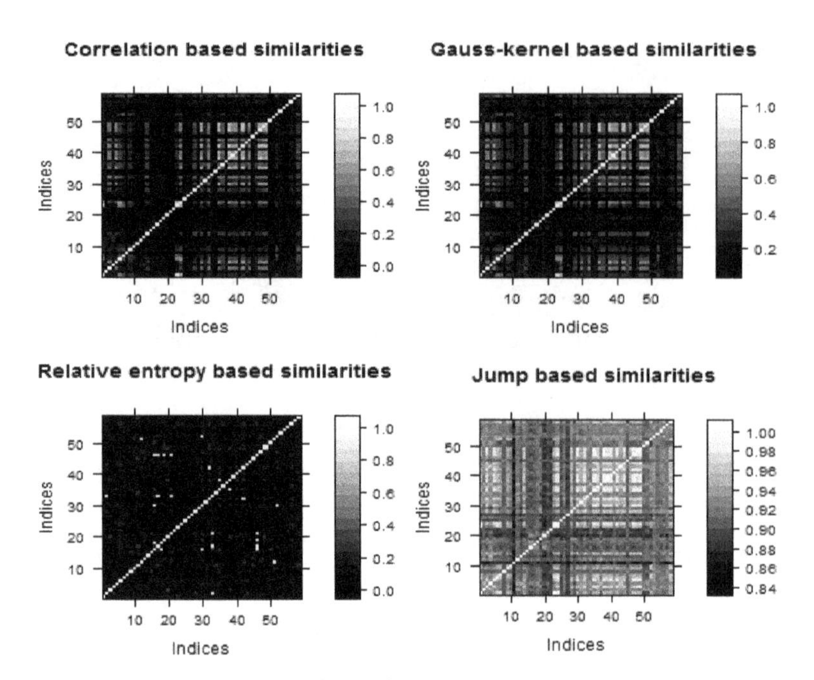

Figure 1. Level plots of daily similarity matrices.

Empirical evidences (Figures 2 and 3) show relative entropy, and Gaussian-kernel can also be used to cluster the stock index network while correlation and jump-based similarities are not promising.

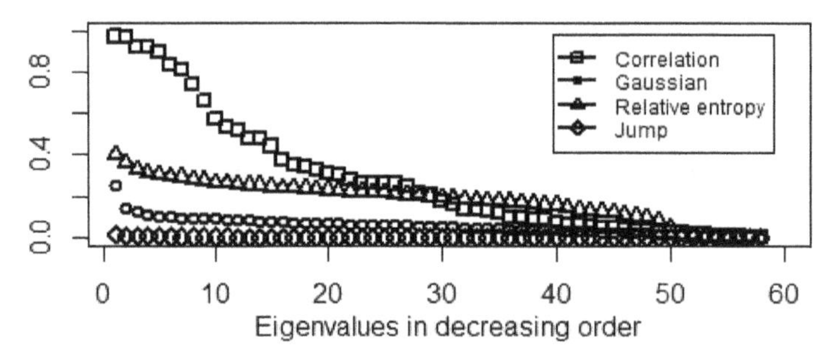

Figure 2. Eigenvalues of normalized modularity matrix in decreasing order.

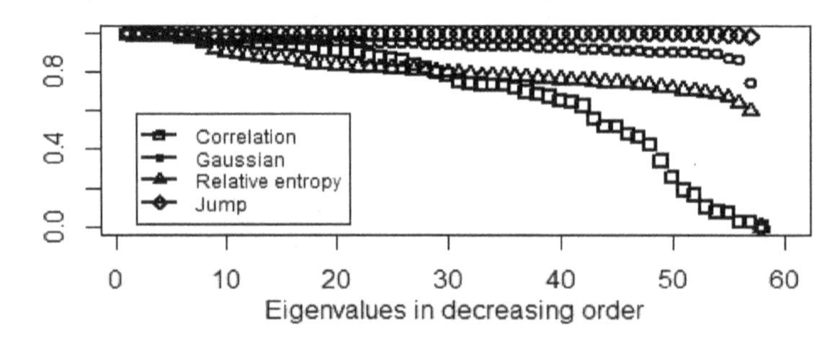

Figure 3. Eigenvalues of normalized Laplacian matrix in decreasing order.

A correlation-based similarity approach implies roughly uniform eigenvalue density on [0, 1]. This means, a lot of gaps appear in the spectrum, hence we could not comment on the optimal number of clusters. Moreover, lower dimensional representations will not contain all the information as some of the large eigenvalues are not considered. These hurdles highlight the problems of squared correlation similarity matrices.

Counting at least two standard deviation jumps results in a small number of eigenvalues with large multiplicity. Therefore, lower dimension representation cannot be used to cluster the data points. Accordingly, jumps are random and do not reflect the network structure; thus we could say all the clusters are exposed to the same systematic risk. Thus, the results provide evidence of spillover effect.

Moreover, we show that shocks and market collapses have a minor effect on the equity index graph i.e., network structure of equity indices.

Gaussian and relative entropy-based similarity matrices infer promising figures, especially in the case of normalized modularity. Here, we get large well separated eigenvalues necessary to transform the data into a lower dimensional space.

Notice that these results are in line with Figure 1 because the normalized Laplacian minimizes the normalized cut (Equation (10)), which in turn, is small if, and only if, the clusters are loosely connected. Whereas, the modularity approach maximizes the information of clustering, hence, it can also be used in a homogeneous network structure as well.

Investigating the spectra, especially the positions of spectral gaps, gives some guidance on the optimal number of clusters. Considering the previous results, the spectra of Gaussian and relative entropy-based normalized modularity matrices are suitable. Figure 4 shows indices could be put into 2, 3, or 5 clusters.

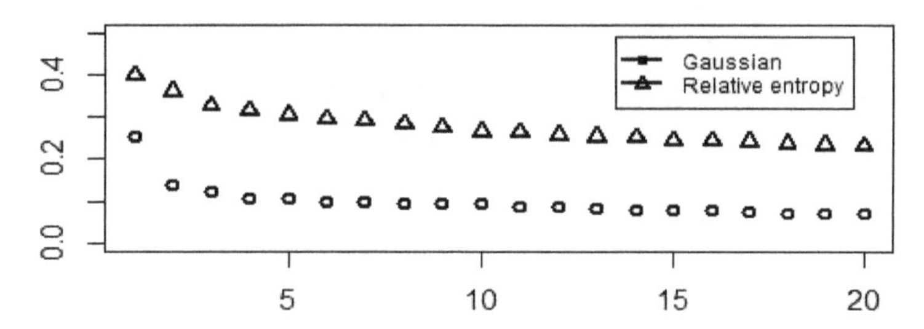

Figure 4. Largest eigenvalues of Gaussian- and relative entropy-based normalized modularity matrices.

In order to identify the spectrum gap, we apply the elbow method to identify the optimal number of clusters. This approach is rather computationally intensive, because of the percentage of variance explained as a function of clusters has to be estimated (Equation (12)); thus, the whole process has to be repeated many times. However, in our case, as we have 59 stock indices, the elbow method can also be used. Figures 5–7 provide evidence for using 2, 3, 4, or 5 clusters.

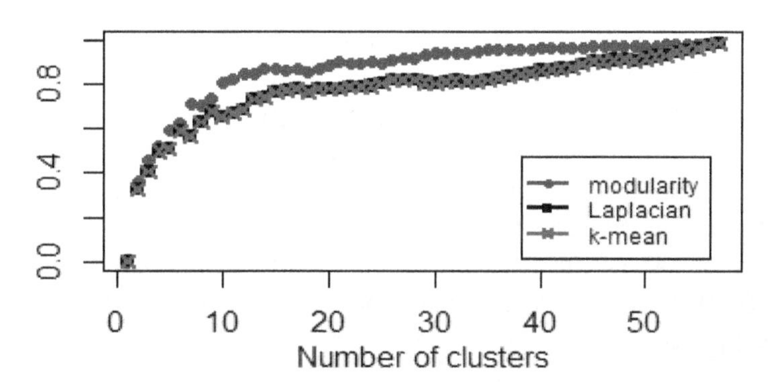

Figure 5. Explained percentage variance of Gaussian-kernel based clusters of representations.

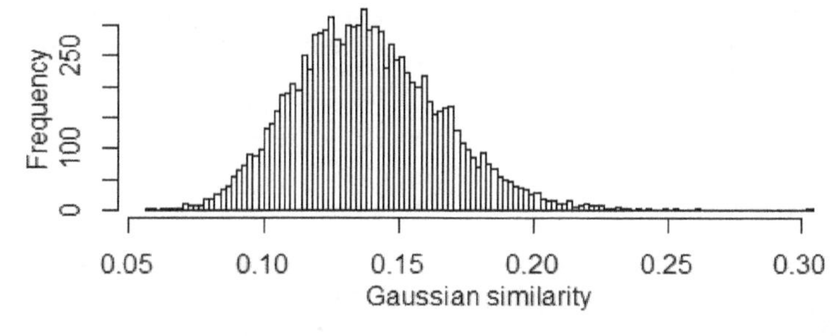

Figure 6. Histogram of 10,000 Gaussian similarities which are generated from i.i.d. 250 dim. standard normal samples.

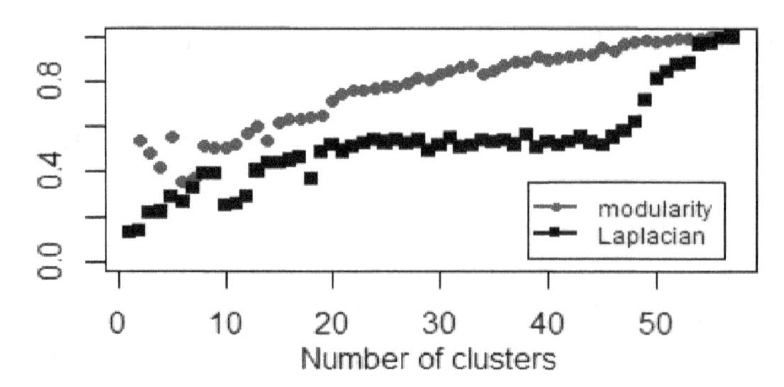

Figure 7. Explained percentage variance of Gaussian-kernel based clusters after zero out similarities less than 0.2.

Analyzing the Gaussian similarity kernel shows that if we randomly generate data, then we would get similarities smaller than 0.2, with probability more than 0.99.

This observation (Figure 6) implies that we have to filter out similarities less than 0.2 from the adjacency matrix.

Figures 2–4 show the Gaussian-kernel infers the clearest spectrum property. The relative entropy-based kernel also gives usable results, whereas, jump and correlation-based approaches are ineffective.

3.2. Comparing Normalized Modularity and Laplacian

We propose the use of an accuracy ratio-based (Engelmann et al. 2003) measure to compare the efficiency of different clustering techniques. Calculating the area between the variance explanation function of the random and the different spectral clustering methods generates an appropriate statistic.

Considering this metric (Zhao 2012), it can be seen that the Gaussian-kernel over-performs relative to the entropy-based approach; this is because in each case its variance explanation function is steeper.

Henceforth, the Gaussian-kernel based normalized modularity matrix is used.

3.3. Equity Index Network Structure

Spectral gap (Figure 4) and variance analyses (Figures 5 and 7) imply equity indices can be studied by using 2, 3, and 5 clusters. The explanatory power of two clusters is 38%. This means roughly one-third of the total variance comes from the sample heterogeneity. If we increase the number of clusters and investigate the three cluster cases, we get a similar explanatory power. However, a spectral gap appears between the third and fourth eigenvalues (Figure 4), so, theoretically, we propose the three clusters. The next gap is between the fifth and sixth eigenvalues. The explanation power of five clusters is 52%. This means, half of the total variance of data can be explained by five clusters.

This result (Figure 8) also suggests that additional clusters have little explanatory power, which is in line with spectrum properties.

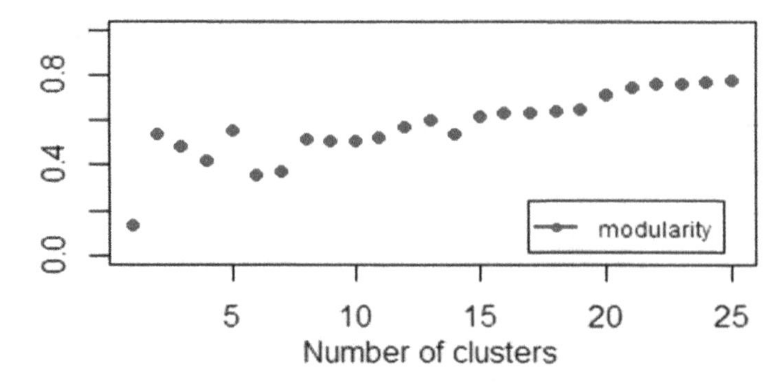

Figure 8. Explained percentage variance of Gaussian-kernel based clusters.

In practice, mean-variance plots can be used to represent risks and rewards. Intuitively, indices with similar risk and return can be believed to be similar. This approach applies a k-means algorithm to cluster the two-dimensional (mean, standard deviation) representation of logarithmic returns.

We have seen this naïve method does not give optimal cuts. However, if we calculate Gaussian similarities and normalized modularity matrix based representation, then we get clusters with a higher variance explanatory power. We have seen stock indices can be put into 2, 3, or 5 clusters. If we plot the mean-variance representation of indices we get Figures 9–11, for 2 and 5 clusters, respectively.

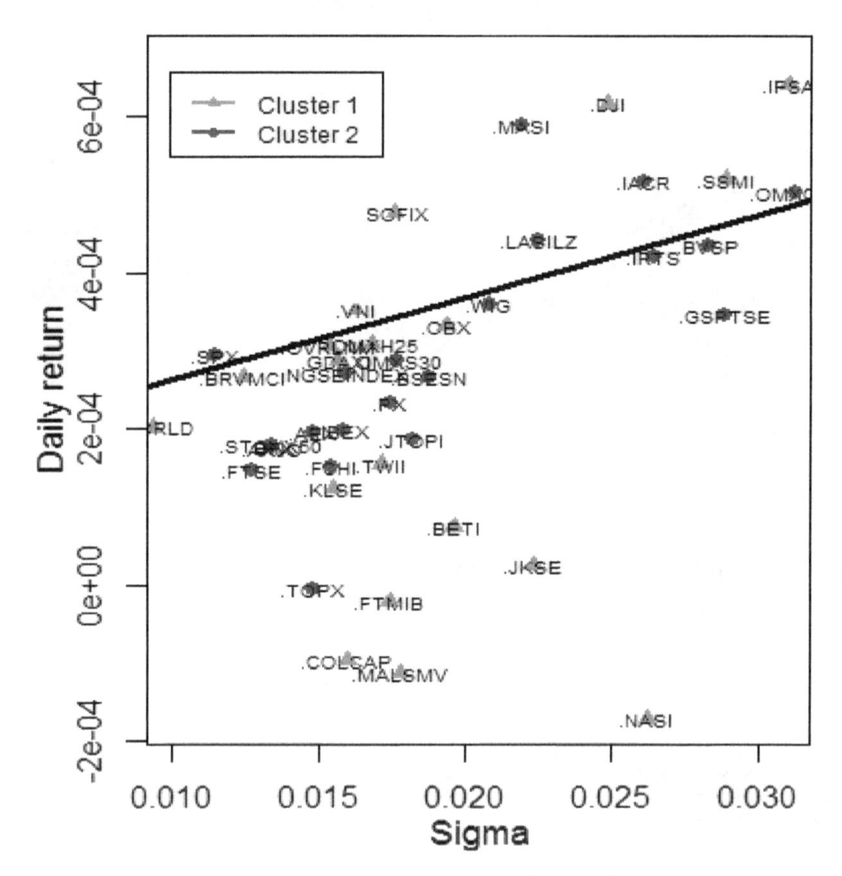

Figure 9. Two Gaussian-kernel based normalized modularity clusters (part of the total graph).

In Figure 9 we can see clusters that are optimizing the modularity cut are concave in a mean-variance framework. If we have a closer look at the indices in Appendix A (Table A1) we could say that a qualitative approach also works, because east-west geographical clustering would imply almost similar results.

Putting the indices into three different clusters (Figure 10) gives a complicated structure, but we could still state that the first cluster is dominated by European countries, the second by American, and the third is a mixture of indices from the rest of the world. Thus, applying geographical diversification is in line with cluster property. The network generated by simple index returns incorporates geographical information.

Calculating five different clusters helps us to gain a deeper understanding of the network. The first surprising result is that despite the penalty of different cluster sizes, the Dhaka Stock Exchange (.DS30) is separated into cluster three. In addition, cluster four contains only two African and two American indices. Another interesting result is the first cluster, which includes the Arabian indices except Morocco. Cluster two primarily comprises developed, while cluster five is dominated by emerging market names. Hence, we could notice that spectral clustering-based classification is similar to qualitative categorizations. However, these results also suggest that a portfolios constructed using only geographical scope can integrate indices which behaves significantly differently compared to real regional regimes.

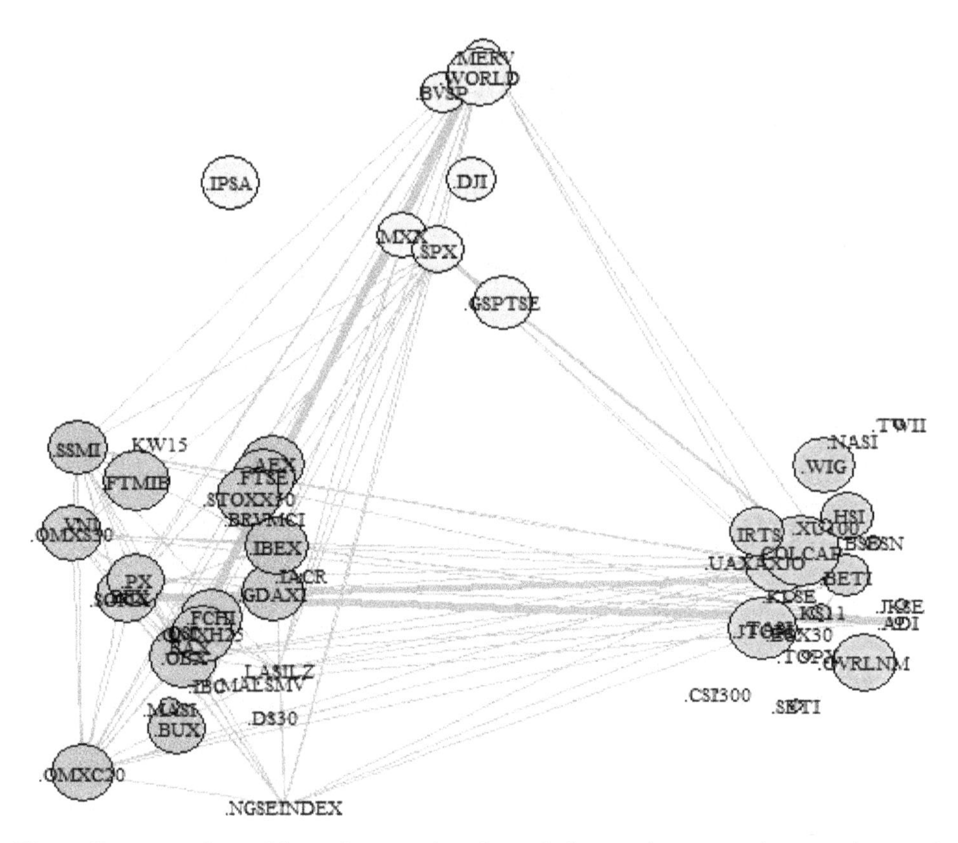

Figure 10. Three Gaussian-kernel based normalized modularity clusters, edges with weights stronger than 0.5.

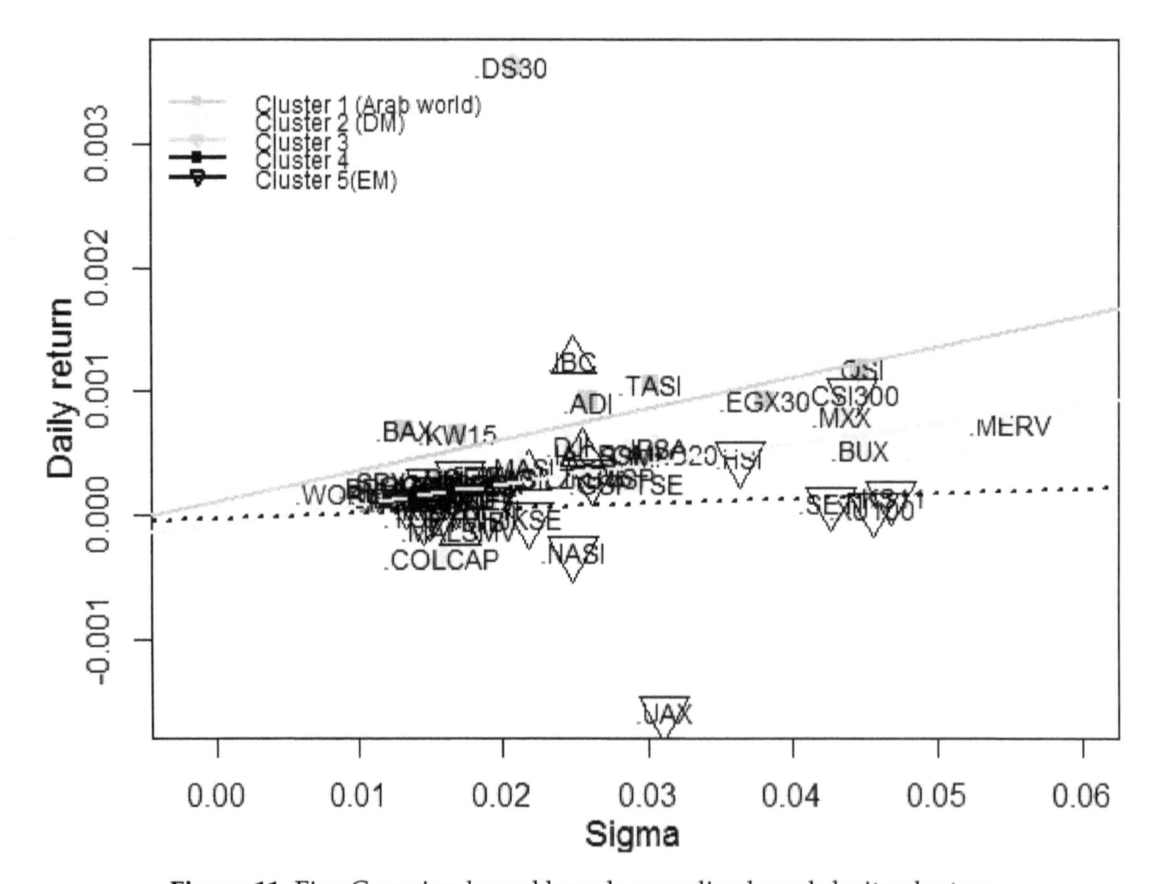

Figure 11. Five Gaussian-kernel based normalized modularity clusters.

In order to compare our quantitative approach with geographical and MSCI classifications, we run the following regressions:

$$r = \beta_0 + \beta_1 \sigma + \beta_2 cluster + \epsilon, \tag{13}$$

The regressions (Table 2) show that spectral clustering provides statistically reliable figures, while geographical- and MSCI-based clusters are not statistically significant.

Table 2. Regression statistics.

Method	Coeff. of Cluster	p-Value
Geographical	−0.000036	0.394
MSCI	−0.000041	0.293
Spectral	−0.000112	0.027

Notes: This table shows the daily linear regression coefficients and p statistics of geographical, MSCI, and spectral clusters. Returns are regressed on standard deviations and clusters.

The outcomes highlight the difficulty of diversification, because the correlation structure of the network is quite homogeneous. Moreover, geographical and other qualitative diversification techniques do not give us statistically significant results. However, indices can be clustered by spectral methods. This means indices in the same cluster are affected by the same risk factor, hence, only cluster wise diversification can be used to eliminate non-systematic global risk.

3.4. Equity Index Graph

Clustering helps us to globally analyze the network. However, the local structure can be better understood by node-specific attributes. Our aim is to find the most influential markets. Hubs can be identified as vertices with the largest vertex weights. Vertex weight of node i can be defined as the sum of the edge weights.

$$V_i := \sum_{j=1}^{N} W_{i,j} \delta(W_{ij} > 0.2), \tag{14}$$

Calculating the histograms, we get Figure 12.

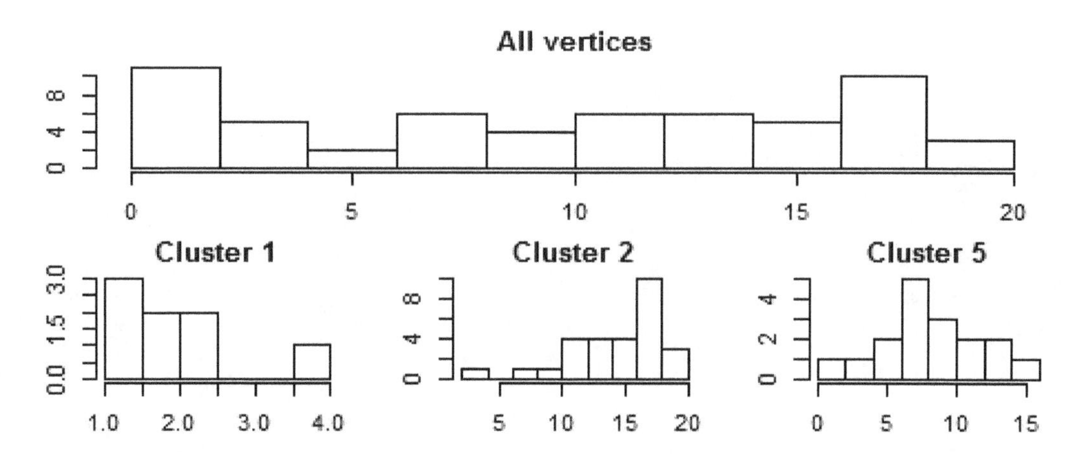

Figure 12. Histogram of vertex weights, five Gaussian cluster, two nodes are connected if their Gaussian similarity is stronger than 0.2.

The outcomes show that essentially cluster wise histograms differ. In each cluster, there are vertices whose connection numbers substantially differ from the cluster wise mean (Figure 12). Note that the vertex connection density of an Erdő-Rényi graph is binomial, hence hubs and separated nodes cannot be generated (Erdős and Alfréd 1960). This implies that preferential attachment processes should be used to model the network structure (Barabási and Réka 1999).

However, the randomness of vertex weights is twofold: one factor is the number of connections, while the other factor is edge weights.

In order to distinguish the effects, we calculate the vertex weights as the sum of connections;

$$V_i^{count} := \sum_{j=1}^{N} \delta(W_{ij} > 0.2), \tag{15}$$

Calculating the histogram of counting-weights we get Figure 13.

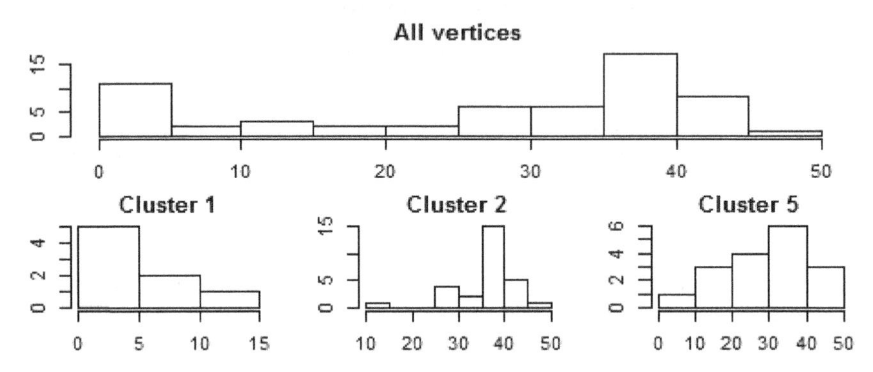

Figure 13. Histogram of vertex count-weights, five Gaussian cluster, two nodes are connected if their similarity is stronger than 0.2.

We could say clusters 1 and 2 contain hubs, whereas, the vertex-count distribution in cluster 5 is more balanced. There is no hub, but there are vertices with more than 40, and less than 10 connections. The results show that the shape of the cluster wise vertex connection differs, hence, the vertex weight distribution is also a mixed distribution.

Comparing Figures 12 and 13 shows that counting implies higher skewness, while having less effect on the shape. When analyzing edge weights, it turns out that they are not uniformly distributed. In addition, different clusters have different edge weight densities.

Moreover, it also can be seen (Figure 14), that if the average connection strength is higher, the vertex has more connections; this is true cluster-wise as well.

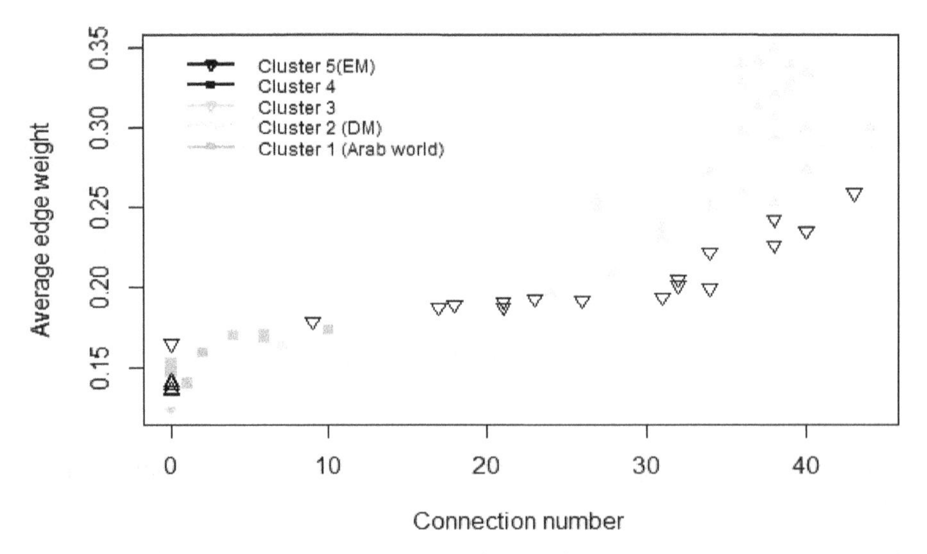

Figure 14. Cluster-wise connection number and average connection strength.

All of this implies that spectral clustering techniques can be used to distinguish subgraphs. Moreover, the number of connections of an index and its average edge weight, follow the preferential attachment process.

3.5. Risk and Reward

To understand the connection between risk and reward, we can use the mean-standard deviation framework. When calculating the regressions we arrive at Table 3. The outcomes imply that the total sample regression does not provide reliable figures, nevertheless, cluster-wise regressions are significant. This points to the conclusion that the relationship between risk and return, cluster wise has different behavior.

Table 3. Descriptive statistics of daily linear regressions.

Clusters	*p*-Value of Intercept	*p*-Value of s.d.	R^2
Total Sample	0.62	0.12	0.05
First cluster	0.62	0.02	0.68
Second cluster	0.29	0.00	0.59
Fifth cluster	0.93	0.71	0.01

Notes: This table shows the p statistics and R^2 values of daily linear regressions. Returns are regressed on standard deviations. Calculation is done for total, only the first, second and fifth clusters.

Figure 11 and Table 3 show higher standard deviations, implying higher returns, because regression lines slope upwards. In addition, it also turns out that connections between returns and standard deviations are strong in Arabian and developed market cases. Nevertheless, emerging markets show different statistics: index returns in the fifth cluster are not linear in standard deviation, hence emerging market returns cannot be estimated in the Markowitz framework.

3.6. Time Stability

Making investment decisions vastly depends on the time stability of our strategy. Therefore, we have to check the stationarity of our clustering method. By splitting the time series by years we get 25 periods. Calculating the stability of explained percentage variance of clustering could be a good proxy of time stability. Stationarity can be analyzed by the augmented Dicky–Fuller (ADF) test.

Note that, the analysis covers 25 years' data, hence we get 25 non-overlapping periods. The t-values (Table 4) show that the variance explanation power process could be stationer, but because of the small sample size the ADF p-value of 0.32. To gain a better understanding of the results, we can compare them with the test statistics of randomly generated 25 long standard normal samples (Figure 15).

Table 4. Augmented Dicky–Fuller (ADF) statistic of explained percentage variance process.

ADF *t*-Value	ADF *p*-Value
−2.67	0.32

Notes: This table shows the ADF t and p statistics of yearly percentage variance process.

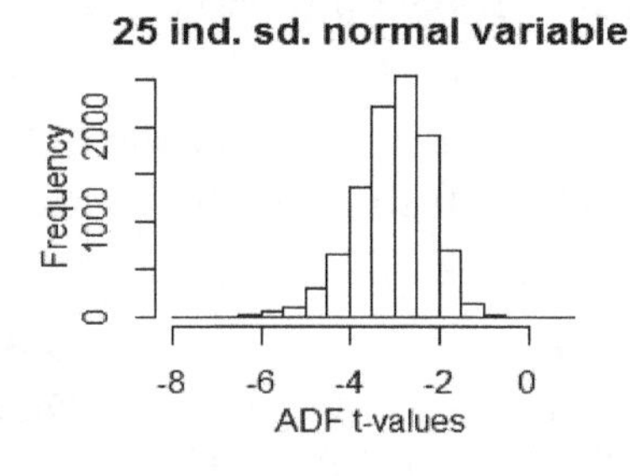

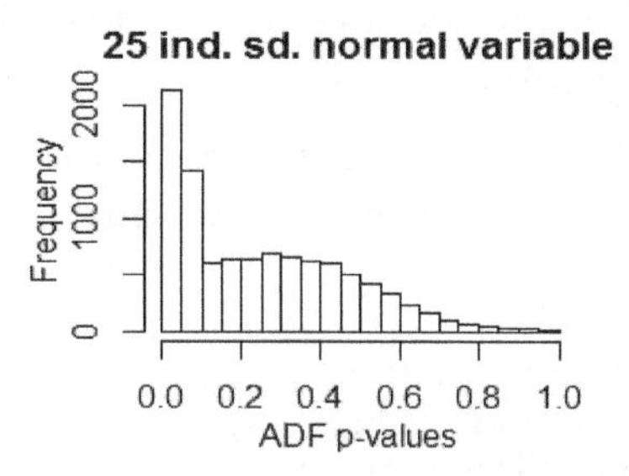

Figure 15. Histogram of ADF statistics of 10,000 independent 25 dim. standard normal sample.

However, we also have to study the time stability of cluster wise mean-standard deviation regressions. Splitting the data into one-year periods, clustering them and calculating regressions shed some light on the robustness of clusters (Figure 16).

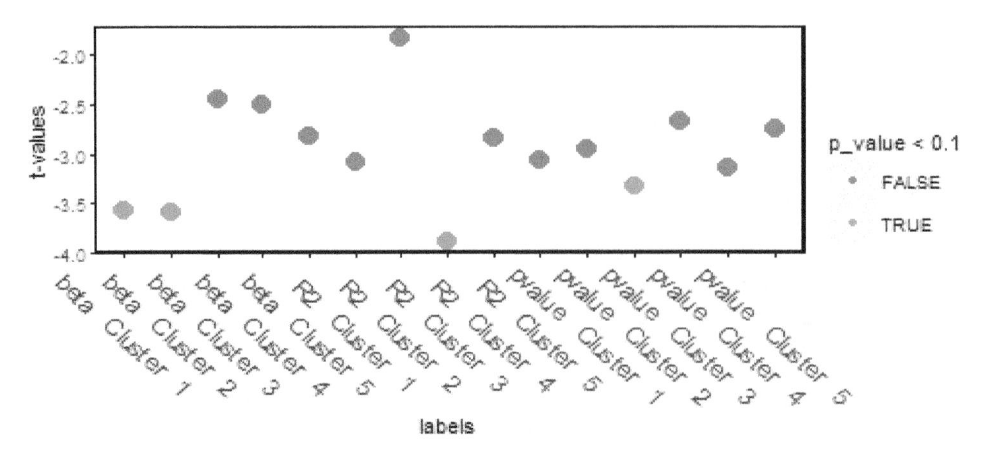

Figure 16. ADF test of cluster wise time shifted regressions.

The results show that cluster wise mean-variance regressions are stationary in cluster 1 and 2. Nevertheless, cluster 3 and 4 are outliers and clusters 5 mostly covers emerging market names. Thus, the Gaussian-based normalized modularity clustering technique can be used to filter out outliers and find robust clusters.

4. Discussion

Spectral clustering techniques can be used to discover the equity index structure. On the one hand, clusters help us to overcome the hardship of heterogeneity and make diversification more efficient. In our paper we shed some light on the relations between spectral, geographical and qualitative clustering. It also turned out that Gaussian-kernel based clusters are more suitable than geographical and qualitative categorizations. In addition, spectral cluster-wise linear regressions give time stationary and significant results.

On the other hand, we stress that correlation does not convey enough information about the network; hence linear dependency-based diversification is not optimal (Sharpe 1964; Maldonado and Anthony 1981). We compared various similarity kernels and spectral clustering methods to demonstrate the inadequacy of a normalized Laplacian approach (Takumasa et al. 2015) and underpin the applicability of the proposed Newman–Girvan cut. Moreover, we highlighted that daily closing prices incorporate the network level information. The results unveiled that tail events have little effect on the dense network structure, in other words, market shocks have no effect on the cluster components; thus, index co-movements are not affected by large price changes.

All of these imply spectral clustering can eliminate non-linear effects, thus regular mean-standard deviation representation gives cluster-wise reliable figures. Instead of qualitative categorization, we suggest that portfolio managers should use Gaussian-based normalized modularity clusters to diversify global non-systematic risk.

An interesting field of further research would be analyzing the evolution of the network to identify patterns that could help us to understand the life cycle of hubs and the vulnerability of the current equity network.

Author Contributions: Conceptualization, L.N. and M.O.; Methodology, L.N. and M.O.; Validation, L.N. and M.O.; Formal Analysis, L.N. and M.O.; Investigation, L.N. and M.O.; Writing-Original Draft Preparation, L.N. and M.O.; Writing-Review & Editing, L.N. and M.O.; Visualization, L.N.; Supervision, M.O.

Acknowledgments: The authors would like to gratefully acknowledge the valuable comments and suggestions of three anonymous referees that contributed to a substantially improved paper. Mihály Ormos acknowledges the support of the János Bolyai Research Scholarship of the Hungarian Academy of Sciences and the support of the Pallas Athéné Domus Educationis Foundation. The views expressed are those of the authors and do not necessarily reflect the official opinion of the Pallas Athéné Domus Educationis Foundation.

Appendix A

Table A1. Clusters of stock indices.

Country	Two Clusters	Three Clusters	Five Clusters
United Arab Emirates	2	3	1
Saudi Arabia	2	3	1
Qatar	2	1	1
Kuwait	2	1	1
Egypt	2	3	1
Bahrain	2	1	1
Vietnam	2	1	1
Nigeria	2	1	1
Dow Jones	1	2	2
Denmark	1	1	2
Switzerland	1	1	2
Canada	1	2	2
Mexico	1	2	2
Chile	1	2	2
Argentina	1	2	2
Hungary	1	1	2
Morocco	2	1	2
S&P 500	1	2	2
MSCI World	1	2	2
Czech Republic	1	1	2
Togo	2	1	2
Spain	1	1	2
Norway	1	1	2
Luxembourg	1	1	2
France	1	1	2
South Africa	1	3	2
Euro Stocks	1	1	2
Sweden	1	1	2
UK	1	1	2
Netherlands	1	1	2
Finland	1	1	2
Poland	1	3	2
Germany	1	1	2
Belgium	1	1	2
Italy	1	1	2
Brazil	1	2	2
Colombia	1	3	2
Bangladesh	2	1	3
Costa Rica	2	1	4
Zambia	2	1	4
Malawi	2	1	4
Venezuela	2	1	4
South Korea	2	3	5
Hong Kong	2	3	5
Thailand	2	3	5
China	2	3	5
Kenya	2	3	5
India	2	3	5
Namibia	2	3	5
Turkey	2	3	5
Indonesia	2	3	5
Malaysia	2	3	5
Russia	2	3	5
Australia	2	3	5
Taiwan	2	3	5
Japan	2	3	5
Ukraine	2	3	5
Bulgaria	2	1	5
Romania	2	3	5

Notes: This table contains the list of indices and clustering results for 2, 3, and 5 clusters.

References

Barabási, Albert L., and Albert Réka. 1999. Emergence of Scaling in Random Networks. *Science* 26: 509–12. [CrossRef]

Berlinet, Alain, and Thomas-Agnan Christine. 2011. *Reproducing Kernel Hilbert Spaces in Probability and Statistics*. Berlin: Springer Science & Business Media, pp. 1–108. ISBN 978-1441990969.

Bolla, Marianna. 2011. Penalized version of Newman-Girvan modularity and their relation to normalized cuts and k-means clustering. *Physical Review E* 84: 016108. [CrossRef] [PubMed]

Chung, Fan R. G. 1997. *Spectral Graph Theory*. Providence: American Mathematical Society, No. 92. pp. 14–81. ISBN 978-0821803158.

Engelmann, Bernd, Evelyn Hayden, and Dirk Tasche. 2003. *Measuring the Discriminative Power of Rating Systems*. Banking and Financial Supervision. Frankfurt: Deutsche Bundesbank.

Erdős, Péter, and Rényi Alfréd. 1960. On the Evolution of Random Graphs. *Acta Mathematica Hungarica* 5: 17–61.

Erdős, Péter, Mihály Ormos, and Dávid Zibriczky. 2011. Non-parametric and semi-parametric asset pricing. *Economic Modelling* 28: 1150–62. [CrossRef]

Fama, Eugene, and Kenneth R. French. 1996. Multifactor explanations of asset pricing anomalies. *The Journal of Finance* 51: 55–84. [CrossRef]

Maurizio, Filippone, Francesco Camastra, Francesco Masulli, and Stefano Rovetta. 2007. A survey of kernel and spectral methods for clustering. *Pattern Recognition* 41: 176–90. [CrossRef]

Heiberger, Raphael H. 2014. Stock network stability in times of crisis. *Physica A: Statistical Mechanics and Its Applications* 393: 376–81. [CrossRef]

Gregory, Leibon, Scott Pauls, Daniel Rockmore, and Robert Savell. 2008. Topological Structures in the Equities Market Network. *PNAS* 105: 20589–94. [CrossRef]

Von Luxburg, Ulrike. 2007. Tutorial on Spectral Clustering. *Statistics and Computing* 17: 395–416. [CrossRef]

Maldonado, Rita, and Saunders Anthony. 1981. International portfolio diversification and the inter-temporal stability of international stock market relationships, 1957–1978. *Financial Management* 10: 54–63. [CrossRef]

MSCI. 2018. Market Classification. Available online: https://www.msci.com/market-classification (accessed on 3 November 2018).

Ormos, Mihály, and Dávid Zibriczky. 2014. Entropy-Based Financial Asset Pricing. *PLoS ONE* 9: E115742. [CrossRef] [PubMed]

Shi, Jianbo, and Jitendra Malik. 2000. Normalized cuts and image segmentation. *IEEE Pattern Analysis and Machine Intelligence* 22: 888–905. [CrossRef]

Sharpe, William F. 1964. Capital asset prices: A theory of market equilibrium under conditions of risk. *Journal of Finance* 19: 425–42.

Song, Dong-Ming, Michele Tumminello, Wei-Xing Zhou, and Rosario N. Mantegna. 2011. Evolution of worldwide stock markets, correlation structure, and correlation-based graphs. *Physical Review E* 84: 026108. [CrossRef] [PubMed]

Takumasa, Sakakibara, Tohgoroh Matsuib, Atsuko Mutoha, and Nobuhiro Inuzuka. 2015. Clustering mutual funds based on investment similarity. *Procedia Computer Science* 60: 881–90. [CrossRef]

Yalamova, Rossitsa. 2009. Correlations in Financial Time Series during Extreme Events-Spectral Clustering and Partition Decoupling Method. Paper presented at World Congress on Engineering, London, UK, July 1–3, Volume 2, pp. 1376–78.

Zhao, Yanchang. 2012. *R and Data Mining: Examples and Case Study*. Cambridge: Academic Press, pp. 49–59.

10

Equity Option Pricing with Systematic and Idiosyncratic Volatility and Jump Risks

Zhe Li

Business School, Nanjing Normal University, Nanjing 210023, China; zheli@njnu.edu.cn

Abstract: Recently, a large number of empirical studies indicated that individual equity options exhibit a strong factor structure. In this paper, the importance of systematic and idiosyncratic volatility and jump risks on individual equity option pricing is analyzed. First, we propose a new factor structure model for pricing the individual equity options with stochastic volatility and jumps, which takes into account four types of risks, i.e., the systematic diffusion, the idiosyncratic diffusion, the systematic jump, and the idiosyncratic jump. Second, we derive the closed-form solutions for the prices of both the market index and individual equity options by utilizing the Fourier inversion. Finally, empirical studies are carried out to show the superiority of our model based on the S&P 500 index and the stock of Apple Inc. on options. The out-of-sample pricing performance of our proposed model outperforms the other three benchmark models especially for short term and deep out-of-the-money options.

Keywords: equity option pricing; factor models; stochastic volatility; jumps

JEL Classification: G13

---

1. Introduction

Most of the existing literature studies on option pricing are for index options, and there are very few about equity options. One approach to modeling equity options is to employ the state-of-the-art model in the index option literature, a stochastic volatility model with jumps (see, for example, Bates 1996, 2000; Bakshi et al. 1997; Duffie et al. 2000; Eraker et al. 2003; Broadie et al. 2007; Christoffersen et al. 2012; Andersen et al. 2015; Bardgett et al. 2019), but to ignore any underlying factor structure.

In Bakshi and Kapadia (2003a), the research results indicated that the volatility risk premium is negative in index options by examining the statistical properties of delta hedged option portfolios, i.e., a portfolio of a long call option position hedged by a short position in the stock. On the one hand, stock returns have a significant market component; the emergence of market volatility risk premiums is bound to have an impact on individual equity option pricing. On the other hand, from the economic point of view, the risk neutral distributions of individual equities are systematically different from the market index. Thus, it is necessary to explore how volatility risk is priced in individual equity options, which also can produce additional insights into the pricing structure of individual equity options (see Bakshi et al. 2003). As is well known, the beta of a stock represents the sensitivity of the risk of the individual equity with respect to the systematic risk of the market and is very useful for portfolio construction in the capital asset pricing model. Therefore, under the assumption that stock returns include a market component and an idiosyncratic component, Bakshi and Kapadia (2003b) developed a factor model for equity option valuation and investigated the pricing of market volatility risk in individual equity options. Their empirical results showed that volatility risk premiums in equity options are smaller than in index options.

Afterwards, Fouque and Kollman (2011) proposed a continuous-time capital asset pricing model (CAPM) where the dynamics of the market index have a stochastic volatility driven by a fast mean reverting process. Moreover, they derived the analytical approximation pricing formulas for both the market index and individual equity call options using a singular perturbation method. Meanwhile, a calibration method for the beta parameter was also presented based on the estimated model parameters of both the market index and individual equity option prices. Subsequently, Fouque and Tashman (2012) extended the constant beta-parameter factor model of Fouque and Kollman (2011) by considering a piecewise-linear relationship between the individual asset and the market index and proposed a regime switching factor model for the pricing market index and individual equity options. Supposing that stock return is linearly related to market index return in terms of the beta parameter, Carr and Madan (2012) developed a factor model for individual equity option pricing under a purely discontinuous Lévy process via fast Fourier transform, in which the variance gamma process for the dynamics of both the market index and stock was taken as an example for illustration. By supposing a continuous-time CAPM with Lévy processes, Wong et al. (2012) also derived analytical solutions to the index and equity options and explored the corresponding static hedging with index futures. Christoffersen et al. (2018) empirically studied the equity volatility levels, skews, and term structures by using equity option prices and principal component analysis. The results indicated that the equity options had a strong factor structure, and then, they developed an equity option pricing model with a CAPM factor structure and stochastic volatility, which allowed for mean reverting stochastic volatility for the dynamics of both the market factor and individual equity.

Recently, Xiao and Zhou (2018) proposed a GARCH-jump model for individual stock returns that took into account four types of risks: the systematic and idiosyncratic jumps and the systematic and idiosyncratic diffusive volatility. By using a dataset consisting of the S&P 500 index and 15 individual stock prices, their empirical results indicated that idiosyncratic jumps were a key determinant of expected stock return.[1] Instead of using only stock returns, Kapadia and Zekhnini (2019) used both stock and option data to decompose the four risk premiums associated with systematic and idiosyncratic diffusive and jump risks and also documented that idiosyncratic jumps are important determinants of the mean returns of a stock from both an ex post and ex ante perspective.

Motivated by the above mentioned insights, we propose to price individual equity options in stochastic volatility jump-diffusion models with a market factor structure, which can be seen as a generalized version of Christoffersen et al. (2018). Specifically, in our proposed model, the individual equity prices are driven by the market factor, as well as an idiosyncratic component that also has stochastic volatility and jump. Due to the model belonging to the affine class, we derive the closed-form solutions for the prices of both the market index and individual equity options by utilizing the Fourier inversion. In addition, we provide the empirical results to test the pricing performance of the proposed factor model based on the S&P 500 index and the stock of Apple Inc. (AAPL) on options. Toward this end, we empirically compare the pricing performance of the proposed model with those of the other three classical two factor stochastic volatility models being taken as benchmark models. Empirical results presented here confirm that the equity option pricing model considering systematic and idiosyncratic volatility and jump risks may offer a good competitor to the models of Bates (2000), Christoffersen et al. (2009), or Christoffersen et al. (2018) for some other option markets.

The remainder of the paper proceeds as follows. In Section 2, we present a novel factor model for equity option valuation and derive the corresponding closed-form solutions. In Section 3,

---

[1] In fact, the work of Xiao and Zhou (2018) is a complement to the recent studies that disentangle the four types of risks in equity premiums, such as Bégin et al. (2020), who developed a GARCH-jump model in which an individual firm's systematic and idiosyncratic risk have both a Gaussian diffusive and a jump component. Their empirical results showed that normal diffusive and jump risks have drastically different effects on the expected return of individual stocks by using 20 years of returns and options on the S&P 500 and 260 stocks.

empirical studies are carried out to show the pricing performance of our proposed model. Finally, some conclusions are stated in Section 4.

2. Equity Option Valuation

In this section, we introduce a general class of stochastic volatility models with jumps for the dynamics of both the market factor and individual equity prices and derive closed form solutions to the prices of the European equity call and put options.

2.1. Model Description

Consider a filtered probability space $(\Omega, \mathcal{F}, \mathbb{Q})$ with information filtration $\{\mathcal{F}_t\}_{0\le t\le T}$ satisfying the usual conditions (increase, right-continuous, and augmented), where $\mathbb{Q}$ is a risk neutral measure. We model an equity market consisting of N firms with a single market factor, I_t (usually approximated by a market index in practice). The individual stock prices are denoted by S_t^i, for $i = 1, 2, \ldots, N$. For the sake of convenience, we ignore the superscript i, and denote $(S_t)_{t\ge 0}$ the pricing process of an individual stock. Investors also have access to a risk free bond that pays a return rate of r. To start, the market factor I_t evolves under a risk neutral measure $\mathbb{Q}$ as:

$$\frac{dI_t}{I_{t-}} = rdt + \sqrt{V_{I,t}}dW_{1,t}^I + \int_R (e^y - 1)\tilde{N}_y(dt, dy), \tag{1}$$

$$dV_{I,t} = \kappa_I(\theta_I - V_{I,t})dt + \sigma_I\sqrt{V_{I,t}}dW_{2,t}^I, \tag{2}$$

where I_{t-} stands for the value of I_t before a possible jump occurs, $y \in R = \mathbb{R} \setminus \{0\}$, $V_{I,t}$ is the variance of market factor, θ_I denotes the long run variance, κ_I captures the mean reversion speed of $V_{I,t}$ to θ_I, σ_I measures the volatility of volatility, $2\kappa_I\theta_I \ge \sigma_I^2$ to ensure that the process $V_{I,t}$ remains strictly positive[2], $W_{1,t}^I$ and $W_{2,t}^I$ are correlated standard Brownian motions, i.e., the innovations to the market return and volatility are correlated with correlation coefficient ρ_I, $\mathrm{Cov}\left(dW_{1,t}^I, dW_{2,t}^I\right) = \rho_I dt$, and $\tilde{N}_y(dt, dy) = N_y(dt, dy) - \nu_y(dy)dt$ is a compensated jump measure, where $N_y(dt, dy)$ is the jump measure and the Lévy kernel (or density) $\nu_y(dy)$ satisfies $\int_R \min(1, y^2)\nu_y(dy) < \infty$.

Furthermore, we separate the effects of the market factor on individual equities' returns into two types of risks: the systematic diffusive volatility and jump. More specifically, the diffusive random variation of individual equities' returns is dependent on the Brownian motion that drives market returns through the coefficient β_{diff}. In addition, the discontinuous movements in the market return can also trigger jumps in individual equities' returns through the coefficient β_{jump}. Therefore, the individual equity prices are driven by the market factor, as well as an idiosyncratic component that also has stochastic volatility and jump, whose process under a risk neutral measure $\mathbb{Q}$ follows:[3]

$$\begin{aligned}\frac{dS_t}{S_{t-}} = rdt + &\underbrace{\beta_{diff}\sqrt{V_{I,t}}dW_{1,t}^I}_{\text{Systematic diffusive}} + \underbrace{\int_R (e^{\beta_{jump}y} - 1)\tilde{N}_y(dt, dy)}_{\text{Systematic jump}} + \underbrace{\sqrt{V_{S,t}}dW_{1,t}^S}_{\text{Idiosyncratic diffusive}} \\ &+ \underbrace{\int_R (e^{\xi} - 1)\tilde{N}_\xi(dt, d\xi)}_{\text{Idiosyncratic jump}},\end{aligned} \tag{3}$$

$$dV_{S,t} = \kappa_S(\theta_S - V_{S,t})dt + \sigma_S\sqrt{V_{S,t}}dW_{2,t}^S, \tag{4}$$

---

2 One can refer to Assumption 2.1 of Cheang et al. (2013) and Cheang and Garces (2019) for a more detailed explanation.

3 Obviously, our proposed model for the dynamics of the market factor and individual equity prices is an extension of Christoffersen et al. (2018). In fact, our model also can be regarded as a further generalization of Cheang et al. (2013) and Cheang and Garces (2019) by taking into account the factor structure.

where S_{t-} stands for the value of S_t before a possible jump occurs, $\xi \in R = \mathbb{R} \setminus \{0\}$, $V_{S,t}$ is the idiosyncratic variance of individual equity, θ_S denotes the long run idiosyncratic variance, κ_S captures the mean reversion speed of $V_{S,t}$ to θ_S, σ_S measures the volatility of idiosyncratic variance, $2\kappa_S\theta_S \geq \sigma_S^2$ to ensure that the process $V_{S,t}$ remains strictly positive[4], $W_{1,t}^S$ and $W_{2,t}^S$ are correlated standard Brownian motions, i.e., the innovations to the idiosyncratic return and volatility are correlated with correlation coefficient ρ_S, $\mathrm{Cov}\left(dW_{1,t}^S, dW_{2,t}^S\right) = \rho_S dt$, but they are independent of Brownian motions in the market factor, i.e., $\mathrm{Cov}\left(dW_{i,t}^S, dW_{j,t}^I\right) = 0$ for $i, j = 1, 2$, and $\tilde{N}_\xi(dt, dy) = N_\xi(dt, d\xi) - \nu_\xi(d\xi)dt$ is a compensated jump measure, where $N_\xi(dt, d\xi)$ is the jump measure and the Lévy kernel (or density) $\nu_\xi(d\xi)$ satisfies $\int_R \min(1, \xi^2)\nu_\xi(d\xi) < \infty$.

2.2. Characteristic Function

In order to be able to derive the pricing formulas for the European call and put equity options, we are particularly interested in the characteristic function of the logarithm asset price. Given the dynamics of the underlying asset price under the $\mathbb{Q}$ measure, we consider the conditional characteristic function of log-asset price $X_T = \ln S_T$ given the market information up to time t, which is denoted by $\varphi(x, v_1, v_2, t, T; \phi)$:

$$\begin{aligned} \varphi(x, v_1, v_2, t, T; \phi) &= \mathrm{E}^{\mathbb{Q}}\left[e^{\mathrm{i}\phi X_T} \middle| X_t = x, V_{I,t} = v_1, V_{S,t} = v_2\right] \\ &\triangleq \mathrm{E}_t^{\mathbb{Q}}\left[e^{\mathrm{i}\phi X_T}\right], \end{aligned} \tag{5}$$

where $\mathrm{E}_t^{\mathbb{Q}}[\cdot]$ denotes the condition expectation under the $\mathbb{Q}$ measure, $t \leq T$, and $\mathrm{i} = \sqrt{-1}$.

Lemma 1. *Suppose that the market factor I_t and individual equity price S_t are driven by Equations (1) and (3), respectively. Then, the conditional characteristic function of log-asset price $X_T = \ln S_T$ is given by:*

$$\varphi(x, v_1, v_2, t, T; \phi) = \exp\{A(\tau)x + B(\tau)v_1 + C(\tau)v_2 + D(\tau)\}, \tag{6}$$

where:

$$\begin{aligned}
A(\tau) &= \mathrm{i}\phi, \\
B(\tau) &= \frac{\kappa_I - \mathrm{i}\phi\beta_{diff}\sigma_I\rho_I - d_1}{\sigma_I^2}\left[\frac{1 - e^{-d_1\tau}}{1 - g_1 e^{-d_1\tau}}\right], \\
C(\tau) &= \frac{\kappa_S - \mathrm{i}\phi\sigma_S\rho_S - d_2}{\sigma_S^2}\left[\frac{1 - e^{-d_2\tau}}{1 - g_2 e^{-d_2\tau}}\right], \\
D(\tau) &= \Bigg[\mathrm{i}\phi r + \underbrace{\int_R \left(e^{\mathrm{i}\phi\beta_{jump}y} - 1 - \mathrm{i}\phi\left(e^{\beta_{jump}y} - 1\right)\right)\nu_y(dy)}_{\mathrm{I}_1} + \underbrace{\int_R \left(e^{\mathrm{i}\phi\xi} - 1 - \mathrm{i}\phi\left(e^{\xi} - 1\right)\right)\nu_\xi(d\xi)}_{\mathrm{I}_2}\Bigg]\tau \\
&\quad + \frac{\kappa_I\theta_I}{\sigma_I^2}\left[\left(\kappa_I - \mathrm{i}\phi\beta_{diff}\sigma_I\rho_I - d_1\right)\tau - 2\ln\frac{1 - g_1 e^{-d_1\tau}}{1 - g_1}\right] \\
&\quad + \frac{\kappa_S\theta_S}{\sigma_S^2}\left[\left(\kappa_S - \mathrm{i}\phi\sigma_S\rho_S - d_2\right)\tau - 2\ln\frac{1 - g_2 e^{-d_2\tau}}{1 - g_2}\right], \\
g_1 &= \frac{\kappa_I - \mathrm{i}\phi\beta_{diff}\sigma_I\rho_I - d_1}{\kappa_I - \mathrm{i}\phi\beta_{diff}\sigma_I\rho_I + d_1},
\end{aligned}$$

---

[4] One can refer to the Assumption 2.1 of Cheang et al. (2013) and Cheang and Garces (2019) for a more detailed explanation.

$$g_2 = \frac{\kappa_S - i\phi\sigma_S\rho_S - d_2}{\kappa_S - i\phi\sigma_S\rho_S + d_2},$$

$$d_1 = \sqrt{\left(i\phi\beta_{diff}\sigma_I\rho_I - \kappa_I\right)^2 + \beta_{diff}^2\sigma_I^2(i\phi + \phi^2)},$$

$$d_2 = \sqrt{(i\phi\sigma_S\rho_S - \kappa_S)^2 + \sigma_S^2(i\phi + \phi^2)},$$

and $\tau = T - t$.

Proof. To obtain the conditional characteristic function of log-asset price $X_T = \ln S_T$, we first take the following transformation by using the Itô lemma for Equation (3):

$$\begin{aligned} d\ln S_t = &\left(r - \frac{1}{2}\beta_{diff}^2 V_{I,t} - \frac{1}{2}V_{S,t} - \int_R \left(e^{\beta_{jump}y} - 1\right)\nu_y(dy) - \int_R \left(e^{\xi} - 1\right)\nu_\xi(d\xi)\right)dt \\ &+ \beta_{diff}\sqrt{V_{I,t}}dW_{1,t}^I + \beta_{jump}\int_R yN_y(dt,dy) + \sqrt{V_{S,t}}dW_{1,t}^S + \int_R \xi N_\xi(dt,d\xi). \end{aligned}$$

The Feynman–Kac formula states that $\varphi(x, v_1, v_2, t, T; \phi)$ is governed by the following partial integro-differential equation (PIDE):

$$\begin{cases} \dfrac{\partial\varphi}{\partial\tau} = \left[r - \dfrac{1}{2}\beta_{diff}^2 V_{I,t} - \dfrac{1}{2}V_{S,t} - \displaystyle\int_R \left(e^{\beta_{jump}y} - 1\right)\nu_y(dy) - \int_R \left(e^{\xi} - 1\right)\nu_\xi(d\xi)\right]\dfrac{\partial\varphi}{\partial x} \\ \quad + \dfrac{1}{2}\left(\beta_{diff}^2 v_1 + v_2\right)\dfrac{\partial^2\varphi}{\partial x^2} + \kappa_I(\theta_I - v_1)\dfrac{\partial\varphi}{\partial v_1} + \dfrac{1}{2}\sigma_I^2 v_1\dfrac{\partial^2\varphi}{\partial v_1^2} \\ \quad + \kappa_S(\theta_S - v_2)\dfrac{\partial\varphi}{\partial v_2} + \dfrac{1}{2}\sigma_S^2 v_2\dfrac{\partial^2\varphi}{\partial v_2^2} + \beta_{diff}\sigma_I\rho_I v_1\dfrac{\partial^2\varphi}{\partial x\partial v_1} + \sigma_S\rho_S v_2\dfrac{\partial^2\varphi}{\partial x\partial v_2} \\ \quad + \displaystyle\int_R \left[\varphi(x + \beta_{jump}y, v_1, v_2, t, T; \phi) - \varphi(x, v_1, v_2, t, T; \phi)\right]\nu_y(dy) \\ \quad + \displaystyle\int_R \left[\varphi(x + \xi, v_1, v_2, t, T; \phi) - \varphi(x, v_1, v_2, t, T; \phi)\right]\nu_\xi(d\xi), \\ \varphi(x, v_1, v_2, t, T; \phi)|_{t=T} = e^{i\phi X_T}. \end{cases} \tag{7}$$

Due to the affine structure of our model, we postulate $\varphi(x, v_1, v_2, t, T; \phi)$ admitting the form of (6). Substituting Equation (6) into the above PIDE (7) gives the following system of ordinary differential equations (ODEs) for $A(\tau)$, $B(\tau)$, $C(\tau)$, and $D(\tau)$:

$$\begin{cases} \dfrac{\partial A(\tau)}{\partial\tau} = 0, \\ \dfrac{\partial B(\tau)}{\partial\tau} = \dfrac{1}{2}\sigma_I^2 B^2(\tau) + \left[\beta_{diff}\sigma_I\rho_I A(\tau) - \kappa_I\right]B(\tau) - \dfrac{1}{2}\beta_{diff}^2\left[A(\tau) - A^2(\tau)\right], \\ \dfrac{\partial C(\tau)}{\partial\tau} = \dfrac{1}{2}\sigma_S^2 C^2(\tau) + \left[\sigma_S\rho_S A(\tau) - \kappa_S\right]C(\tau) - \dfrac{1}{2}\left[A(\tau) - A^2(\tau)\right], \\ \dfrac{\partial D(\tau)}{\partial\tau} = rA(\tau) + \kappa_I\theta_I B(\tau) + \kappa_S\theta_S C(\tau) + \displaystyle\int_R \left[e^{A(\tau)\beta_{jump}y} - 1 - A(\tau)\left(e^{\beta_{jump}y} - 1\right)\right]\nu_y(dy) \\ \quad + \displaystyle\int_R \left[e^{A(\tau)\xi} - 1 - A(\tau)\left(e^{\xi} - 1\right)\right]\nu_\xi(d\xi), \end{cases}$$

where the boundary conditions are given as $A(0) = i\phi$ and $B(0) = C(0) = D(0) = 0$.

By solving the above ODEs, we can obtain the characteristic function (6). □

Lemma 2. *Suppose that the market factor I_t is driven by Equation (1). Then, the conditional characteristic function of log-market factor $Z_T = \ln I_T$ is given by:*

$$\begin{aligned}\psi(z, v_1, t, T; \phi) &= \mathrm{E}^{\mathbb{Q}}\left[e^{\mathrm{i}\phi Z_T} | Z_t = z, V_{I,t} = v_1\right] \\ &= \exp\left\{\tilde{A}(\tau)z + \tilde{B}(\tau)v_1 + \tilde{D}(\tau)\right\},\end{aligned} \tag{8}$$

where:

$$\tilde{A}(\tau) = \mathrm{i}\phi,$$

$$\tilde{B}(\tau) = \frac{\kappa_I - \mathrm{i}\phi\sigma_I\rho_I - d}{\sigma_I^2}\left[\frac{1 - e^{-d\tau}}{1 - ge^{-d\tau}}\right],$$

$$\tilde{D}(\tau) = \left[\mathrm{i}\phi r + \underbrace{\int_R \left(e^{\mathrm{i}\phi y} - 1 - \mathrm{i}\phi\left(e^y - 1\right)\right) \nu_y(dy)}_{\mathrm{I}_3}\right]\tau + \frac{\kappa_I\theta_I}{\sigma_I^2}\left[\left(\kappa_I - \mathrm{i}\phi\sigma_I\rho_I - d\right)\tau - 2\ln\frac{1 - ge^{-d\tau}}{1 - g}\right],$$

$$g = \frac{\kappa_I - \mathrm{i}\phi\sigma_I\rho_I - d}{\kappa_I - \mathrm{i}\phi\sigma_I\rho_I + d},$$

$$d = \sqrt{(\mathrm{i}\phi\sigma_I\rho_I - \kappa_I)^2 + \sigma_I^2(\mathrm{i}\phi + \phi^2)},$$

and $\tau = T - t$.

Proof. Similar to the proof of Lemma 1, we can easily verify the above results. □

2.3. Valuation of the European Index and Equity Options

Once the characteristic function is found, it is straightforward to calculate the prices of European options by using Fourier inversion. Let $C(S_t, T, K)$ and $P(S_t, T, K)$ be the prices of the European equity call and put options at time t with strike price K and maturity T under the risk neutral measure $\mathbb{Q}$, respectively. Then, these option prices are determined by:

$$C(S_t, T, K) = e^{-r\tau}\mathrm{E}_t^{\mathbb{Q}}\left[\max(S_T - K, 0)\right]$$

and:

$$P(S_t, T, K) = e^{-r\tau}\mathrm{E}_t^{\mathbb{Q}}\left[\max(K - S_T, 0)\right]$$

where $\tau = T - t$ is the time to maturity.

Theorem 1. *Suppose that the market factor I_t and the individual equity price S_t are driven by Equations (1) and (3), respectively. Then, the prices of the European equity call and put options with strike price K and maturity $\tau = T - t$ are given by:*

$$C(S_t, T, K) = S_t\Pi_1\left(S_t, T, K; \beta_{diff}, \beta_{jump}\right) - Ke^{-r\tau}\Pi_2\left(S_t, T, K; \beta_{diff}, \beta_{jump}\right) \tag{9}$$

and:

$$P(S_t, T, K) = Ke^{-r\tau}\left[1 - \Pi_2\left(S_t, T, K; \beta_{diff}, \beta_{jump}\right)\right] - S_t\left[1 - \Pi_1\left(S_t, T, K; \beta_{diff}, \beta_{jump}\right)\right] \tag{10}$$

where the risk neutral probability distribution functions Π_1 and Π_2 are defined by:

$$\Pi_1\left(S_t, T, K; \beta_{diff}, \beta_{jump}\right) = \frac{1}{2} + \frac{e^{-r\tau}}{\pi S_t}\int_0^{+\infty} \Re\left[\frac{e^{-\mathrm{i}\phi\ln K}\varphi(x, v_1, v_2, t, T; \phi - \mathrm{i})}{\mathrm{i}\phi}\right] d\phi$$

and:

$$\Pi_2\left(S_t,T,K;\beta_{diff},\beta_{jump}\right)=\frac{1}{2}+\frac{1}{\pi}\int_0^{+\infty}\Re\left[\frac{e^{-\mathrm{i}\phi\ln K}\varphi(x,v_1,v_2,t,T;\phi)}{\mathrm{i}\phi}\right]d\phi,$$

where $\varphi(x,v_1,v_2,t,T;\phi)$ is the conditional characteristic function of $\ln S_T$, which can be seen in Equation (6), and $\Re[\cdot]$ indicates the real part of a complex number.

Proof. In order to get the pricing formulas of the European equity call and put options, let us first introduce a change of measure from $\mathbb{Q}$ to $\tilde{\mathbb{Q}}$ by the following Radon–Nikodym derivative:

$$\frac{d\tilde{\mathbb{Q}}}{d\mathbb{Q}}=e^{-r(T-t)}\frac{S_T}{S_t}.$$

We denote the conditional characteristic function of $X_T=\ln S_T$ under the $\tilde{\mathbb{Q}}$ measure by $\tilde{\varphi}(x,v_1,v_2,t,T;\phi)$. Then, $\tilde{\varphi}(x,v_1,v_2,t,T;\phi)$ can be expressed as:

$$\begin{aligned}\tilde{\varphi}(x,v_1,v_2,t,T;\phi)&=\mathrm{E}_t^{\tilde{\mathbb{Q}}}\left[e^{\mathrm{i}\phi X_T}\right]\\&=\mathrm{E}_t^{\mathbb{Q}}\left[e^{-r(T-t)}\frac{S_T}{S_t}e^{\mathrm{i}\phi X_T}\right]\\&=e^{-r(T-t)-x}\mathrm{E}_t^{\mathbb{Q}}\left[e^{\mathrm{i}(\phi-\mathrm{i})X_T}\right]\\&=e^{-r(T-t)-x}\varphi(x,v_1,v_2,t,T;\phi-\mathrm{i}).\end{aligned}$$

Thus, the price of a European equity call option $C(S_t,T,K)$ can be calculated by utilizing $\varphi(x,v_1,v_2,t,T;\phi)$ and $\tilde{\varphi}(x,v_1,v_2,t,T;\phi)$:

$$\begin{aligned}C(S_t,T,K)&=e^{-r\tau}\mathrm{E}_t^{\mathbb{Q}}\left[\max(S_T-K,0)\right]\\&=e^{-r\tau}\mathrm{E}_t^{\mathbb{Q}}\left[S_T1_{\{S_T\geq K\}}\right]-Ke^{-r\tau}\mathrm{E}_t^{\mathbb{Q}}\left[1_{\{S_T\geq K\}}\right]\\&=S_t\mathrm{E}_t^{\tilde{\mathbb{Q}}}\left[1_{\{S_T\geq K\}}\right]-Ke^{-r\tau}\mathrm{E}_t^{\mathbb{Q}}\left[1_{\{S_T\geq K\}}\right]\\&=S_t\mathrm{E}_t^{\tilde{\mathbb{Q}}}\left[1_{\{X_T\geq\ln K\}}\right]-Ke^{-r\tau}\mathrm{E}_t^{\mathbb{Q}}\left[1_{\{X_T\geq\ln K\}}\right]\\&=S_t\Pi_1\left(S_t,T,K;\beta_{diff},\beta_{jump}\right)-Ke^{-r\tau}\Pi_2\left(S_t,T,K;\beta_{diff},\beta_{jump}\right).\end{aligned}$$

Once the conditional characteristic function $\varphi(x,v_1,v_2,t,T;\phi)$ is obtained, we can easily calculate the probability distribution functions $\Pi_1\left(S_t,T,K;\beta_{diff},\beta_{jump}\right)$ and $\Pi_2\left(S_t,T,K;\beta_{diff},\beta_{jump}\right)$ according to the Lévy inversion formula:

$$\Pi_1\left(S_t,T,K;\beta_{diff},\beta_{jump}\right)=\frac{1}{2}+\frac{1}{\pi}\int_0^{+\infty}\Re\left[\frac{e^{-\mathrm{i}\phi\ln K}\tilde{\varphi}(x,v_1,v_2,t,T;\phi)}{\mathrm{i}\phi}\right]d\phi$$

and:

$$\Pi_2\left(S_t,T,K;\beta_{diff},\beta_{jump}\right)=\frac{1}{2}+\frac{1}{\pi}\int_0^{+\infty}\Re\left[\frac{e^{-\mathrm{i}\phi\ln K}\varphi(x,v_1,v_2,t,T;\phi)}{\mathrm{i}\phi}\right]d\phi,$$

A similar approach can be used to derive the pricing formula for the European equity put option. □

In a similar way, we also can present the pricing formulas for the European index call and put options.

Theorem 2. *Suppose that the market factor I_t is driven by Equation (1). Then, the time t prices of the European index call and put options with strike price K and maturity $\tau = T - t$ are given by:*

$$C(I_t, T, K) = I_t \tilde{\Pi}_1 (I_t, T, K) - Ke^{-r\tau} \tilde{\Pi}_2 (I_t, T, K) \tag{11}$$

and:

$$P(I_t, T, K) = Ke^{-r\tau} \left[1 - \tilde{\Pi}_2 (I_t, T, K)\right] - I_t \left[1 - \tilde{\Pi}_1 (I_t, T, K)\right] \tag{12}$$

where the risk neutral probability distribution functions Π_1 and Π_2 are defined by:

$$\tilde{\Pi}_1 (I_t, T, K) = \frac{1}{2} + \frac{e^{-r\tau}}{\pi I_t} \int_0^{+\infty} \Re \left[\frac{e^{-\mathrm{i}\phi \ln K} \psi(z, v_1, t, T; \phi - \mathrm{i})}{\mathrm{i}\phi} \right] d\phi$$

and:

$$\tilde{\Pi}_2 (I_t, T, K) = \frac{1}{2} + \frac{1}{\pi} \int_0^{+\infty} \Re \left[\frac{e^{-\mathrm{i}\phi \ln K} \psi(z, v_1, t, T; \phi)}{\mathrm{i}\phi} \right] d\phi,$$

where $\psi(z, v_1, t, T; \phi)$ is the conditional characteristic function of $\ln I_T$, which can be seen in Equation (8).

3. Empirical Studies

In this section, we empirically compare the pricing performance of our proposed model with those of the classical two factor stochastic volatility models, such as Bates (2000) (two variance SVmodel with price jumps, 2-SVJmodel), Christoffersen et al. (2009) (two-variance SV model, 2-SV model), and Christoffersen et al. (2018) (two-variance SV model with a single market factor, 2-FSVmodel), being taken as benchmark models.

3.1. Data Description

We used the S&P 500 index (SPX) to proxy for the market factor and AAPL as the individual equity. We employed the delayed market quotes on arbitrary date 8 May 2019, which was the last date available at the time of writing, as the in-sample data to calibrate the risk neutral parameters, and those on 9 May 2019 were used for the out-of-sample test. We used mid-quotes to represent the option prices. To eliminate the sample noise in raw option data, we adopted some filtering rules commonly used within the related literature: (i) we omitted those options with fewer than seven days and more than 365 days to maturity; (ii) all observations with zero trading volume were discarded; (iii) all options with implied volatility equal to zero and larger than 1.0 were discarded. In addition, for the convenience of the empirical analysis in the following, we only considered the sample data of the index call options and individual equity call options with the same expiration date. Thus, we focused only on ten maturities slices, namely on the maturities of 24 May 2019, 31 May 2019, 7 June 2019, 14 June 2019, 21 June 2019, 19 July 2019, 16 August 2019, 20 September 2019, 18 October 2019, and 17 January 2020.

After these filters, we had a total of 401 observations for the S&P 500 index call option on 8 May 2019. The individual equity option sample contained 233 call options on 8 May 2019 and 264 call options on 9 May 2019, respectively. Due to the life of an option being usually less than one year, we chose the three month U.S. Treasury Bill Rate to substitute for the risk free interest rate. All of the data were downloaded from the Chicago Board Options Exchange (http://www.cboe.com/).

3.2. Parameter Estimation

Our proposed model allowed a general distribution for jump components of the market factor and individual equity price and thus could be easily introduced to the special cases such that the jump components follow the compound Poisson process of Merton (1976) and Kou (2002), etc. For different types of Lévy kernels, different forms of our model can be presented. In order to keep consistent with

Bates (2000) for comparative analysis, in the following, we assumed that the jump components of the dynamics for the market factor and individual equity followed compound Poisson processes and the jump magnitude was drawn from the log-normal distribution of Merton (1976). Thus, the Lévy kernels for the market factor and individual equity, respectively, are given by:

$$\nu_y(dy) = \lambda_I \frac{1}{\sqrt{2\pi\delta_I^2}} \exp\left\{-\frac{(y-\mu_I)^2}{2\delta_I^2}\right\} dy \tag{13}$$

and:

$$\nu_\xi(d\xi) = \lambda_S \frac{1}{\sqrt{2\pi\delta_S^2}} \exp\left\{-\frac{(\xi-\mu_S)^2}{2\delta_S^2}\right\} d\xi, \tag{14}$$

where λ_j, for $j = I, S$, denotes the jump intensity, μ_j is the mean of the jump size, and δ_j is the variance of the jump size. Then, the integrals I_i, for $i = 1, 2, 3$, in Lemmas 1 and 2 can be calculated as follows:

$$I_1 = \lambda_I \left[e^{i\phi\beta_{jump}\mu_I - \frac{1}{2}\phi^2\beta_{jump}^2\delta_I^2} - 1 - i\phi\left(e^{\beta_{jump}\mu_I + \frac{1}{2}\beta_{jump}^2\delta_I^2} - 1\right)\right],$$

$$I_2 = \lambda_S \left[e^{i\phi\mu_S - \frac{1}{2}\phi^2\delta_S^2} - 1 - i\phi\left(e^{\mu_S + \frac{1}{2}\delta_S^2} - 1\right)\right],$$

and

$$I_3 = \lambda_I \left[e^{i\phi\mu_I - \frac{1}{2}\phi^2\delta_I^2} - 1 - i\phi\left(e^{\mu_I + \frac{1}{2}\delta_I^2} - 1\right)\right].$$

Based on Theorems 1 and 2, we employed a two step calibration procedure (see, for example, Wong et al. 2012; Christoffersen et al. 2018) to estimate the model parameters. First, we calibrated the market index dynamic Θ_I based on the S&P 500 index option price alone. Second, we used the equity option price to calibrate the individual equity dynamic Θ_S conditional on estimates of Θ_I. Consider the situation in which an investor wants to hedge his or her equity position with index options and hedging horizon T. For brevity, we further suppose that the investor observes index option prices and equity option prices both with maturity T, the same as hedging horizon. Specifically, the dataset contains M_t index option prices $C(I_t, T, K_i)$, for $i = 1, 2, \ldots, M_t$, and N_t equity option prices $C(S_t, T, K_j)$, for $j = 1, 2, \ldots, N_t$.

In the calibration process, the risk neutral model parameters were backed out by minimizing a loss function capturing the fit between the theoretical model and market prices. We employed the root mean squared errors (RMSE) as the objective function. The first step calibrated the risk neutral parameters for the index process, which are calibrated by:

$$\mathrm{RMSE}(I) = \arg\min_{\Theta_I} \sqrt{\frac{1}{M_t}\sum_{i=1}^{M_t}\left[C_{i,market}(I_t, T, K_i) - C_{i,model}^{\Theta_I}(I_t, T, K_i)\right]^2}, \tag{15}$$

where $C_{i,market}(I_t, T, K_i)$ is the market price of the index call option contract from the sample and $C_{i,model}^{\Theta_I}(I_t, T, K_i)$ represents the model price calculated using Equation (15) and the vector of model input parameters Θ_I.

The second calibrated the beta and the parameters for the idiosyncratic risk:

$$\mathrm{RMSE}(S) = \arg\min_{\Theta_S} \sqrt{\frac{1}{N_t}\sum_{j=1}^{N_t}\left[C_{j,market}(S_t, T, K_j) - C_{j,model}^{\Theta_S}(S_t, T, K_j)\right]^2}, \tag{16}$$

where $C_{j,market}(S_t, T, K_j)$ is the market price of the equity call option contract from the sample and $C^{\Theta_S}_{j,model}(S_t, T, K_j)$ represents the model price calculated using Equation (13) and the vector of model input parameters Θ_S.

On the basis of the above calibration method, Table 1 presents the risk neutral parameter estimates across various model specifications. Note that the values of the diffusive beta β_{diff} and jump beta β_{jump} for our proposed model were 0.3891 and 0.8429, respectively. The corresponding value of β_{diff} for the 2-FSV model was 0.2457. Obviously, both our proposed model and the 2-FSV model showed that AAPL tended to have a relatively low exposure to diffusive market movements. However, the jump exposure coefficient $\beta_{jump} = 0.8429$ indicated that the AAPL had a strong exposure to market jumps, which meant that the factor structure of the jumps was much stronger than the one of the diffusive movements. The reason for this result may be related to the sample data we selected. If we can get more sample data in the future, we will do an in-depth analysis. Moreover, we also can see that the values of correlation ρ were strongly negative for four models, capturing the so-called leverage effect both in the index and individual equity.

Table 1. Estimated parameters. Note: This table shows the average of the estimated parameters obtained by minimizing the root mean squared pricing errors between the market price and the model price for each option on 8 May 2019. Standard errors are reported in parentheses .

Parameters	**Our**		**2-FSV**		**2-SV**	**2-SVJ**
	SPX	**AAPL**	**SPX**	**AAPL**	**AAPL**	**AAPL**
$V_{I,0}/V_{1,0}$	0.0133 (0.0000)		0.0119 (0.0000)		0.0239 (0.0002)	0.0181 (0.0001)
$V_{S,0}/V_{2,0}$		0.0470 (0.0000)		0.0514 (0.0000)	0.0197 (0.0002)	0.0176 (0.0002)
κ_I/κ_1	0.2496 (0.0212)		0.2929 (0.0148)		0.3489 (0.0118)	0.4064 (0.0311)
κ_S/κ_2		0.2454 (0.0288)		0.1504 (0.0797)	0.4131 (0.0729)	0.4108 (0.0171)
θ_I/θ_1	0.2820 (0.0181)		0.3066 (0.0317)		0.3314 (0.0534)	0.2817 (0.0348)
θ_S/θ_2		0.2303 (0.0190)		0.3683 (0.0590)	0.2447 (0.0365)	0.3415 (0.0423)
σ_I/σ_1	0.3472 (0.0127)		0.3932 (0.0137)		0.1615 (0.0081)	0.1898 (0.0106)
σ_S/σ_2		0.1496 (0.0056)		0.1640 (0.0135)	0.2206 (0.0386)	0.1970 (0.0059)
λ_I	0.0450 (0.0017)					
λ_S		0.3413 (0.2463)				0.3065 (0.1194)
μ_I	0.1657 (0.0599)					
μ_S		0.0889 (0.0391)				0.0333 (0.0042)
δ_I	0.0850 (0.0113)					
δ_S		0.0679 (0.0078)				0.0534 (0.0013)
β_{diff}		0.3891 (0.0381)		0.2457 (0.0983)		
β_{jump}		0.8429 (0.8091)				
ρ_I/ρ_1	−0.9290 (0.0063)		−0.8498 (0.0080)		−0.9222 (0.0096)	−0.7445 (0.0297)
ρ_S/ρ_2		−0.9926 (0.0001)		−0.8938 (0.0469)	−0.7673 (0.1632)	−0.7817 (0.0549)

3.3. Pricing Performance

In this subsection, we present the empirical results for the calibrated models. In order to investigate the impacts of the systematic and idiosyncratic volatility and jump risks on equity option pricing, we took the 2-FSV, 2-SV, and 2-SVJ models as benchmark models to evaluate the pricing performance of our proposed model.

Figures 1–10 exhibit the predicted prices of the four model specifications and market prices listed on 9 May 2019, with 11, 16, 21, 26, 31, 51, 71, 96, 116, and 181 trading days to expiry, respectively. Here, the predicted prices (out-of-sample pricing) were calculated by the in-sample calibration parameters reported in Table 1. One can clearly observe from the left panels of Figures 1–10 that the option prices obtained by theoretical models were generally closer to the market prices for different strike prices. To further investigate the pricing performance of the four models, the right panels of Figures 1–10 show the relative price differences (relative errors) between the theoretical model prices and market prices.[5] For simplicity, we refer to a call option as deep out-of-the-money (DOTM) if $S/K \leq 0.90$; out-of-the-money (OTM) if $0.90 < S/K \leq 0.97$; at-the-money (ATM) if $0.97 < S/K \leq 1.03$; in-the-money (ITM) if $0.97 < S/K \leq 1.10$; and deep in-the-money (ITM) if $1.10 < S/K$. Moreover, we considered options less than 60 days to expiration as short term; options with 60–120 days to expiration as medium term; and options larger than 120 days to expiration as long term. For the options with 11, 16, 21, 26, 31, and 51 trading days to expiry, the relative pricing errors produced by our proposed model were all significantly lower than those of 2-FSV, 2-SV, and 2-SVJ models in the case of DOTM options, while the relative errors of all models were slightly higher.

It is also worth noting that the pricing performance of the stochastic model with jump behavior was much better than that of the model without jump in the case of deep out-of-money. For the options with 71, 96, 116, and 181 trading days to expiry, we did not find the same conclusions as the above short term options. In conclusion, the pricing performance of equity option valuation model considering market and idiosyncratic volatility and jump risks was significantly improved for short term and DOTM options.

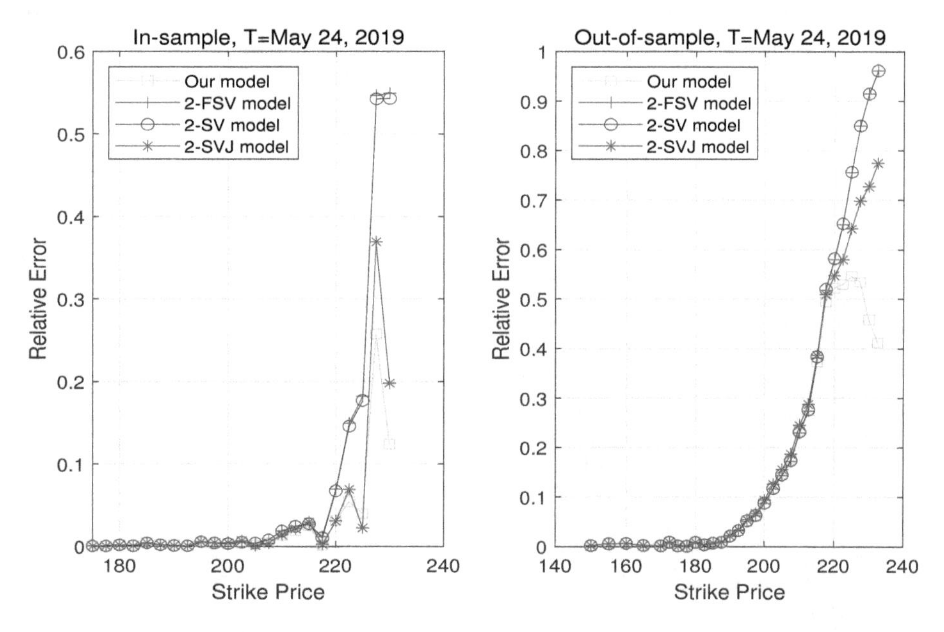

Figure 1. The comparison of predicted prices of four model specifications and market prices on 9 May 2019, with maturity T = 24 May 2019.

[5] The relative error is defined by $\frac{|C_{model} - C_{market}|}{C_{market}} \times 100\%$, where C_{model} and C_{market} denote the theoretical model option prices and the real market prices, respectively.

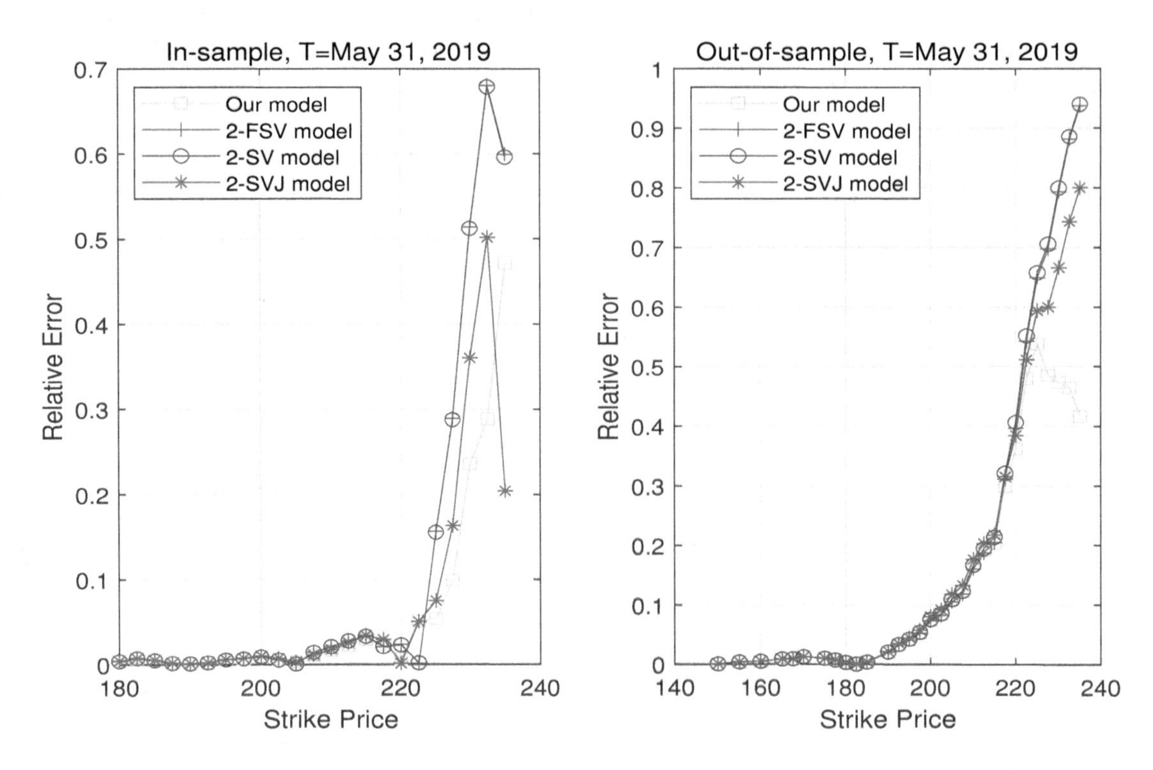

Figure 2. The comparison of predicted prices of four model specifications and market prices on 9 May 2019, with maturity T = 31 May 2019.

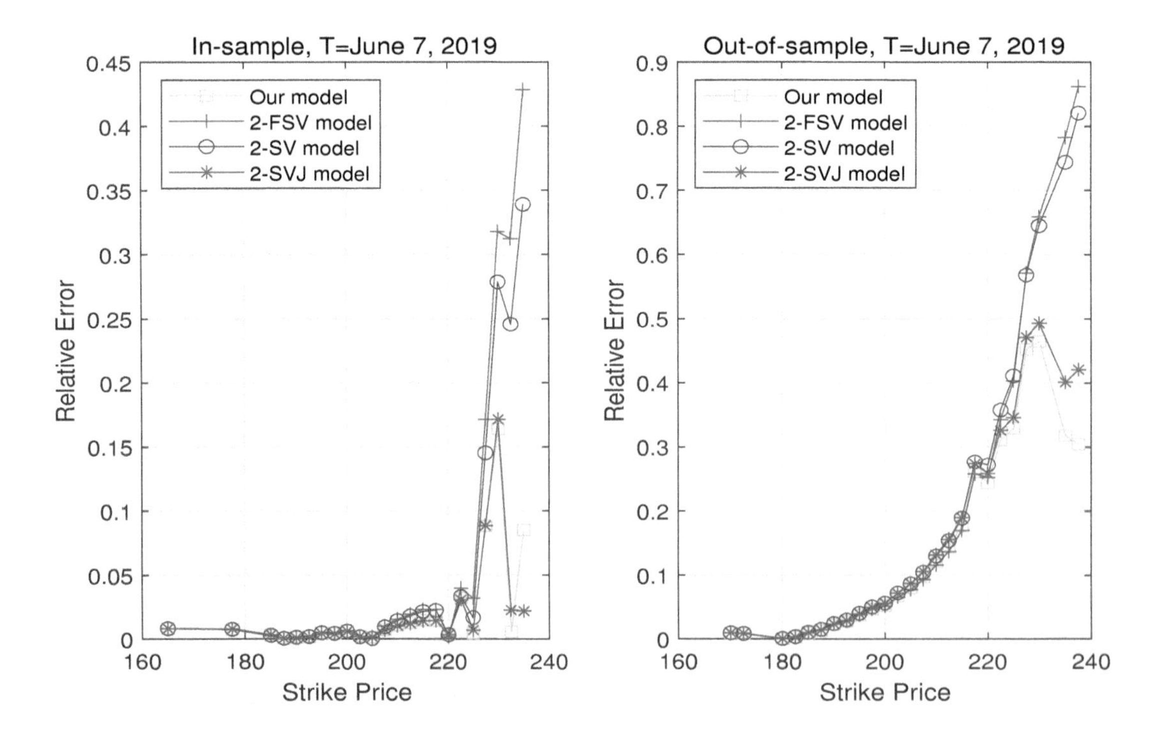

Figure 3. The comparison of predicted prices of four model specifications and market prices on 9 May 2019, with maturity T = 7 June 2019.

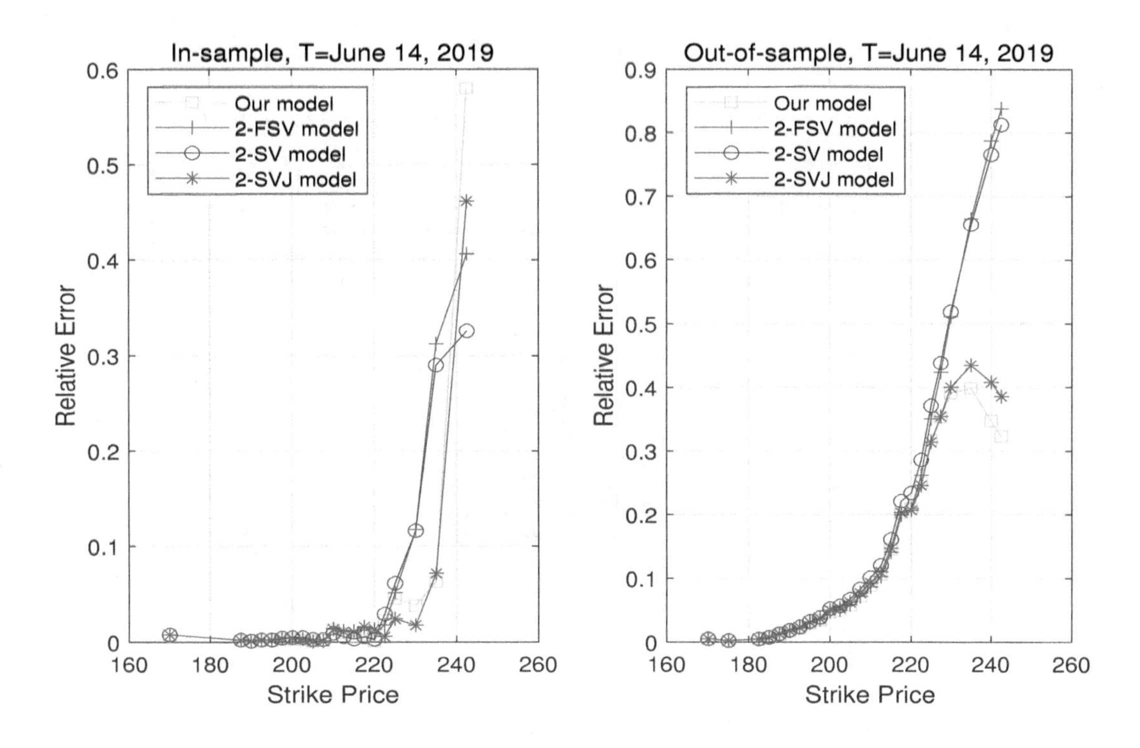

Figure 4. The comparison of predicted prices of four model specifications and market prices on 9 May 2019, with maturity T = 14 June 2019.

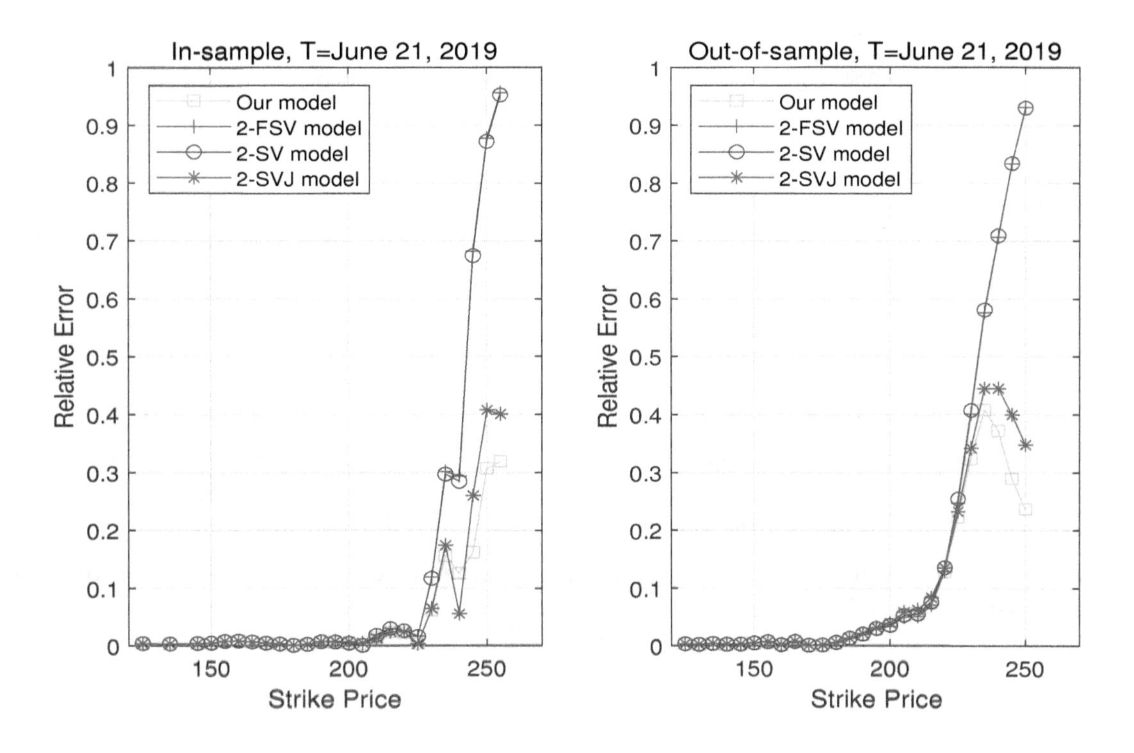

Figure 5. The comparison of predicted prices of four model specifications and market prices on 9 May 2019, with maturity T = 21 June 2019.

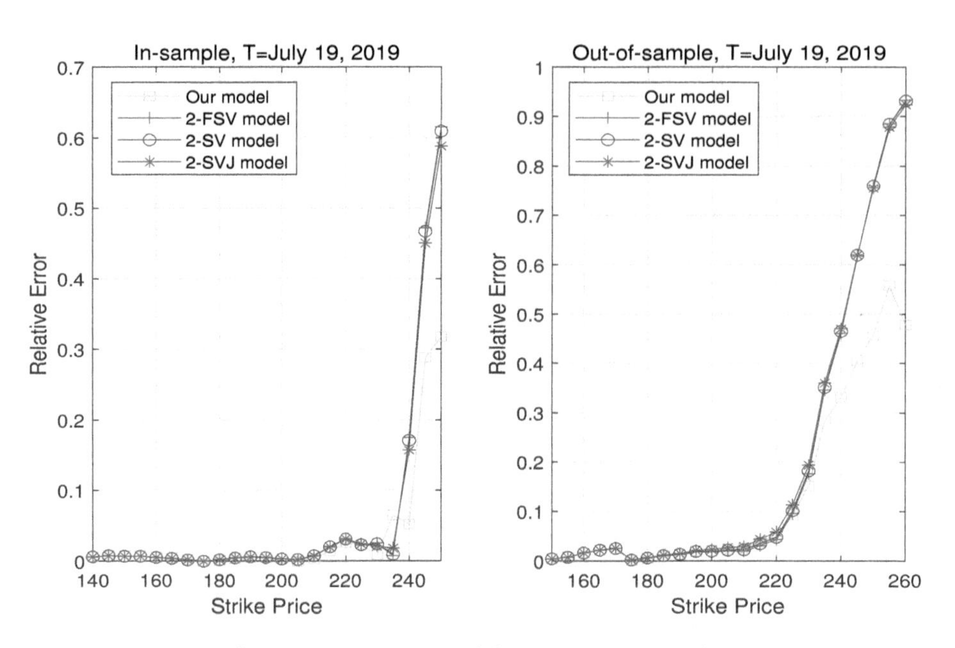

Figure 6. The comparison of predicted prices of four model specifications and market prices on 9 May 2019, with maturity T = 19 July 2019.

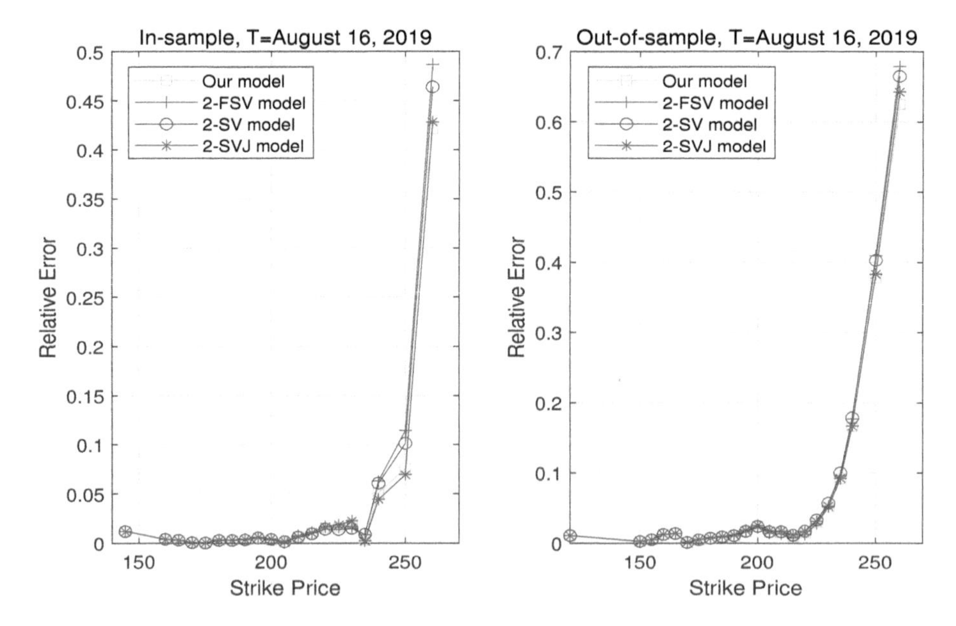

Figure 7. The comparison of predicted prices of four model specifications and market prices on 9 May 2019, with maturity T = 16 August 2019.

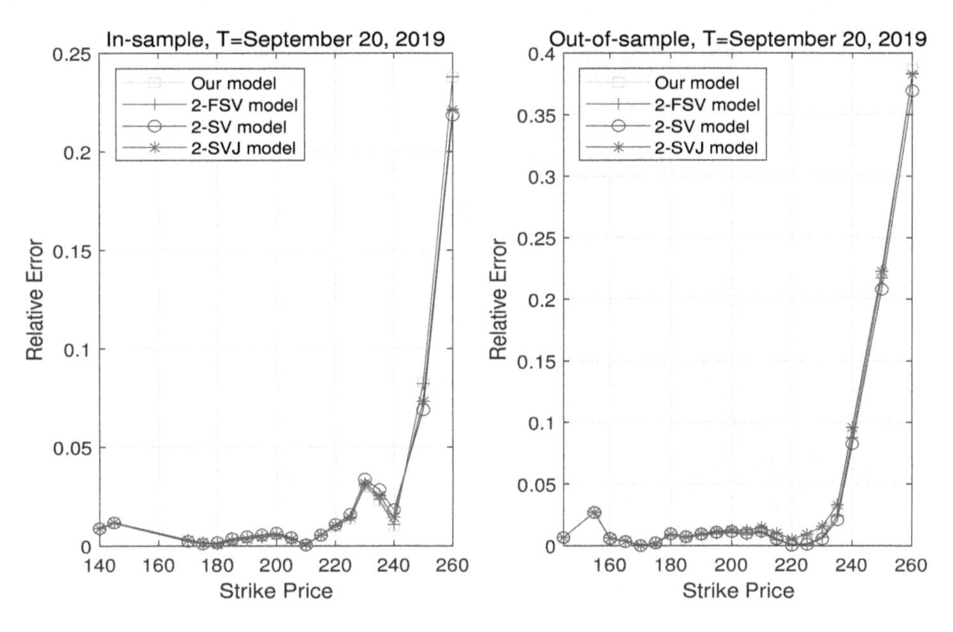

Figure 8. The comparison of predicted prices of four model specifications and market prices on 9 May 2019, with maturity $T = 20$ September 2019.

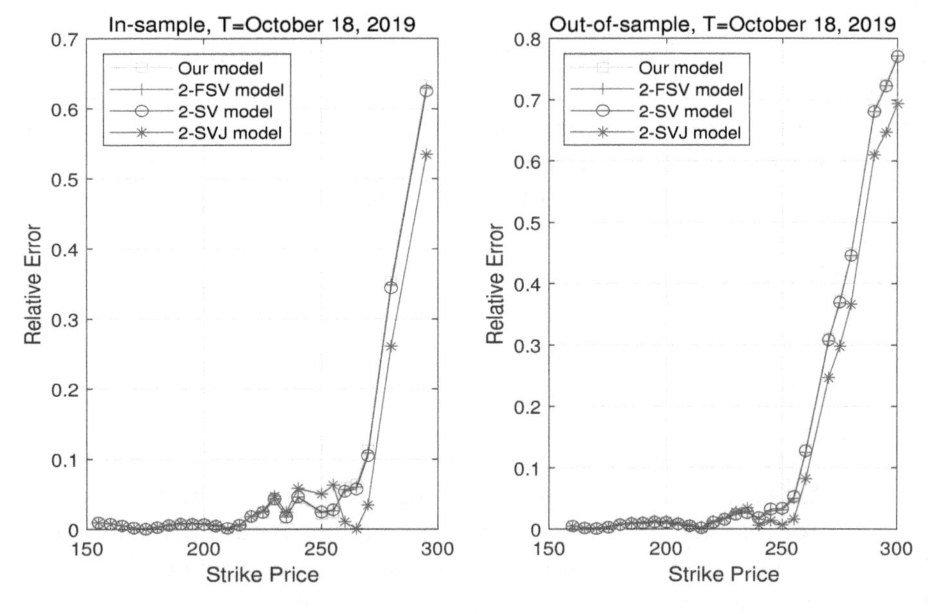

Figure 9. The comparison of predicted prices of four model specifications and market prices on 9 May 2019, with maturity $T = 18$ October 2019.

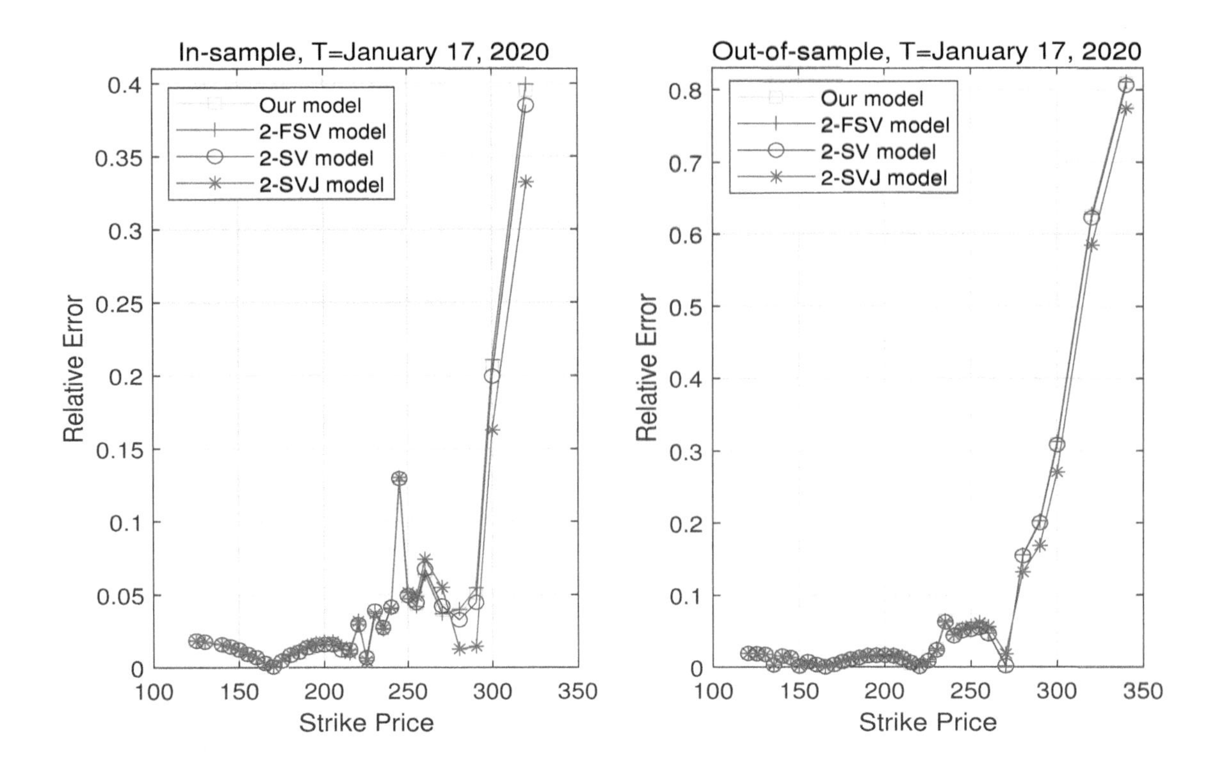

Figure 10. The comparison of predicted prices of four model specifications and market prices on 9 May 2019, with maturity T = 17 January 2019.

To summarize the model calibration results, we also adopted the RMSE as a measure of the goodness of fit. Table 2 reports the out-of-sample pricing errors for the four models across different maturities. Note from Table 2 that our proposed model generally outperformed the other three models in terms of out-of-sample pricing errors. In fact, the same was true for in-sample, whose pricing errors were generally lower than those of the out-of-sample. We will not repeat them here. To measure the extent to which a model was better or worse than another, we defined the improvement rate as the relative differences between the pricing errors from the benchmark model and our proposed model, i.e.,

$$\text{Improvement rate} = \frac{\text{RMSE}_{benchmark} - \text{RMSE}_{our}}{\text{RMSE}_{benchmark}} \times 100\%$$

where RMSE_{our} and $\text{RMSE}_{benchmark}$ denote the RMSE implied by our model and benchmark model, respectively. A positive (or negative) value of improvement rate meant that our model yielded lower (or higher) pricing errors than benchmark model, implying that the pricing performance of the former was better (or worse) than that of the latter by a percentage of that value.

From the last column of Table 2, we can see that our model was superior to the 2-SVJ model across different maturities, which meant that it was necessary to consider the market factor structure in equity option pricing. From the third last column of Table 2, the improvement rate indicated that our model slightly outperformed the 2-FSV model in terms of short term options, but was worse than that of both medium and long term. In spite of this, our empirical study presented here could at least illustrate that the equity option pricing model considering systematic and idiosyncratic volatility and jump risks may offer a good competitor of the models of Bates (2000), Christoffersen et al. (2009), or Christoffersen et al. (2018) for some other equity option markets.

Table 2. Out-of-sample pricing errors. Note: This table shows the out-of-sample pricing errors across different maturities. Pricing errors are reported as the root mean squared errors (RMSE) of option prices for four models.

RMSE	Our	2-FSV	2-SV	2-SVJ	Improvement Rate		
Maturity					Our vs. 2-FSV	Our vs. 2-SV	Our vs. 2-SVJ
24 May 2019	0.2573	0.2574	0.2596	0.2707	0.0373%	0.8803%	4.9568%
31 May 2019	0.2507	0.2508	0.2564	0.2652	0.0392%	2.2499%	5.4846%
7 June 2019	0.2343	0.2347	0.2527	0.2474	0.1764%	7.2947%	5.3044%
14 June 2019	0.1992	0.2041	0.2261	0.2099	2.4278%	11.9155%	5.0858%
21 June 2019	0.1824	0.1827	0.1873	0.1916	0.1399%	2.5963%	4.7934%
19 July 2019	0.3256	0.3301	0.3326	0.3383	1.3434%	2.0948%	3.7368%
16 August 2019	0.2856	0.2835	0.2879	0.2922	−0.7573%	0.7946%	2.2384%
20 September 2019	0.3177	0.3159	0.3162	0.3222	−0.5932%	-0.4851%	1.4002%
18 October 2019	0.1185	0.1180	0.1215	0.1272	−0.4458%	2.4886%	6.8593%
17 January 2020	0.4882	0.4882	0.4893	0.4943	−0.0071%	0.2182%	1.2201%

4. Conclusions

In Christoffersen et al. (2018), the issues of the equity volatility levels, skews, and term structures were investigated by using equity option prices and the principal component analysis method. Their empirical results indicated that the equity options had a strong factor structure, and then, they developed an equity option pricing model with a CAPM factor structure and stochastic volatility. In addition, jumps in stock returns of individual firms were triggered by either systematic events or idiosyncratic shocks. Some recent studies indicated that idiosyncratic jumps were a key important determinant of expected stock; see, for example, Xiao and Zhou (2018), Kapadia and Zekhnini (2019) and Bégin et al. (2020).

Motivated by these insights, we developed a novel model for pricing individual equity options that incorporated a market factor structure, which could be seen as a generalized version of the work by Christoffersen et al. (2018). Specifically, in our model, the individual equity prices were driven by the market factor, as well as an idiosyncratic component that also had stochastic volatility and jump. Due to our model belonging to the affine class, we derived the closed-form solutions for the prices of both the market index and individual equity options by utilizing the Fourier inversion. In addition, we provided the empirical results to test the pricing performance of our proposed factor model based on the S&P 500 index and the AAPL stock on options. Toward this end, we empirically compared the pricing performance of our proposed model with those of the other three classical two factor stochastic volatility models being taken as benchmark models. The out-of-sample pricing performance of equity option valuation model considering market and idiosyncratic volatility and jump risks was significantly improved for short term and DOTM options. In conclusion, the empirical results presented here at least confirmed that the equity option pricing model considering systematic and idiosyncratic volatility and jump risks may offer as good competitor of the models of Bates (2000), Christoffersen et al. (2009), or Christoffersen et al. (2018) for some other option markets.

References

Andersen, Torben G., Nicola Fusari, and Viktor Todorov. 2015. The risk premia embedded in index options. *Journal of Financial Economics* 117: 558–84. [CrossRef]

Bakshi, Gurdip, Charles Cao, and Zhiwu Chen. 1997. Empirical performance of alternative option pricing models. *Journal of Finance* 52: 2003–49. [CrossRef]

Bakshi, Gurdip, Nikunj Kapadia, and Dilip Madan. 2003. Stock return characteristics, skew laws, and the differential pricing of individual equity options. *Review of Financial Studies* 16: 101–43. [CrossRef]

Bakshi, Gurdip, and Nikunj Kapadia. 2003a. Delta-hedged gains and the negative market volatility risk premium. *Review of Financial Studies* 16: 527–66.

Bakshi, Gurdip, and Nikunj Kapadia. 2003b. Volatility risk premiums embedded in individual equity options: Some new insights. *Journal of Derivatives* 11: 45–54. [CrossRef]

Bardgett, Chris, Elise Gourier, and Markus Leippold. 2019. Inferring volatility dynamics and risk premia from the S&P 500 and VIX markets. *Journal of Financial Economics* 131: 593–618. [CrossRef]

Bates, David S. 1996. Jumps and stochastic volatility: Exchange rate processes implicit in Deutsche mark options. *Review of Financial Studies* 9: 69–107. [CrossRef]

Bates, David S. 2000. Post-'87 crash fears in the S&P 500 futures option market. *Journal of Econometrics* 94: 181–238. [CrossRef]

Broadie, Mark, Mikhail Chernov, and Michael Johannes. 2007. Model specification and risk premia: Evidence from futures options. *Journal of Finance* 62: 1453–90. [CrossRef]

Bégin, Jean-François, Christian Dorion, and Geneviève Gauthier. 2020. Idiosyncratic jump risk matters: Evidence from equity returns and options. *Review of Financial Studies* 33: 155–211.

Carr, Peter, and Dilip B. Madan. 2012. Factor models for option pricing. *Asia-Pacific Financial Markets* 19: 319–29. [CrossRef]

Cheang, Gerald H. L., Carl Chiarella, and Andrew Ziogas. 2013. The representation of American options prices under stochastic volatility and jump-diffusion dynamics. *Quantitative Finance* 13: 241–53. [CrossRef]

Cheang, Gerald H. L., and Len Patrick Dominic M. Garces. 2019. Representation of exchange option prices under stochastic volatility jump-diffusion dynamics. *Quantitative Finance*. [CrossRef]

Christoffersen, Peter, Kris Jacobs, and Chayawat Ornthanalai. 2012. Dynamic jump intensities and risk premiums: Evidence from S&P 500 returns and options. *Journal of Financial Economics* 106: 447–72.

Christoffersen, Peter, Mathieu Fournier, and Kris Jacobs. 2018. The factor structure in equity options. *Review of Financial Studies* 31: 595–637. [CrossRef]

Christoffersen, Peter, Steven Heston, and Kris Jacobs. 2009. The shape and term structure of the index option smirk: Why multifactor stochastic volatility models work so well. *Management Science* 55: 1914–32. [CrossRef]

Duffie, Darrell, Jun Pan, and Kenneth Singleton. 2000. Transform analysis and asset pricing for affine jump diffusions. *Econometrica* 68: 1343–76. [CrossRef]

Eraker, Biørn, Miichael Johannes, and Nicholas Polson. 2003. The Impact of Jumps in Volatility and Returns. *Journal of Finance* 58: 1269–1300. [CrossRef]

Fouque, Jean-Pierre, and Adam P. Tashman. 2012. Option pricing under a stressed-beta model. *Annals of Finance* 8: 183–203. [CrossRef]

Fouque, Jean-Pierre, and Eli Kollman. 2011. Calibration of stock betas from skews of implied volatilities. *Applied Mathematical Finance* 18: 119–37. [CrossRef]

Kapadia, Nishad, and Morad Zekhnini. 2019. Do idiosyncratic jumps matter? *Journal of Financial Economics* 131: 666–92. [CrossRef]

Kou, Steven G. 2002. A jump-diffusion model for option pricing. *Management Science* 48: 1086–101. [CrossRef]

Merton, Robert C. 1976. Option pricing when underlying stock returns are discontinuous. *Journal of Financial Economics* 3: 125–44. [CrossRef]

Wong, Hoi Ying, Edwin Kwan Hung Cheung, and Shiu Fung Wong. 2012. Lévy betas: Static hedging with index futures. *Journal of Futures Markets* 32:1034–59. [CrossRef]

Xiao, Xiao, and Chen Zhou. 2018. The decomposition of jump risks in individual stock returns. *Journal of Empirical Finance* 47: 207–28.

11

Comparison of Financial Models for Stock Price Prediction

Mohammad Rafiqul Islam and Nguyet Nguyen *

Department of Mathematics and Statistics, Youngstown State University, Youngstown, OH 44555, USA; mislam02@student.ysu.edu
* Correspondence: ntnguyen01@ysu.edu

Abstract: Time series analysis of daily stock data and building predictive models are complicated. This paper presents a comparative study for stock price prediction using three different methods, namely autoregressive integrated moving average, artificial neural network, and stochastic process-geometric Brownian motion. Each of the methods is used to build predictive models using historical stock data collected from Yahoo Finance. Finally, output from each of the models is compared to the actual stock price. Empirical results show that the conventional statistical model and the stochastic model provide better approximation for next-day stock price prediction compared to the neural network model.

Keywords: stock price prediction; auto-regressive integrated moving average; artificial neural network; stochastic process-geometric Brownian motion; financial models

---

1. Introduction

Predicting modeling is one of the most popular mathematical methods in many fields such as business, social science, engineering, and finance. In business, predictive modeling is also known as predictive analytics. Among many, one of the most important applications of predictive modeling is to predict the stock price. Modern predictive modeling can be categorized into two basic categories such as statistical and soft computing techniques (Adebiyi et al. 2014). Autoregressive integrated moving average (ARIMA) is one of the most popular and widely used statistical techniques for making predictions using past observations (Meyler et al. 1998). In spite of having great popularity in making predictions, this method has some limitations such as seasonality, non-stationarity, and other factors (Tambi 2005). In contrast, as a machine learning method or soft computing technique, artificial neural networks (ANNs) are one of the most accurate and widely used forecasting models for forecasting, pattern recognition, and image processing (Khashei and Bijari 2010). Neural network models have become more popular in forecasting over the last decade in business, economics, and finance (Avcı 2007). According to Khashei and Bijari (2010), ANNs are distinguished and most effective for predictive modeling because of their data-driven self-adaptive nature and they are universal function approximators. The network can generalize, this means that once the network learns the data, it can predict the unseen or future part of the data even if the given data is not smooth.

In addition to the above two methods, stochastic modeling that uses geometric Brownian motion to predict the stock price is very popular. Brownian motion is a special type of motion of molecular particles, first observed and described by the British-Scottish botanist in 1827. However, Louis Bachelier, a French mathematician named this *Brownian motion* and proposed a model to predict stock prices using Brownian motion in 1900. According to the geometric Brownian motion model, the returns on a certain stock in successive, equal periods of time are independents and normally distributed (Dmouj 2006). The equation of geometric Brownian motion has a constant volatility and drift, but in real-world scenario these are not constant and vary over time (Estember and Maraña 2016). Hence, we consider time variant volatility and drift in our analysis.

There are many researchers using the three basic techniques: ARIMA, ANN, and stochastic models to predict stock prices, which will be reviewed in the next section. However, in the literature, there are no comparisons of using each of the three models to predict prices of one stock. Most of the researchers compared performances of the two models ARIMA and ANN in stock price predictions, but not all of the three methods. Therefore, in this paper, we build predictive models using all of the above three modeling techniques and compare the models' performance for stock price predictions, which are discussed in the subsequent sections. Section 2 represents the literature review and related works. In Section 3, we describe the general theories for each of the methods and then build the models specifically for S & P 500 index. In Section 4, we describe the results from each of the three models and model diagnostics. Section 5 contains the conclusions.

2. Literature Review

Prediction has long been a popular field in mathematical science, so there is plenty of related research in the field. The first significant study of neural network models for stock price prediction was done by (White 1988). His predictive model was based on IBM's daily common stock and the training predictions were very optimistic. Thereafter, a lot of research was performed to check the neural networks' accuracy of prediction to forecast the stock market. Hassan et al. (2007) proposed a fusion model by combining the hidden Markov model (HMM), artificial neural network (ANN), and genetic algorithms (GA) to forecast financial market behavior. They found that the performance of the fusion tool is better than that of the basic model (Hassan and Nath 2005) where they used only a single HMM. They also indicated that the performance of the fusion model is similar to that of the ARIMA model. Zhang and Wu (2009) proposed an integrated model improved bacterial chemotaxis optimization (IBCO) and back propagation artificial neural network to predict the S & P 500 index. The IBCO based back propagation (or IBCO-BP) model is less computationally complex and has better accuracy. Khashei and Bijari (2010) found that the performance of a neural network for some real time series is not satisfactory. Hence, using ARIMA models, they suggested a novel hybrid type of artificial neural network. The proposed model provided better predictions for three separate actual datasets than just the neural artificial network model. Yao et al. (1999) compared the back propagation neural network model and ARIMA model stock index forecasting. They found that the neural network results in better accuracy in forecasting than the traditional ARIMA models. Adebiyi et al. (2014) compared the forecasting performance by ARIMA and artificial neural network for stock data. They analyzed daily stock prices for the Dell Incorporation and found a superiority of the neural network model over the ARIMA model.

Merh et al. (2010) developed a three-layer feed-forward neural network model and auto-regressive integrated moving average model to predict the future value of the stock price and revealed that the ARIMA models perform better over ANN models. Lee et al. (2007) did a comparative study of the forecasting performance by neural network models and the time series model (SARIMA) for the Korean Stock Index data. They also found ARIMA models outperforming ANN models for the stock price prediction. Agustini et al. (2018) used several stock indexes under the Jakarta Corporate Index to build a predictive model with Brownian motion. They found a higher accuracy for prediction with a mean absolute percentage error (MAPE) less than 20%. Rathnayaka et al. (2014) developed a forecasting model using the geometric Brownian motion model and compared the predictions with the results from the traditional time series model ARIMA. They used the Colombo Stock Exchange (CSE), Sri Lanka data to build their models and found that the stochastic model prediction is more significant than the traditional model.

The literature shows different opinions on the relative performances of the two of the three models depending on data. Hence, further comparative studies of all the three models can assemble a consistent methodology for stock price prediction. In this paper, we study the comparative performances of the three models in predicting next-day stock prices for S & P 500 index data from Yahoo Finance.

3. Methodology

The methodology section contains the basic four subsections. The first subsection describes the data that were used to build the models. Then, each of the subsections describes general theories and procedures to build the models and then how the models were fitted for a particular dataset. The overall performance of each of the models was checked by the analysis of the residuals and four different error measures, namely the absolute percentage error (APE), the average absolute error (AAE), the average relative percentage error (ARPE), and the root-mean-square error (RMSE) (Nguyet Nguyen and Wakefield 2018). The formula to calculate these errors are as follows:

$$\begin{aligned} APE &= \frac{1}{\bar{r}} \sum_{i=1}^{N} \frac{r_i - \hat{r}_i}{N}, \\ AEE &= \sum_{i=1}^{N} \frac{r_i - \hat{r}_i}{N}, \\ ARPE &= \frac{1}{N} \sum_{i=1}^{N} \frac{r_i - \hat{r}_i}{N}, \\ RMSE &= \sqrt{\frac{1}{N} \sum_{i=1}^{N} \frac{r_i - \hat{r}_i}{N}}. \end{aligned} \tag{1}$$

3.1. Data

S & P 500 daily stock for the period 1 January 2015 to 31 December 2019 was used in this research. We used the quantmod package (Ryan et al. 2020) in statistical software R, version 1.2.1335 to collect the data directly from *Yahoo Finance*. Initially, the dataset contains six variables, namely daily *Open*, *High*, *Low*, *Close*, *Volume*, and *Adjusted Close* price. Addition to the six variables, we created two more variables, i.e., *Average* and *Return*. The *Average* variable is the average of daily *Open*, *High*, *Low*, and *Close* price. All the predictive models were built to predict the *Adjusted Close* price for the next day on the basis of the present day's predictor variables. There were 63 trading days per quarter in 2019. All the models were used to predict the next-day stock price for the last quarter of 2019. A total of 63 predictions were made.

3.2. Autoregressive Integrated Moving Average Process

3.2.1. ARIMA(p,d,q) Models

The time series analysis requires the stationarity of the data, meaning the statistical properties such as mean, variance, and so on do not change over time. However, most of the real-world data, like stock data, are non-stationary by nature. This non-stationarity can be taken care of by using the Box–Jenkins *ARIMA(p,d,q)* approach (Makridakis and Hibon 1997). A time series $\{Y_t\}$ is said to follow an *Autoregessive Integrated Moving Average ARIMA(p,d,q)* if the dth difference $W_t = \nabla^d Y_t$ is a stationary $ARMA(p,q)$ process (Cryer and Kellet 1991). A generalised $ARIMA(p,d,q)$ model can be written as

$$W_t = \phi_1 W_{t-1} + \phi_2 W_{t-2} + \ldots + \phi_p W_{t-p} + e_t + \theta_1 e_{t-1} + \theta_2 e_{t-2} + \ldots + \theta_q e_{t-q},$$

where $\phi_1, \phi_2, \ldots, \phi_p$ and $\theta_1, \theta_2, \ldots, \theta_q$ are the autoregressive and moving average parameters, respectively, and $e's$ are the white noise. The autoregression $AR(p)$, order p, and moving average $MA(q)$, order q, are determined from the analysis of the autocorrelation function. The number d indicates the number of differences applied to the time series to remove the trend. The autoregressive parameters $\phi's$ and moving average parameters $\theta's$ are estimated from the model based on $p, d,$ and q.

3.2.2. ARIMA(p,d,q) Model for S & P 500 Index

The adjusted close price of S & P 500 is a time series process $\{X_t\}$ that has been analysed to build the model. The process $\{X_t\}$ is not a stationary process, if we see the following graphs.

Figure 1 shows an upward trend in the data. Inspecting the sample autocorrelation plot from Figure 2, it is clear that the Auto Correlation Function (ACF) dies down very slowly and the Partial Autocorrelation Function (PACF) cuts off at lag 1 with correlation one.

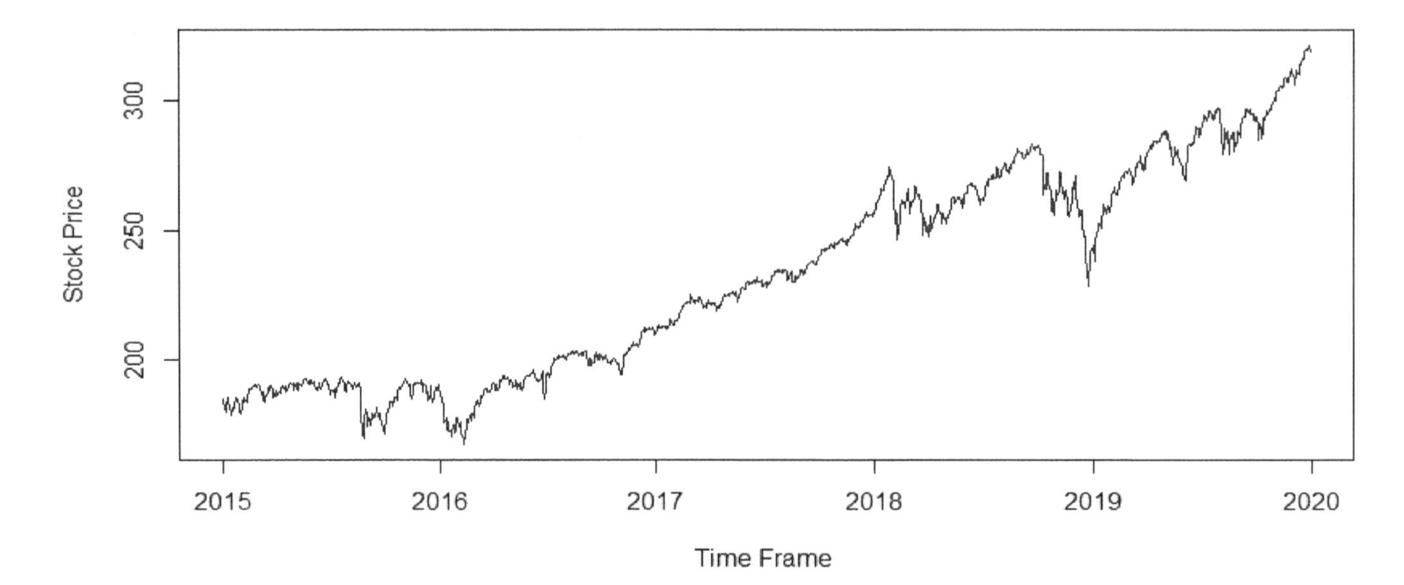

Figure 1. Time plot of the raw data.

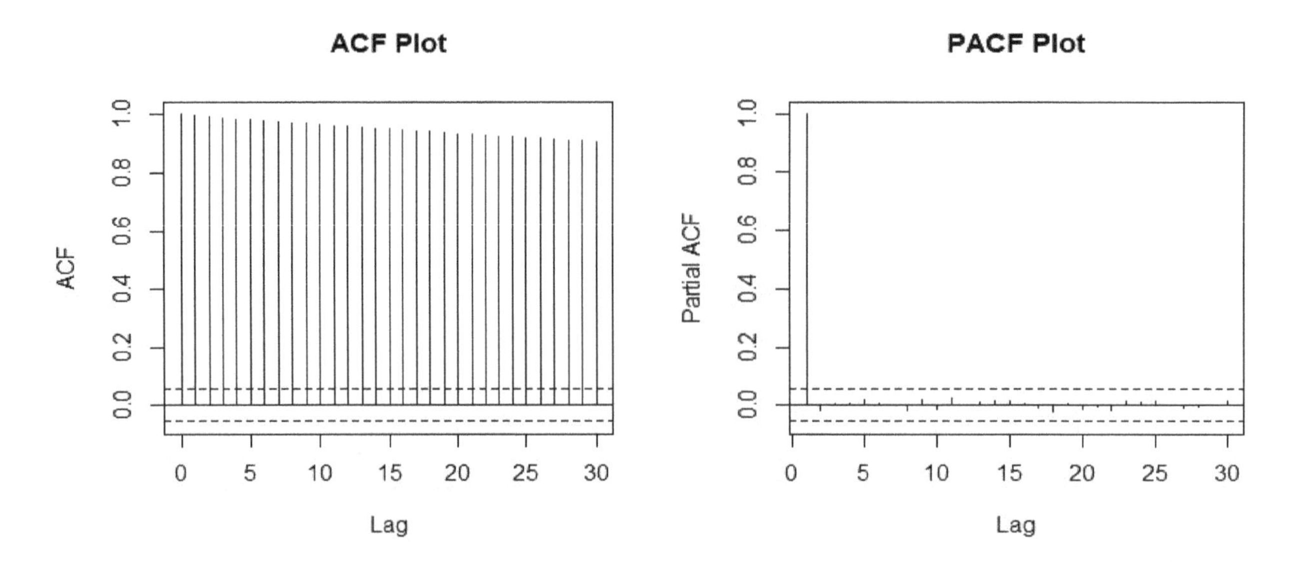

Figure 2. Sample autocorrelation plot.

The slow dying-down nature of the ACF indicates that the process is non-volatile. That is, the current value is relating with all the past values. These facts ensure the process is non stationary. To make the process stationary, we transform the $\{X_t\}$ series to the $\{Y_t\} = \{\log X_t\}$ series and then to a new series $\{W_t\} = \{\nabla^2 Y_t\}$, where $W_t = Y_t - 2Y_{t-1} + Y_{t-2}$.

Figure 3 is the window plot of the second differenced log transformed stock price. From this plot, the data looks stationary and randomized. Stationarity has been confirmed from the augmented Dickey–Fuller test (Cheung and Lai 1995) with a p value of 0.01, where the alternative hypothesis was stationary. Next, the autoregressive and moving average orders p and q were determined from the PACF and ACF plot from Figure 4.

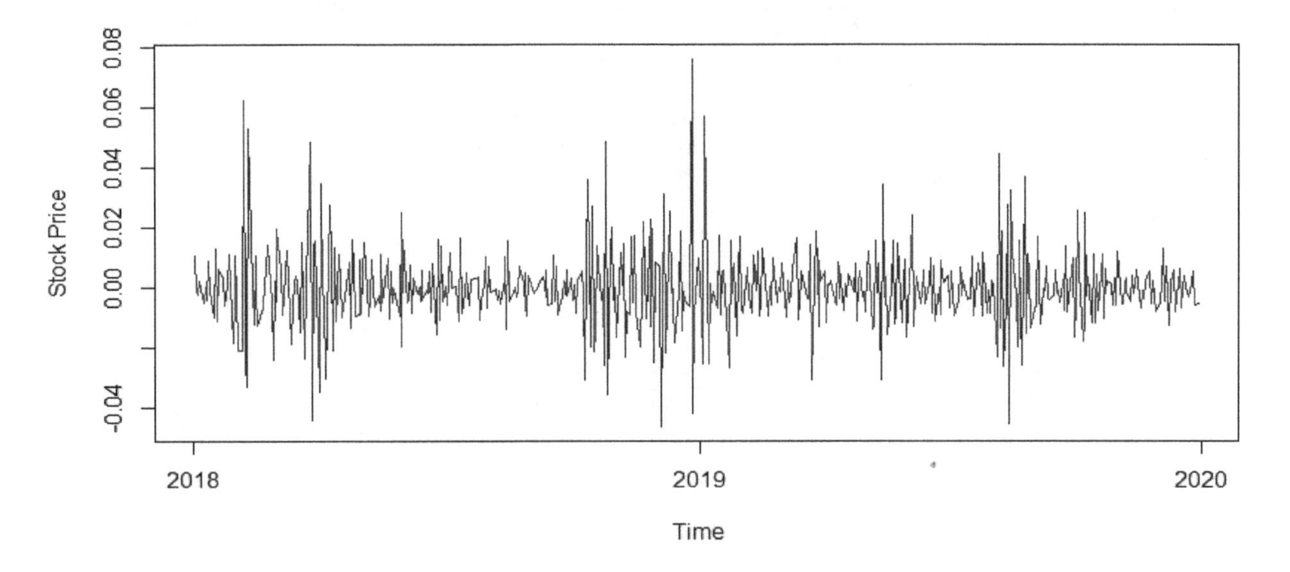

Figure 3. Window plot of the second differenced log transformed stock price.

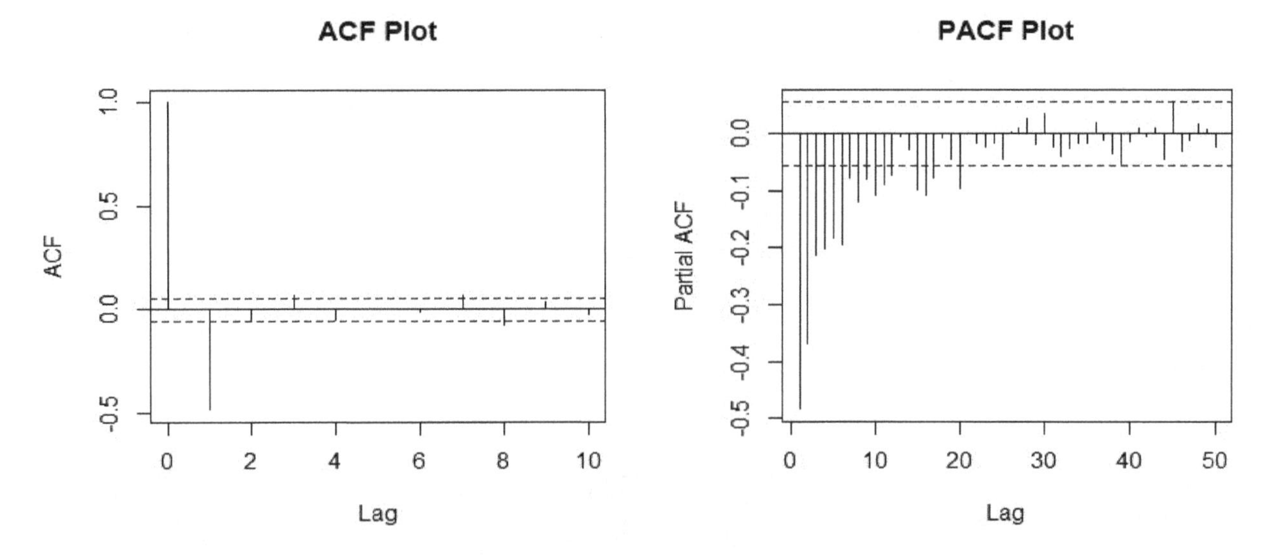

Figure 4. ACF and PACF of the second differenced log transformed stock price.

The ACF cuts of at lag 1 which indicates that the process incorporates an MA process of order $q = 1$ whereas the PACF gradually dies down. Therefore, the series W_t follows an $MA(1)$ process or the series Y_t follows an $IMA(2,1)$ process i.e. $Y_t \sim ARIMA(0,2,1)$. Other $ARIMA(p,d,q)$ models were also considered in this research, as shown in Table 1. The best model has been chosen based on the *Schwarz Bayesian Information Criterion* (BIC) (Neath and Cavanaugh 2012) criteria, the more negative, the more accurate model. The reason of not choosing *Akaike Information Criterion* (AIC) or *Bias Corrected Akaike Information Criterion* (AICc) is that those models provide over-fitting and non significant parameters.

From Table 1, the $ARIMA(0,2,1)$ model has the most negative BIC value which fits the data most perfectly. The *Arima* function with order $(p,d,q) = (0,2,1)$ was run in RStudio and the summary of the model is displayed in Table 2.

Table 1. ARIMA (p,d,q) model comparison.

Model	AIC	BIC	AICc
ARIMA(0,2,0)	−7144.11	−7139.03	−7144.11
ARIMA(0,2,1)	−7981.48	−7971.31	−7981.47
ARIMA(0,2,2)	−7979.84	−7964.59	−7979.82
ARIMA(0,2,3)	−7981.77	−7961.43	−7981.73
ARIMA(0,2,4)	−7980.02	−7954.61	−7979.97
ARIMA(1,2,0)	−7455.29	−7445.12	−7455.28
ARIMA(1,2,1)	−7979.81	−7964.56	−7979.79
ARIMA(1,2,2)	−7982.16	−7961.83	−7982.13
ARIMA(1,2,3)	−7983.04	−7957.62	−7982.98
ARIMA(1,2,4)	−7981.53	−7951.03	−7981.45
ARIMA(2,2,0)	−7631.73	−7616.48	−7631.71
ARIMA(2,2,1)	−7981.50	−7961.16	−7981.46
ARIMA(2,2,2)	−7982.88	−7957.46	−7982.83
ARIMA(2,2,3)	−7982.62	−7952.12	−7982.55
ARIMA(2,2,4)	−7979.91	−7944.32	−7979.81
ARIMA(3,2,0)	−7692.63	−7672.30	−7692.60
ARIMA(3,2,1)	−7979.84	−7954.42	−7979.79
ARIMA(3,2,2)	−7981.50	−7951.00	−7981.43
ARIMA(3,2,3)	−7977.84	−7942.25	−7977.74
ARIMA(3,2,4)	−7978.07	−7937.40	−7977.95
ARIMA(4,2,0)	−7738.80	−7713.39	−7738.75
ARIMA(4,2,1)	−7980.13	−7949.63	−7980.06
ARIMA(4,2,2)	−7978.69	−7943.11	−7978.60
ARIMA(4,2,3)	−7978.28	−7937.61	−7978.15
ARIMA(4,2,4)	−7984.84	−7939.09	−7984.69

Table 2. ARIMA(0,2,1) model summary.

	Model	**Arima(x = tr.stock, order = c(0, 2, 1))**					
MA(1) Coefficient	−1.00						
Standard Error	0.0027						
Sigma-squared estimated as	0.0000718						
Log likelihood	3992.74						
AIC	−7981.48						
AICc	−7981.47						
BIC	−7971.31						
Training set error measures	ME	RMSE	MAE	MPE	MAPE	MASE	ACF1
	0.00013	0.00846	0.00573	0.00220	0.10561	0.99516	−0.01682

The model in difference equation is given as

$$
\begin{aligned}
W_t &= e_t + \theta e_{t-1} \\
\nabla^2 Y_t &= e_t + \theta e_{t-1} \\
Y_t - 2Y_{t-1} + Y_{t-2} &= e_t + \theta e_{t-1} \\
Y_t &= 2Y_{t-1} - Y_{t-2} + e_t + \theta e_{t-1}. \qquad (2)
\end{aligned}
$$

Finally, substituting the MA parameter $\theta = -1$ in Equation (2), the model for $Y_t = \log X_t$ is given as

$$Y_t = 2Y_{t-1} - Y_{t-2} + e_t - e_{t-1}. \qquad (3)$$

A fixed window of 1194 past observed stock prices have been used to predict each of the next-day prices using the model in Equation (3). Hence, the training dataset moved and the end price of the window was updated with the actual price. The results and diagnostics of this model are discussed in Section 4.1.

3.3. *Stochastic Model Geometric Brownian Motion*

A process that generates some outcomes which are time-dependent but can not be said ahead of time is known as a *stochastic process*. A stochastic process $\{W(t) : 0 \leq t \leq T\}$ is a standard Brownian motion on $[0, T]$ if

1. $W(0) = 0$
2. It has independent increments. That is, for any $t_1, t_2, \ldots, t_n$, $W(t_2) - W(t_1), W(t_3) - W(t_4) \ldots, W(t_n) - W(t_{n-})$ are independent random variables.
3. For every $0 \leq s < t \leq T, W(t) - W(s) \sim \mathbb{N}(0, t-s)$.

A stochastic process $\{X(t) : 0 < t < T\}$ is said to be a general Brownian motion with a drift parameter μ and diffusion coefficient σ^2 if $\frac{X(t)-\mu t}{\sigma}$ is a standard Brownian motion, written as $X(t) \sim BM(\mu, \sigma^2)$.

The general Brownian motion still follows first two properties of the standard Brownian motion. However, the third property is modified as $X(t) - X(s) \sim \mathbb{N}(\mu(t-s), \sigma^2(t-s))$ for any $0 \leq s < t < T$.

3.3.1. Geometric Brownian Motion (GBM) Model

If $X(t) \sim BM(\mu, \sigma^2)$ then $X(t)$ satisfies the stochastic differential equation (Yang and Aldous 2015)

$$dX(t) = \mu t + \sigma dW(t), \tag{4}$$

where, $W(t)$ is the standard Brownian motion or Wiener process. If the stochastic process is defined as $X(t) = \log S(t)$ then $dS(t) = \mu S(t)dt + \sigma S(t)dW(t)$ is the stochastic differential equation for the stock price random process.

For a given time $t > 0$, the standard model for stock prediction can be given from the stochastic differential equation by integration

$$S(t) = S(0) + \mu \int_0^t S(r)dr + \sigma \int_0^t S(r)dW(r). \tag{5}$$

A more explicit formula can be derived using Ito's formula (Ševcovic et al. 2011) to the function $F(\log S(t), t)$

$$dF = \left[\frac{\partial F}{\partial t} + \mu \frac{\partial F}{\partial S(t)} + \frac{1}{2}\sigma^2 \frac{\partial^2 F}{\partial^2 S(t)}\right] dt + \left(\sigma \frac{\partial F}{\partial S(t)}\right) dW(t),$$

which results

$$\begin{aligned} d \log S(t) &= \frac{1}{S(t)} dS(t) + \frac{1}{2} \frac{-1}{S^2(t)} (dS(t))^2 \\ &= \mu dt + \sigma dW(t) + \frac{1}{2} \frac{-1}{S^2(t)} (\mu S(t)dt + \sigma S(t)dW(t))^2 \\ &= (\mu - \frac{1}{2}\sigma^2)dt + \sigma dW(t). \end{aligned}$$

For any time $t > 0$, the differential can be written as

$$\begin{aligned} \log S(t) &= \log S(0) + (\mu - \frac{1}{2}\sigma^2)t + \sigma W(t) \\ \text{Or,} \quad S(t) &= S(0)e^{(\mu - \frac{1}{2}\sigma^2)t + \sigma W(t)}. \end{aligned} \tag{6}$$

3.3.2. Geometric Brownian Motion Model for S & P 500 Index: GBM(μ, σ^2) Simulation

For a given time set, $t_0 = 0 < t_1 < t_2 < \ldots < t_n$, the stock price $S(t)$ at time $t_0, t_1, \ldots, t_n$ can be generated by

$$S(t_{i+1}) = S(t_i)e^{(\mu-\frac{1}{2}\sigma^2)(t_{i+1}-t_i)+\sigma\sqrt{(t_{i+1}-t_i)}Z_{i+1}}, \tag{7}$$

where $Z_1, Z_2, \ldots Z_n$ are independent and identically distributed standard normals and $i = \overline{0, (n-1)}$. In our case, the time interval $t_{i+1} - t_i = 1$ for all $i = \overline{0, (n-1)}$, since we are predicting the next-day price. Hence, the model becomes

$$S(t_{i+1}) = S(t_i)e^{(\mu-\frac{1}{2}\sigma^2)+\sigma Z_{i+1}}. \tag{8}$$

Using the model in Equation (8), we simulate a large number of prices, and from that we take the average to predict the next-day price. For our data, this large number is 100,000. A total of 63 predictions have been made using this model. A fixed window of 1194 past observed stock prices have been used to predict each of the next-day prices. Hence, the training dataset moved, and the end price of the window was updated with the actual price. The results and diagnostics of this model are discussed in Section 4.2.

3.4. *Artificial Neural Network*

This section describes how an artificial neural network can be used to predict the stock price and how to build a model based on the stock data for S & P 500 index.

3.4.1. Model Descriptions

Artificial neural network is one of the most popular machine learning techniques for nonlinear approximations because of its ability to deal with a large number of functions with a high degree of accuracy (Chen et al. 2003). The idea of ANN came from the structure of the animal brain, more specifically, from the human neural system. It is based on the idea of how brain works, how the neurons in the brain receive information from the input neurons, analyse it, and finally identify the object or pattern. Fundamentally, the mechanism has three layers—input layer, hidden layers, and output layer. Each layer consists of neurons or nodes. The hidden part may consist of many layers, however, for the time series analysis and forecasting, the single hidden layer feed forward network is the most widely used model structure (Zhang et al. 1998). A simple three layer neural network has the following mathematical form

$$Y_t = W_0 + \sum_{j=1}^{q} W_j.g(W_{0,j} + \sum_{i=1}^{p} W_{i,j}.Y_{t-i}) + \epsilon_t, \tag{9}$$

where, $W_{i,j}$ and W_j for $i = 1, 2, \ldots, p,\ \ j = 1, 2, \ldots, q$ are known as connection weights. The parameter p and q are the number of input and output nodes respectively. The network involves an activation function which plays a very important role because it converts the input signals to be used for the neurons or nodes in the next layer, eventually the output neuron. The most widely used activation functions are the logistic and hyperbolic functions (Khashei and Bijari 2010), which are shown in Equations (10) and (11)

$$sig(x) = \frac{1}{1 - e^{-x}} \tag{10}$$

$$\tan^{-1}(x) = \frac{1 - e^{-2x}}{1 + e^{-2x}}. \tag{11}$$

Most of the modelers prefer the hyperbolic tangent function as the activation functions because of its faster convergence, and it makes the optimization easier. Hence, we used this activation function in our model. There is no systematic rule of choosing the number of neurons or nodes, q in the hidden layer (Khashei and Bijari 2010). In most of the cases it is data-dependent and chosen on the basis of trial and error.

3.4.2. Artificial Neural Network for S & P 500 Index

The model proposed for S & P 500 in this section is a three layer model—input, hidden, and the output layer. The input layer consists of a total of seven nodes which are daily *Open, Close, High, Low, Average, Volume,* and *Return*. The variable *Average* is the average of daily *Open, Close, High,* and *Low*. The *Volume* was converted to million units. Daily return was calculated by this formula $r_t = \log \frac{S_t}{S_{t-1}}$, where S_t is the adjusted closed price and day one return, r_0 was considered zero. The output layer has only one node that corresponds to the predicting variable *Adjusted Close* price. The number of the nodes in the hidden layer was chosen based on the error measures in Equation (1) for different combinations of the hidden nodes, which are displayed in Table 3. From this Table 3, we see that model ANN(7-15-1) has the lowest APE, AAE, ARPE, and RMSE and the highest adjusted R^2 value.

Table 3. Error measures for different network structures.

MODEL	R^2	APE	AAE	ARPE	RMSE
7-2-1	0.3137	0.0231	7.0435	0.1118	0.3344
7-3-1	0.4158	0.0211	6.412	0.1018	0.319
7-4-1	0.0422	0.028	8.5128	0.1351	0.3676
7-5-1	0.0368	0.0292	8.8796	0.1409	0.3754
7-6-1	0.3716	0.0215	6.5496	0.104	0.3224
7-7-1	0.3431	0.0226	6.8687	0.109	0.3302
7-8-1	0.3907	0.0219	6.6781	0.106	0.3256
7-9-1	0.4036	0.0212	6.4713	0.1027	0.3205
7-10-1	0.4108	0.0214	6.5237	0.1036	0.3218
7-11-1	0.5155	0.0195	5.9284	0.0941	0.3068
7-12-1	0.5392	0.0187	5.6802	0.0902	0.3003
7-13-1	0.4777	0.0194	5.8945	0.0936	0.3059
7-14-1	0.4676	0.0196	5.9702	0.0948	0.3078
7-15-1	0.6216	0.0167	5.0928	0.0808	0.2843
7-16-1	0.5701	0.0182	5.5382	0.0879	0.2965

The original dataset had 1257 observations, but the dataset used in this method was modified in this way—all the predictor and predicting variables have the same length of 1256, however, the predictor variables started from day 1 to the 1256th day and the predicting variable day 2 to the 1257th day. Then, the dataset was divided into two parts to run the model. The test dataset contained the last 63 actual stock prices (adjusted close) which were compared to the predicted prices. The best model was selected on the basis of the adjusted R^2 and four error measures (Table 3). The model architecture is shown in Figure 5 and the result of this model is discussed in Section 4.3.

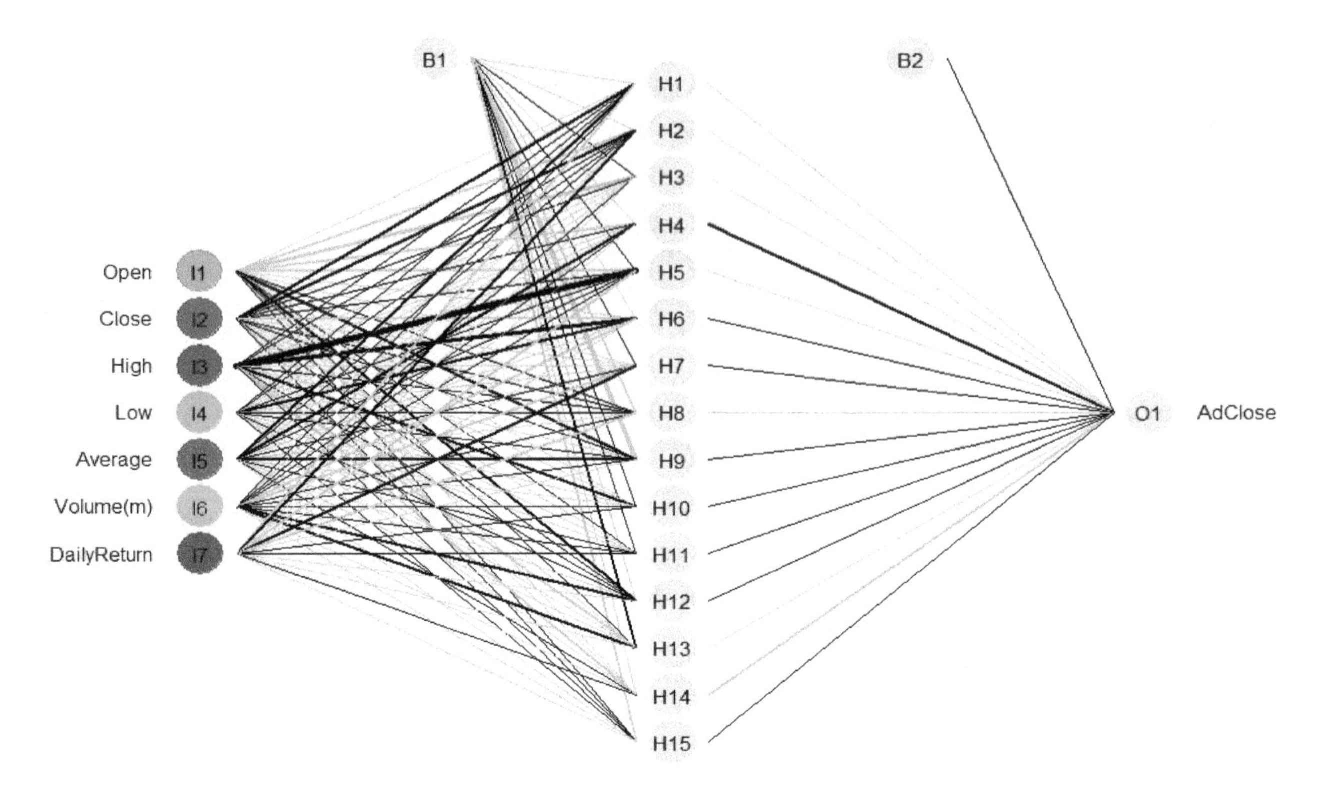

Figure 5. Artificial neural network architecture.

4. Results

In this section, the result of from the above three models is discussed and a window of the predicted and actual price is shown together with a graphical presentation. Finally, we assess how our model is performing by model diagnostics.

4.1. Autoregressive Integrated Moving Average

4.1.1. ARIMA Model Result

The model $ARIMA(0,2,1)$ in Equation (3) produces the prediction in a logarithmic scale, which is then converted back to the original scale by the formula Predicted Price $= e^{prediction}$. Total 63 trading days have been predicted by the model and compared with the actual prices which has been shown in Table 4 with individual prediction error calculated by the formula,

$$\text{error} = \frac{\text{actual} - \text{predicted}}{\text{actual}} \times 100. \tag{12}$$

From Table 4, we see that the forecast errors are less than one dollar for the daily period from 12 November, 2019 to 30 December 2019, with the relative errors within the range of 0.00003 to 0.00292. Figure 6 shows the graphical representation of the actual and predicted stock price by the model. The black line represents the actual stock price and the red line represents the predicted stock price for S & P 500. Figure 6 also shows that the ARIMA (0,2,1) predicted prices follow closely to the trend of the actual prices.

Table 4. Prediction by ARIMA(0,2,1) model.

Date	Actual	Predicted	Error
11-12-2019	305.69	305.17	0.17
11-13-2019	305.79	305.81	0.01
11-14-2019	306.23	305.91	0.10
11-15-2019	308.45	306.35	0.68
11-18-2019	308.68	308.57	0.04
11-19-2019	308.59	308.80	0.07
11-20-2019	307.44	308.71	0.41
11-21-2019	306.95	307.56	0.20
11-22-2019	307.63	307.07	0.18
11-25-2019	310.01	307.75	0.73
11-26-2019	310.72	310.14	0.19
11-27-2019	312.10	310.85	0.40
11-29-2019	310.94	312.23	0.41
12-2-2019	308.30	311.07	0.90
12-3-2019	306.23	308.43	0.72
12-4-2019	308.12	306.36	0.57
12-5-2019	308.68	308.25	0.14
12-6-2019	311.50	308.81	0.86
12-9-2019	310.52	311.64	0.36
12-10-2019	310.17	310.65	0.15
12-11-2019	311.05	310.30	0.24
12-12-2019	313.73	311.18	0.81
12-13-2019	313.92	313.86	0.02
12-16-2019	316.08	314.05	0.64
12-17-2019	316.15	316.21	0.02
12-18-2019	316.17	316.29	0.04
12-19-2019	317.46	316.31	0.36
12-20-2019	318.86	317.60	0.40
12-23-2019	319.34	319.00	0.11
12-24-2019	319.35	319.48	0.04
12-26-2019	321.05	319.49	0.49
12-27-2019	320.97	321.20	0.07
12-30-2019	319.20	321.12	0.60

4.1.2. ARIMA Model Diagnostics

The performance of the *ARIMA(0,2,1)* model was assessed by the analysis of the four error measures state in Equation (1) and the residuals plot, which is depicted in Figure 7 and those four error measures are tabulated in Table 5.

Table 5. Prediction error by ARIMA(0,2,1) model.

MODEL	APE	AAE	ARPE	RMSE
ARIMA(0,2,1)	0.0044	1.3651	0.0217	0.1472

From the results in Table 5, we see that all the error measures are comparatively very low to the actual prices, this indicates that the model is performing better in its prediction. From Figure 7 it is clear that the residuals do not follow any special pattern, they are a randomized plot. Correlations in the few lags are significant. Overall, the model fits very well to predict the stock price.

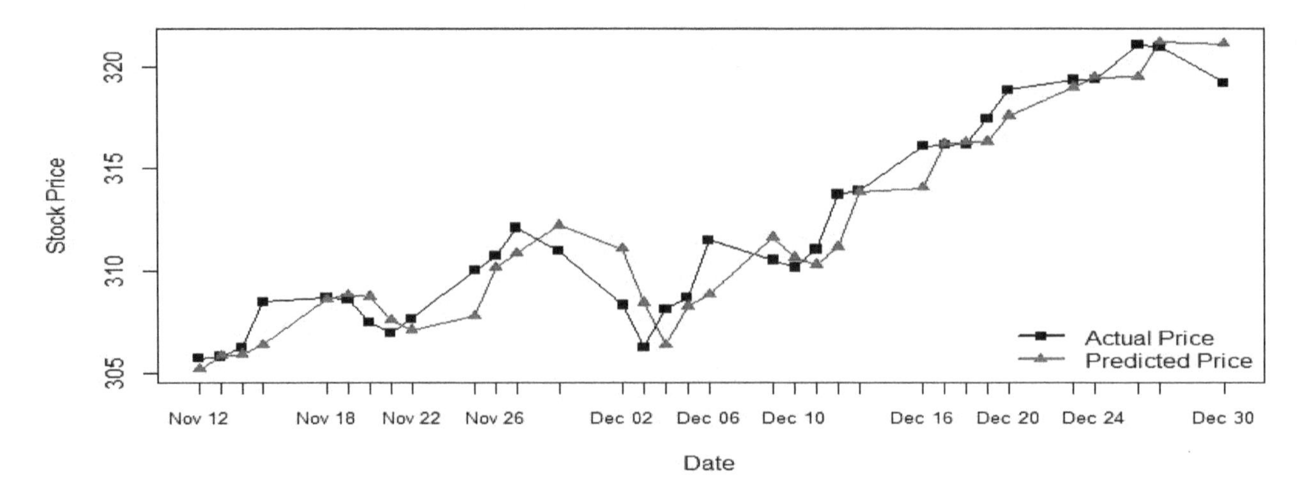

Figure 6. ARIMA (0,2,1) model prediction.

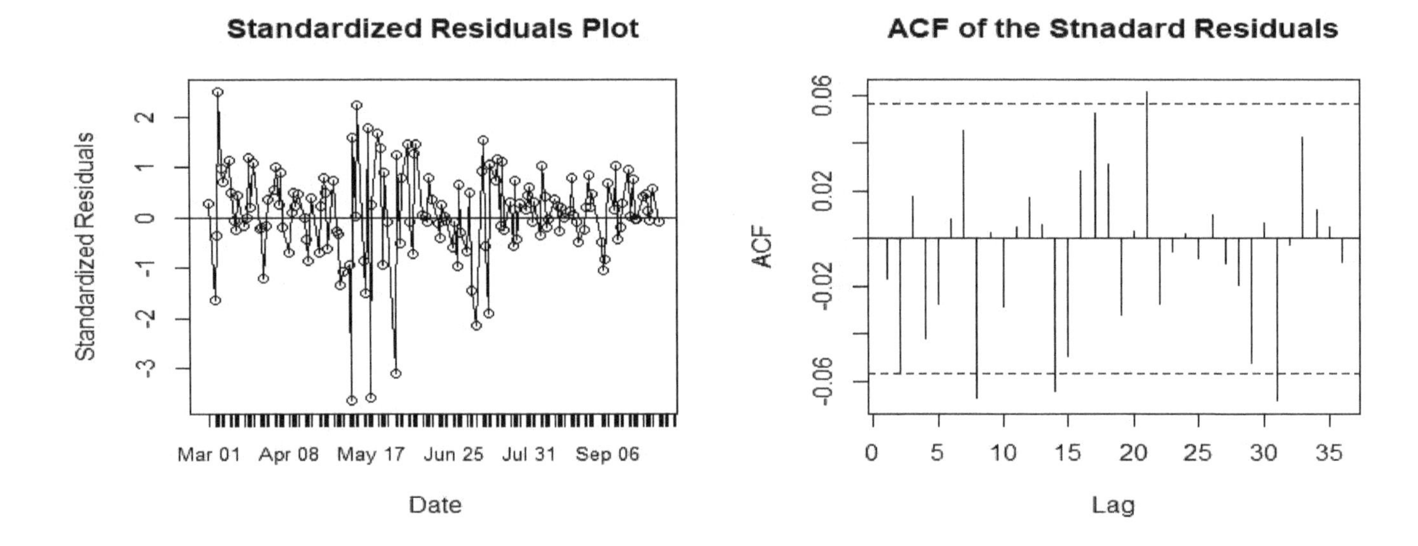

Figure 7. ARIMA(0,2,1) model residual analysis.

4.2. Stochastic Model

4.2.1. Stochastic Model Result

The model proposed in Equation (8) requires the calculation of 63 distinct values of the means μ and standard deviations σ of the daily returns. Both of the parameters were calculated on the basis of the same number of returns each time. Predicted values, actual values, and individual errors are shown in Table 6, and the errors were calculated by the same formula (12) used in the previous model. The results in Tables 5 and 6 are almost the same except at some points. Thus, Figures 7 and 8 are almost identical.

Figure 8 displays the graphical representation of the actual and predicted stock prices from the stochastic model. The black line represents the actual stock price and the green line represents the predicted stock price for S & P 500 index.

Table 6. Prediction by geometric Brownian motion.

Date	Actual	Predicted	Error
11-12-2019	305.69	305.17	0.17
11-13-2019	305.79	305.81	0.01
11-14-2019	306.23	305.91	0.10
11-15-2019	308.45	306.35	0.68
11-18-2019	308.68	308.59	0.03
11-19-2019	308.59	308.80	0.07
11-20-2019	307.44	308.71	0.41
11-21-2019	306.95	307.58	0.21
11-22-2019	307.63	307.06	0.19
11-25-2019	310.01	307.74	0.73
11-26-2019	310.72	310.15	0.18
11-27-2019	312.10	310.86	0.40
11-29-2019	310.94	312.23	0.41
12-2-2019	308.30	311.09	0.90
12-3-2019	306.23	308.43	0.72
12-4-2019	308.12	306.36	0.57
12-5-2019	308.68	308.25	0.14
12-6-2019	311.50	308.82	0.86
12-9-2019	310.52	311.63	0.36
12-10-2019	310.17	310.64	0.15
12-11-2019	311.05	310.30	0.24
12-12-2019	313.73	311.21	0.80
12-13-2019	313.92	313.88	0.01
12-16-2019	316.08	314.06	0.64
12-17-2019	316.15	316.21	0.02
12-18-2019	316.17	316.27	0.03
12-19-2019	317.46	316.31	0.36
12-20-2019	318.86	317.59	0.40
12-23-2019	319.34	319.00	0.11
12-24-2019	319.35	319.47	0.04
12-26-2019	321.05	319.51	0.48
12-27-2019	320.97	321.20	0.07
12-30-2019	319.20	321.11	0.60

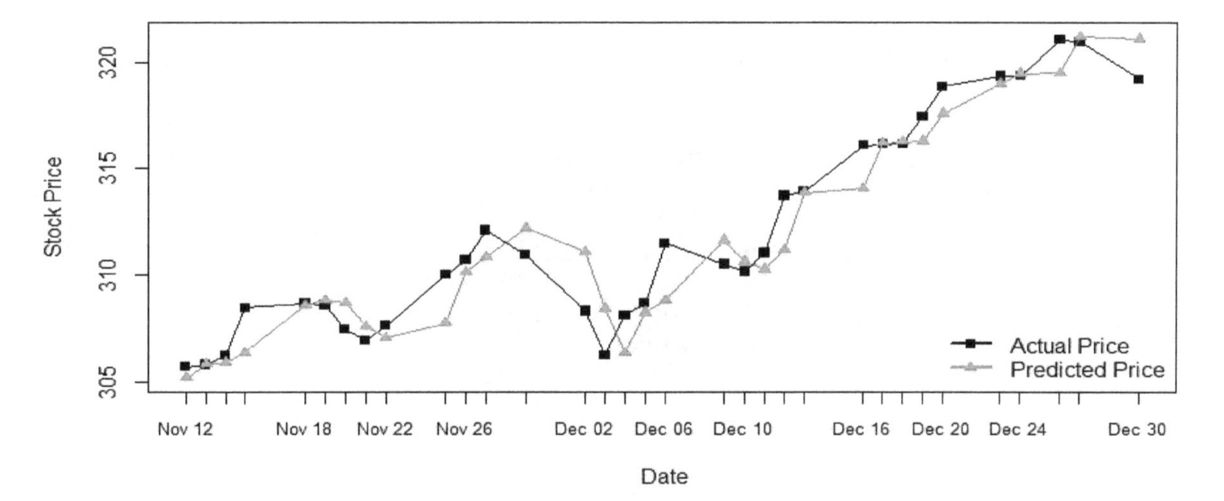

Figure 8. Stochastic model geometric Brownian motion prediction.

4.2.2. Stochastic Model Diagnostics

The performance of the stochastic model was assessed by the analysis of the four error measures stated in Equation (1) and the residual plot which is depicted in Figure 9 and the calculation of the four different error measures, as shown in Table 7.

Table 7. Prediction error by geometric Brownian motion.

MODEL	APE	AAE	ARPE	RMSE
GBM	0.0044	1.3341	0.0212	0.1455

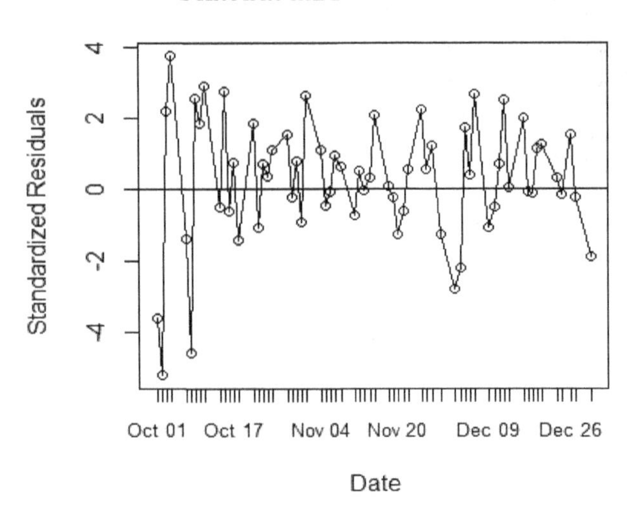

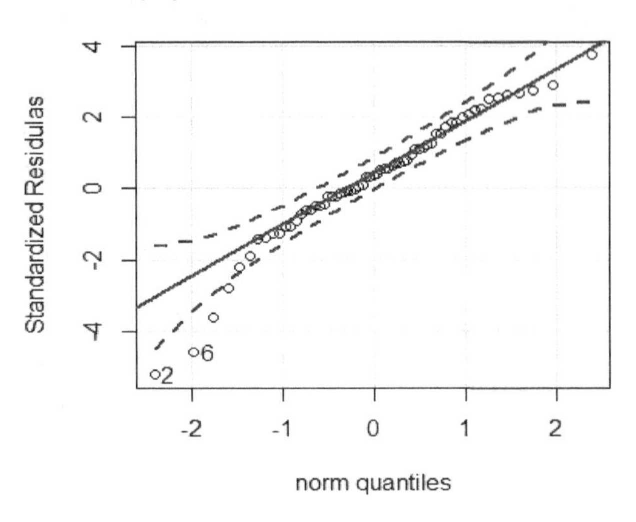

Figure 9. GBM model diagnostics.

The standardized residual plot is random and the mean passes through the zero line. A few of the residuals at the lower end are outside of the band in the Q-Q plot of the residuals. Still, both of the plots depict the approximate normal behavior of the residuals.

4.3. *Artificial Neural Network*

4.3.1. ANN(7-15-1) Results

Both of the training and test datasets were converted to normal and the prediction was converted back to the original scale by inverse transformation. The model required approximately 7000 steps with an error of 0.1972. Actual prices, predicted prices, and the corresponding errors are displayed in Table 8.

Table 8. Prediction by ANN (7-15-1) model.

Date	Actual	Predicted	Error
11-12-2019	305.69	302.13	1.17
11-13-2019	305.79	302.52	1.07
11-14-2019	306.23	302.26	1.30
11-15-2019	308.45	302.39	1.97
11-18-2019	308.68	303.56	1.66
11-19-2019	308.59	304.19	1.43
11-20-2019	307.44	303.54	1.27
11-21-2019	306.95	302.52	1.44
11-22-2019	307.63	302.86	1.55
11-25-2019	310.72	305.06	1.82
11-27-2019	312.10	305.81	2.02
11-29-2019	310.94	306.38	1.47
12-2-2019	308.30	306.02	0.74
12-3-2019	306.23	303.59	0.86
12-4-2019	308.12	301.63	2.11
12-5-2019	308.68	304.03	1.51

Table 8. *Cont.*

Date	Actual	Predicted	Error
12-6-2019	311.50	304.28	2.32
12-9-2019	310.52	306.05	1.44
12-10-2019	310.17	306.07	1.33
12-11-2019	311.05	305.23	1.87
12-12-2019	313.73	305.55	2.61
12-13-2019	313.92	306.48	2.37
12-16-2019	316.08	306.73	2.96
12-17-2019	316.15	307.56	2.72
12-18-2019	316.17	308.05	2.57
12-19-2019	317.46	308.40	2.85
12-20-2019	318.86	308.04	3.39
12-23-2019	319.34	306.20	4.11
12-24-2019	319.35	308.94	3.26
12-26-2019	321.05	309.71	3.53
12-27-2019	320.97	310.31	3.32
12-30-2019	319.20	309.93	2.90

The predicted errors in Table 8 are much higher than those in Tables 5 and 6. Precise comparisons of the three models are given in the next section. The graph associated with this result is displayed in Figure 10. The black line represents the actual stock price and the blue line represents the predicted stock price for the S & P 500 index. From the graph, it is clear that the model is working better at the beginning of the prediction interval.

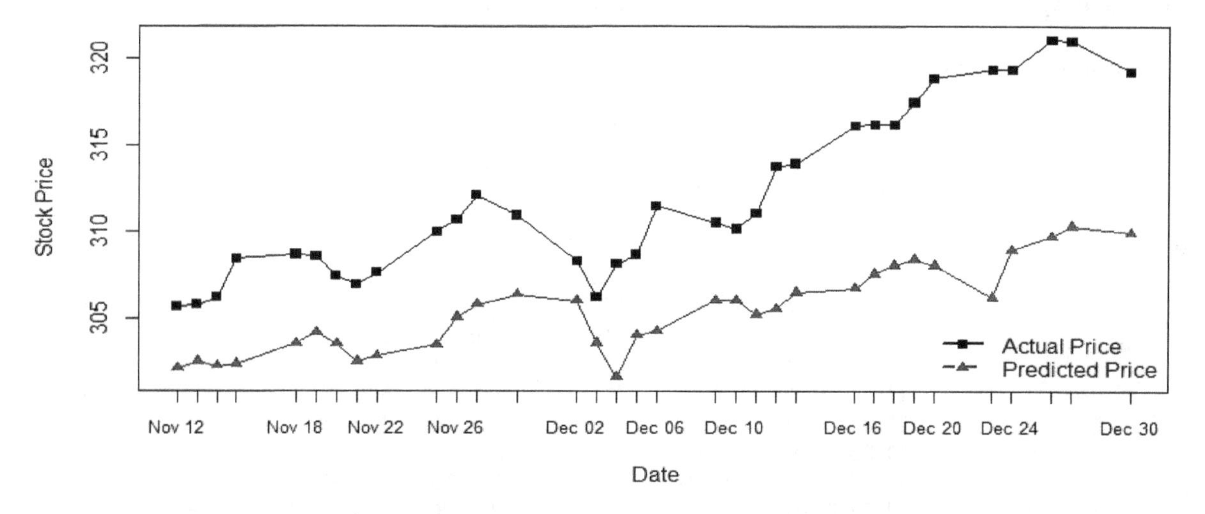

Figure 10. ANN(7-15-1) prediction.

4.3.2. ANN(7-15-1) Model Diagnostics

The performance of the neural network ANN(7-15-1) was assessed by the analysis of the four error measures stated in Equation (1) and the standardized residuals plot, which is depicted in Figure 11, and the calculation of the four different error measures are shown in Table 9.

Table 9. Prediction error by ANN(7-5-1) model.

MODEL	APE	AAE	ARPE	RMSE
ANN(7-15-1)	0.0167	5.09279	0.08084	0.28432

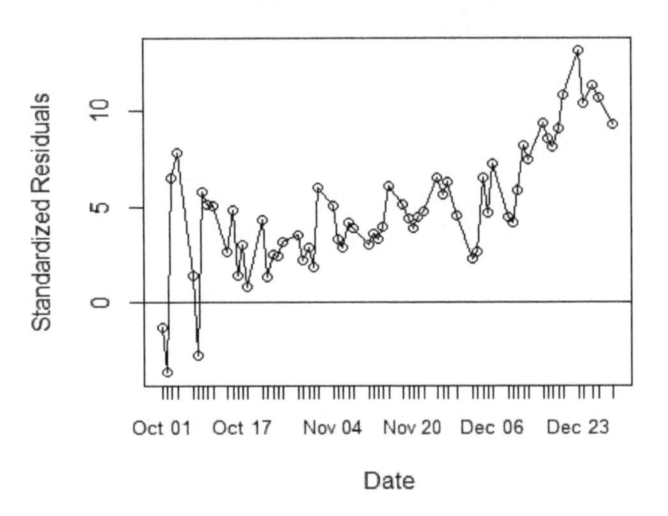

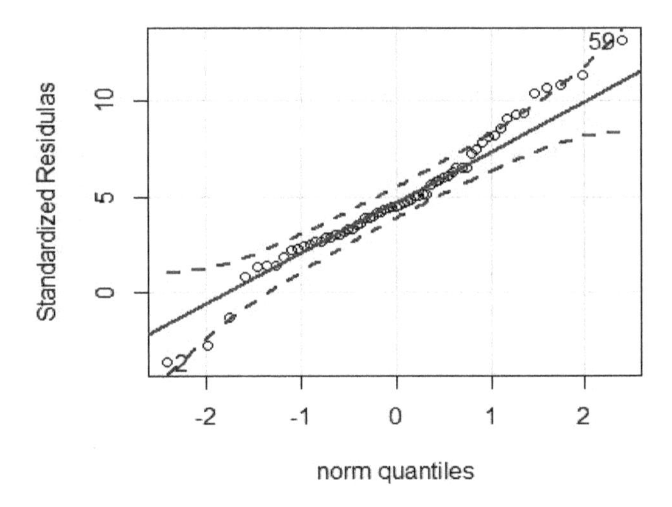

Figure 11. ANN(7-15-1) model diagnostics.

The standardized residual plot does not show normal behavior. The error increases in an exponential shape as the predicting interval increases.

4.4. Comparison

In this section, the combined output from the three models above is discussed. Table 10 shows a sample of the empirical results obtained from the models and Figure 12 displays the result graphically.

Table 10. Sample results from the models—ARIMA(0,2,1), $GBM(\mu, \sigma^2)$, and ANN(7-15-1).

Date	Actual	ARIMA	GBM	ANN
11-12-2019	305.69	305.17	305.17	302.13
11-13-2019	305.79	305.81	305.81	302.52
11-14-2019	306.23	305.91	305.91	302.26
11-15-2019	308.45	306.35	306.35	302.39
11-18-2019	308.68	308.57	308.59	303.56
11-19-2019	308.59	308.80	308.80	304.19
11-20-2019	307.44	308.71	308.71	303.54
11-21-2019	306.95	307.56	307.58	302.52
11-22-2019	307.63	307.07	307.06	302.86
11-25-2019	310.01	307.75	307.74	303.50
11-26-2019	310.72	310.14	310.15	305.06
11-27-2019	312.10	310.85	310.86	305.81
11-29-2019	310.94	312.23	312.23	306.38
12-2-2019	308.30	311.07	311.09	306.02
12-3-2019	306.23	308.43	308.43	303.59
12-4-2019	308.12	306.36	306.36	301.63
12-5-2019	308.68	308.25	308.25	304.03
12-6-2019	311.50	308.81	308.82	304.28
12-9-2019	310.52	311.64	311.63	306.05
12-10-2019	310.17	310.65	310.64	306.04
12-11-2019	311.05	310.30	310.30	305.23
12-12-2019	313.73	311.18	311.21	305.55
12-13-2019	313.92	313.86	313.88	306.48
12-16-2019	316.08	314.05	314.06	306.73
12-17-2019	316.15	316.21	316.21	307.56

Table 10. *Cont.*

Date	Actual	ARIMA	GBM	ANN
12-18-2019	316.17	316.29	316.27	308.05
12-19-2019	317.46	316.31	316.31	308.40
12-20-2019	318.86	317.60	317.59	308.04
12-23-2019	319.34	319.00	319.00	306.20
12-24-2019	319.35	319.48	319.47	308.94
12-26-2019	321.05	319.49	319.51	309.71
12-27-2019	320.97	321.20	321.20	310.31
12-30-2019	319.20	321.12	321.11	309.93

From Figure 12 it clear that that ARIMA(0,2,1) model's output and GBM model's output are very close, sometimes they coincide, whereas the output from the ANN(7-15-1) model gets far from the actual points as time increases.

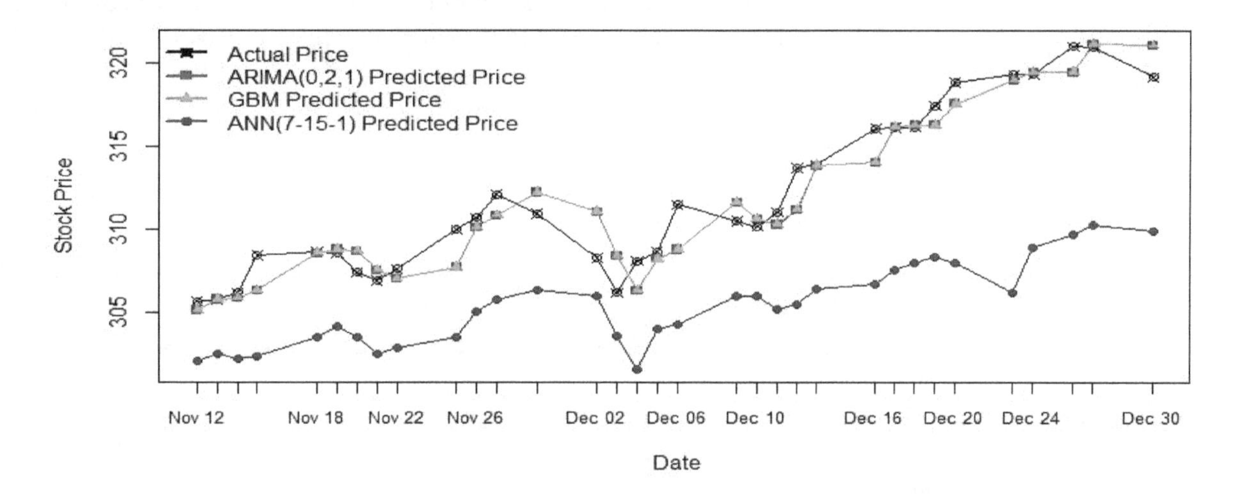

Figure 12. Prediction by all three models against the actual stock price.

Looking at the error measures in Table 11, it is clear that the ARIMA model and stochastic model perform better than the neural network model for predicting the next-day stock price.

Table 11. Error measures comparison from the three models.

MODEL	APE	AAE	ARPE	RMSE
ARIMA(0,2,1)	0.00438	1.33476	0.02119	0.14556
GBM	0.00438	1.33426	0.02118	0.14553
ANN(7-15-1)	0.01672	5.09279	0.08084	0.28432

5. Conclusions

This study represents a comparative study of three financial models ARIMA, ANN, and Geometric Brownian Motion to predict the next-day stock prices. Results obtained from the analysis of the S & P 500 index show that the conventional statistical model ARIMA and the stochastic model-geometric Brownian motion model perform better than the artificial neural network models for short term next-day stock price prediction. The results are in contradiction with the results in Khashei and Bijari (2010), which concluded that the ARIMA was no better than the ANN model in time series predictions. However, their proposed hybrid ANN model outperformed the traditional ANN and the ARIMA models. Furthermore, our results are similar to the conclusions in Merh et al. (2010) and Lee et al. (2007) which stated that ARIMA models outperform ANN models for stock price predictions. On the other hand, Rathnayaka et al. (2014) found that the stochastic model prediction is more significant than the traditional ARIMA model. In fact, on the basis of our results, the ARIMA model and the stochastic model produce almost the same results. Thus, for short term prediction using the time series data, the ARIMA model and the stochastic model can be used interchangeably. For the ANN models, further studies, hybridization of existing models, and adding more independent variables can improve the neural network models in predicting stock prices. One model can work better than other models with particular time series data. Therefore, researchers or investors should examine some different models to predict the prices of each stock to find the best prediction model.

Author Contributions: M.R.I. setup and ran models, processed data, and wrote the first draft. N.N. introduced the methodology, refined the manuscript, and supervised the project. All authors have read and agreed to the published version of the manuscript.

References

Adebiyi, Ayodele Ariyo, Aderemi Oluyinka Adewumi, and Charles Korede Ayo. 2014. Comparison of arima and artificial neural networks models for stock price prediction. *Journal of Applied Mathematics* 2014: 614342. [CrossRef]

Agustini, W. Farida, Ika Restu Affianti, and Endah R. M. Putri. 2018. Stock price prediction using geometric brownian motion. *Journal of Physics: Conference Series* 974: 012047.

Avcı, Emin. 2007. Forecasting daily and sessional returns of the ise-100 index with neural network models. *Dogus Universitesi Dergisi* 8: 128–42. [CrossRef]

Chen, An-Sing, Mark T. Leung, and Hazem Daouk. 2003. Application of neural networks to an emerging financial market: forecasting and trading the taiwan stock index. *Computers & Operations Research* 30: 901–23.

Cheung, Yin-Wong, and Kon S. Lai. 1995. Lag order and critical values of the augmented dickey–fuller test. *Journal of Business & Economic Statistics* 13: 277–80.

Cryer, Jonathan D., and Natalie Kellet. 1991. *Time Series Analysis*. Berlin and Heidelberg: Springer.

Dmouj, Abdelmoula. 2006. Stock Price Modelling: Theory and Practice. Masters's thesis, Vrije Universiteit, Amsterdam, The Netherlands.

Estember, Rene D., and Michael John R. Maraña. 2016. Forecasting of stock prices using brownian motion–monte carlo simulation. Paper presented at the 2016 International Conference on Industrial Engineering and Operations Management Kuala Lumpur, Kuala Lumpur, Malaysia, March 8–10, pp. 704–13.

Hassan, Md Rafiul, and Baikunth Nath. 2005. Stock market forecasting using hidden markov model: A new approach. Paper presented at the International Conference on Intelligent Systems Design and Applications (ISDA'05), Pretoria, South Africa, December 3–5, Piscataway: IEEE, pp. 192–96.

Hassan, Md Rafiul, Baikunth Nath, and Michael Kirley. 2007. A fusion model of hmm, ann and ga for stock market forecasting. *Expert Systems with Applications* 33: 71–80. [CrossRef]

Khashei, Mehdi, and Mehdi Bijari. 2010. An artificial neural network (p, d, q) model for timeseries forecasting. *Expert Systems with Applications* 37: 479–89. [CrossRef]

Lee, Kyungjoo, Sehwan Yoo, and John Jongdae. 2007. Neural network model versus sarima model in forecasting korean stock price index (kospi). *Issues in Information System* 8: 372–8.

Makridakis, Spyros, and Michele Hibon. 1997. Arma models and the box–jenkins methodology. *Journal of Forecasting* 16: 147–63. [CrossRef]

Merh, Nitin, Vinod P. Saxena, and Kamal Raj Pardasani. 2010. A comparison between hybrid approaches of ann and arima for indian stock trend forecasting. *Business Intelligence Journal* 3: 23–43.

Meyler, Aidan, Geoff Kenny, and Terry Quinn. 1998. *Forecasting Irish Inflation Using Arima Models*. Dublin: Central Bank of Ireland.

Neath, Andrew A., and Joseph E. Cavanaugh. 2012. The bayesian information criterion: Background, derivation, and applications. *Wiley Interdisciplinary Reviews: Computational Statistics* 4: 199–203. [CrossRef]

Nguyet Nguyen, Dung Nguyen, and Thomas P. Wakefield. 2018. Using the hidden markov model to improve the hull-white model for short rate. *International Journal of Trade, Economics and Finance* 9. [CrossRef] [CrossRef]

Rathnayaka, R. M. Kapila Tharanga, Wei Jianguo, and DMK N. Seneviratna. 2014. Geometric brownian motion with ito's lemma approach to evaluate market fluctuations: A case study on colombo stock exchange. Paper presented at the 2014 International Conference on Behavioral, Economic, and Socio-Cultural Computing (BESC2014), Shanghai, China, October 30–November 2, Piscataway: IEEE, pp. 1–6.

Ryan, Jeffrey A., Joshua M. Ulrich, Wouter Thielen, Paul Teetor, Steve Bronder, and Maintainer Joshua M. Ulrich. 2020. Package 'quantmod'. Available online: https://cran.r-project.org/web/packages/quantmod/quantmod.pdf (accessed on 14 August 2020).

Ševcovic, Daniel, B. Stehlıková, and K. Mikula. 2011. *Analytical and Numerical Methods for Pricing Financial Derivatives*. Hauppauge: Nova Science.

Tambi, Mahesh Kumar. 2005. Forecasting exchange rate: A univariate out of sample approach. *The IUP Journal of Bank Management* 4: 60–74.

White, Halbert. 1988. Economic prediction using neural networks: The case of ibm daily stock returns. Paper presented at the IEEE 1988 International Conference on Neural Networks, San Diego, CA, USA, July 24–27, vol. 2, pp. 451–58.

Yang, Zhijun, and D. Aldous. 2015. *Geometric Brownian Motion Model in Financial Market*. Berkeley: University of California.

Yao, Jingtao, Chew Lim Tan, and Hean-Lee Poh. 1999. Neural networks for technical analysis: A study on klci. *International Journal of Theoretical and Applied Finance* 2: 221–41. [CrossRef]

Zhang, Guoqiang, B. Eddy Patuwo, and Michael Y. Hu. 1998. Forecasting with artificial neural networks: The state of the art. *International Journal of Forecasting* 14: 35–62. [CrossRef]

Zhang, Yudong, and Lenan Wu. 2009. Stock market prediction of S & P 500 via combination of improved bco approach and bp neural network. *Expert Systems with Applications* 36: 8849–54.

Permissions

All chapters in this book were first published in MDPI; hereby published with permission under the Creative Commons Attribution License or equivalent. Every chapter published in this book has been scrutinized by our experts. Their significance has been extensively debated. The topics covered herein carry significant findings which will fuel the growth of the discipline. They may even be implemented as practical applications or may be referred to as a beginning point for another development.

The contributors of this book come from diverse backgrounds, making this book a truly international effort. This book will bring forth new frontiers with its revolutionizing research information and detailed analysis of the nascent developments around the world.

We would like to thank all the contributing authors for lending their expertise to make the book truly unique. They have played a crucial role in the development of this book. Without their invaluable contributions this book wouldn't have been possible. They have made vital efforts to compile up to date information on the varied aspects of this subject to make this book a valuable addition to the collection of many professionals and students.

This book was conceptualized with the vision of imparting up-to-date information and advanced data in this field. To ensure the same, a matchless editorial board was set up. Every individual on the board went through rigorous rounds of assessment to prove their worth. After which they invested a large part of their time researching and compiling the most relevant data for our readers.

The editorial board has been involved in producing this book since its inception. They have spent rigorous hours researching and exploring the diverse topics which have resulted in the successful publishing of this book. They have passed on their knowledge of decades through this book. To expedite this challenging task, the publisher supported the team at every step. A small team of assistant editors was also appointed to further simplify the editing procedure and attain best results for the readers.

Apart from the editorial board, the designing team has also invested a significant amount of their time in understanding the subject and creating the most relevant covers. They scrutinized every image to scout for the most suitable representation of the subject and create an appropriate cover for the book.

The publishing team has been an ardent support to the editorial, designing and production team. Their endless efforts to recruit the best for this project, has resulted in the accomplishment of this book. They are a veteran in the field of academics and their pool of knowledge is as vast as their experience in printing. Their expertise and guidance has proved useful at every step. Their uncompromising quality standards have made this book an exceptional effort. Their encouragement from time to time has been an inspiration for everyone.

The publisher and the editorial board hope that this book will prove to be a valuable piece of knowledge for researchers, students, practitioners and scholars across the globe.

List of Contributors

Oscar V. De la Torre-Torres and Evaristo Galeana-Figueroa
Faculty of Accounting and Management, Saint Nicholas and Hidalgo Michoacán State University (UMSNH) 58030 Morelia, Mexico

José Álvarez-García
Financial Economy and Accounting Department, Faculty of Business, Finance and Tourism, University of Extremadura, 10071 Cáceres, Spain

Hoang Viet Long
Division of Computational Mathematics and Engineering, Institute for Computational Science, Ton Duc Thang University, Ho Chi Minh City 70000, Vietnam
Faculty of Mathematics and Statistics, Ton Duc Thang University, Ho Chi Minh City 758307, Vietnam

Haifa Bin Jebreen
Department of Mathematics, College of Science, King Saud University, Riyadh 11451, Saudi Arabia

Y. Chalco-Cano
Departamento de Matemática, Universidad de Tarapacá, Casilla 7D, Arica 09010, Chile

Imran Yousaf and Shoaib Ali
Air University School of Management, Air University, Islamabad 44000, Pakistan

Wing-Keung Wong
Department of Finance, Fintech Center and Big Data Research Center, Asia University, Taichung 41354, Taiwan
Department of Medical Research, China Medical University Hospital, Taichung 40402, Taiwan
Department of Economics and Finance, The Hang Seng University of Hong Kong, Hong Kong 999077, China

Alex Golodnikov and Viktor Kuzmenko
V.M. Glushkov Institute of Cybernetics, 40, pr. Akademika Glushkova, 03187 Kyiv, Ukraine

Stan Uryasev
Applied Mathematics & Statistics, Stony Brook University, B-148 Math Tower, Stony Brook, NY 11794, USA

Rafiuddin Ahmed
Program of Accounting and Finance, James Cook University, Douglas, QLD 4814, Australia

Rafiqul Bhuyan
Department of Accounting and Finance, Alabama A&M University, Normal, AL 35762, USA

Sel Ly, Kim-Hung Pho and Sal Ly
Faculty of Mathematics and Statistics, Ton Duc Thang University, Ho Chi Minh City 756636, Vietnam

Jian Huang and Huazhang Liu
Division of Business Management, Beijing Normal University-HongKong Baptist University United International College, Zhuhai 519087, China

Lászl6 Nagy
Department of Finance, Budapest University of Technology and Economics, Magyar tudosok krt. 2., 1117 Budapest, Hungary

Mihály Ormos
Department of Economics, Janos Selye University, Hradná ul. 21., 94501 Komarno, Slovakia

Zhe Li
Business School, Nanjing Normal University, Nanjing 210023, China

Mohammad Rafiqul Islam and Nguyet Nguyen
Department of Mathematics and Statistics, Youngstown State University, Youngstown, OH 44555, USA

Index

Printed in the USA
CPSIA information can be obtained
at www.ICGtesting.com
JSHW051625061123
51533JS00005B/110

9 781647 285258